INTERACTIONS

BOSTON STUDIES IN THE PHILOSOPHY OF SCIENCE

VOLUME 251

INTERACTIONS
Mathematics, Physics and Philosophy, 1860-1930

Edited by

VINCENT F. HENDRICKS
Department of Philosophy and Science Studies,
Roskilde University, Denmark

KLAUS FROVIN JØRGENSEN
Department of Philosophy and Science Studies,
Roskilde University, Denmark

JESPER LÜTZEN
Institute for Mathematical Sciences,
University of Copenhagen, Denmark

and

STIG ANDUR PEDERSEN
Department of Philosophy and Science Studies,
Roskilde University, Denmark

 Springer

A C.I.P. Catalogue record for this book is available from the Library of Congress.

ISBN-10 1-4020-5194-8 (HB)
ISBN-13 978-1-4020-5194-4 (HB)
ISBN-10 1-4020-5195-6 (e-book)
ISBN-13 978-1-4020-5195-1 (e-book)

Published by Springer,
P.O. Box 17, 3300 AA Dordrecht, The Netherlands.

www.springer.com

Printed on acid-free paper

REVIEWS

This stimulating volume covers a wide range of topics which are of direct interest to anyone who thinks about the curious relation between mathematics and the natural world. Philosophers often pose interesting questions about the "dispensability" of mathematics to science. But they too often overlook the wealth of philosophical perplexities that can arise in detailed examples and case studies, both contemporary and historical. This volume refocuses our attention by addressing a number of topics connected to applied mathematics, any one of which is worthy of every philosopherís attention.

—James Robert Brown, University of Toronto

What to make of neo-Kantianism in its hey-day, from 1840–1940? It was the most prolific of times and the most seminal, it was the most muddled and confused, it is philosophy working at its hardest with science and most damagingly against science.

It is examined here episodically, as it engaged individual scientists: Helmholtz, Hertz, Poincare, Minkowski, Hilbert, Eddington and Weyl. If Einstein is not in their number, he had to contend with their influence, and anyway he transformed their agenda. The essays on these figures are glinting in their focus and scholarship. Whatever one thinks of neo-Kantianism, this book is history and philosophy of science at its best: mathematically and physically informed, historically engaged, and philosophically driven.

—Simon Saunders, University of Oxford

Ten first-rate philosopher-historians probe insightfully into key conceptual questions of pre-quantum mathematical physics, from Helmholtz and Boltzmann, through Hertz and Lorentz, to Einstein, Weyl and Eddington, with an interesting aside on the rarely studied philosophy of Federigo Enriques. A rich and effective display of what the critical history of science can do for our understanding of scientific thought and its achievements.

—Roberto Torretti, University of Puerto Rico

CONTENTS

PREFACE

The main theme of this book is the unique interaction between mathematics, physics and philosophy during the beginning of the 20th century. In this period seminal theories of modern physics and new fundamental mathematical structures were discovered or formed. Significant physicists such as Lorentz and Einstein as well as mathematicians such as Poincaré, Minkowski, Hilbert and Weyl contributed to this development. They created the new physical theories and the mathematical disciplines that play such a paramount role in their mathematical formulation—and in other areas of mathematics. These physicists and mathematicians were also key figures in the philosophical discussions of nature and science—from philosophical tendencies like logical empiricism via critical rationalism to various neo-Kantian trends.

During the first half of the 20th century relativity theory and quantum mechanics raised different kinds of philosophical problems. This compilation primarily deals with the philosophical issues related to relativity theory. Their root in the 19th century considerations of the nature of space, time, mechanics and electromagnetism is the subject of the first papers. Hyder analyzes Helmholtz's proof of the centrality of forces and its equivalence to the conservation of energy. He connects it to Helmholtz's changing adaptation of Kantian epistemology and shows that these physico-philosophical considerations form the basis of his later papers on geometry. Lützen argues that Hertz's philosophical theory of images, presented in his book on mechanics, had their origin in Hertz's work on electromagnetism and was aimed at presenting a mechanistic foundation of all of physics. After Hertz several physicists tried to replace this mechanistic philosophy by an electromagnetic program. This is at the heart of Janssen's and Mecklenburg's paper on the electromagnetic models of the electron in which they give a detailed account of the transition from classical to relativistic mechanics.

Also Pulte's and DiSalle's papers connect post relativity ways of thinking with 19th century ideas. Pulte discusses Schlick's changing philosophy of spatial intuition and the foundation of geometry relating it to his reading of Helmholtz and DiSalle argues that the creation of a mathematical relativistic world structure was a continuation of the epistemological reflections of Helmholtz and Poincaré.

Hilbert's axiomatic program has often had a bad press in so far as it deals with physical theories, but Majer defends Hilbert's views while explaining in some detail how Hilbert imagined one should go about such an axiomatization.

Many of the mathematicians and physicists of the early 20th century published their epistemological reflections in semi-popular works on science. In his paper Gray analyzes a less well-known such work namely Enriques's somewhat anti-Kantian positivist book that may be considered as a reaction to Poincaré's views. Rowe's paper in turn deals with reactions of a different kind, namely the public and scientific reactions to Einstein's theory of relativity.

The central part of the book ends with Scholz's thorough analysis of Weyl's changing concept of matter 1918-30. It is followed by the discussion of one of the two central philosophical problems of statistical mechanics: Why does the standard measure work?

The papers of this anthology are revised versions (in some cases greatly revised versions) of talks given at the meeting "The Interaction between Mathematics, Physics and Philosophy from 1850 to 1940" held on September 26–28, 2002 at the Carlsberg Academy, Copenhagen. The meeting was arranged by MATHNET – The Danish Network for the History and Philosophy of Mathematics and was sponsored by The Danish Natural Science Research Council.

We would like to extend our gratitude to Springer, and in particular senior publishing editor Charles Erkelens for taking on this project.

Vincent F. Hendricks   Klaus Frovin Jørgensen

Jesper Lützen   Stig Andur Pedersen

Copenhagen, March 2006

CONTRIBUTING AUTHORS

In order of appearance:

David Hyder studied philosophy at Yale, Göttingen, and Toronto. His research interests include the philosophies of Wittgenstein and Russel, Kant's philosophy of science, and the history and philosophy of science. His is currently Assistant Professor of Philosophy at the University of Ottawa.

Jesper Lützen is professor of the History of Mathematics at the University of Copenhagen. His main interests are in mathematics and mechanics in the 19th and early 20th century. He is the author of *Joseph Liouville 1809–1882. Master of Pure and Applied Mathematics* (Springer 1990) and *Mechanistic Images in Geometric Form: Heinrich Hertz's Principles of Mechanics* (Oxford University Press 2005).

Michel Janssen used to be an editor for the Einstein Papers Project. He is now in the Program in History of Science and Technology at the University of Minnesota.

Matthew Mecklenburg earned his B.S. in physics from the University of Minnesota. He is currently a physics graduate student at UCLA.

Jeremy Gray is a Professor of the History of Mathematics and Director of the Centre for the History of the Mathematical Sciences at the Open University in England, and is an Honorary Professor in the Mathematics Department at the University of Warwick. He is the author, co-author, or editor of 13 books on the history of mathematics in the 19th and 20th Centuries, including issues in the philosophy and cultural significance of mathematics. His most recent book is *Janos Bolyai, non-Euclidean geometry, and Nature of Space* (MIT Press 2004). He has recently finished a book on the history of geometry in the 19th Century, and is now working on one on mathematical modernism and the philosophy of mathematics.

Ulrich Majer is apl. Professor at the Georg-August-Universität Göttingen, Germany. His main interests are the philosophy and history of mathematics and the exact sciences in the 20th century. In particular he is interested in the works of H. Hertz, B. Russell, and H. Poincaré and their influence on the philosophy of science and its history in the 20th century. His research, however, is focused primarily on the works of E. Husserl, H. Weyl, and D. Hilbert. He is co-editor of F.P. Ramsey's work *On Truth* (Kluwer 1991), *Semantical Aspects of Spacetime Theories* (BI 1994) and *Reflections on Spacetime* (Kluwer

1995) as well as chief editor of the *Nachlass* of David Hilbert in Göttingen (Springer 2004).

Helmut Pulte is full professor of Philosophy and History of Science at Ruhr-Universität Bochum. His main interests are in history and philosophy of mathematics and physics, modern analytical philosophy and the history of philosophy of science in general. Inter alia, he is the author of *Axiomatik und Empirie* (WB Darmstadt, 2005), which deals with the philosophical development of mathematical physics from the late 17th century (I. Newton and others) to the lat 19th century (C. Neumann and others).

Robert DiSalle is Associate Professor of Philosophy at the University of Western Ontario. His work concerns the history and philosophy of modern physics, especially the foundations of space-time theories. He is the author of *Understanding Space-Time: The Philosophical Development of Physics from Newton to Einstein* (Cambridge University Press, 2006).

David E. Rowe is professor of history of mathematics and science at Mainz University. His research interest focus on mathematics and physics in Germany during the period 1800–1945. He is coeditor with Robert Schulmann of a sourcebook entitled *The Political Einstein: Private Thoughts and Public Passions* scheduled for publication by Princeton University Press in 2006.

Erhard Scholz is (associate) professor for the history of mathematics at the University of Wuppertal. His main interest lie in the exchanges between mathematics and physics in the 19th and 20th centuries and their philosophical implications. He is author of *Geschichte des Mannigfaltiglkeitsbegriffs von Riemann bis Poincaré* (Birkhäuser 1980), *Symmetrie – Gruppe – Dualität. Zur Beziehung zwischen theoretischer Mathematik und Anwendungen in Kristallographie und Baustatik des 19. Jahrhunderts* (Birkhäuser 1989). Moreover he has edited books on the history of algebra and on Hermann Weyl's *Raum–Zeit–Materie* and is coeditor of F. Hausdorff's *Gesammelte Schriften*.

Lawrence Sklar is the Carl G. Hempel and William K. Frankena Distinguished University Professor of Philosophy at the University of Michigan. His main interests are in the philosophy of physics and methodological problems of theories in physical science. He is the author of *Space, Time and Spacetime* (California 1974), *Philosophy and Spacetime Physics* (California 1985), *Physics and Chance* (Cambridge 1993) and *Theory and Truth* (OUP 2000).

DAVID HYDER

KANT, HELMHOLTZ AND THE DETERMINACY OF PHYSICAL THEORY[0]

ABSTRACT

In this paper, I analyse Helmholtz's arguments for the centrality of forces in the opening sections of his *Erhaltung der Kraft*, paying special regard to Helmholtz's insistence on the "determinacy" of physical science. Helmholtz applies this Kantian principle to prove that the intensity of forces must be a function of position, and that since position is definable only in relative space, force functions must be definable with regard only to the relative positions of the mass-points comprising a physical system. I claim that this argument is an adaptation of Kant's criticisms of Newton's parallelogram law of force additivity in the *Metaphysical Foundations*: for both Helmholtz and Kant, forces can only be well defined if they are conceived as connections between material points, thus all such definitions may take account only of the relative or "empirical" spaces determined by these points. I conclude by arguing that Helmholtz's 1854 defence of the *Erhaltung* against Clausius's objections lays special emphasis on the empirical conditions which are required to establish the congruence relations holding between pairs of points. I suggest that Helmholtz's observation in that paper that such conditions are necessary for the "empirical application" of physical principles forms the point of departure for his much later papers on geometry.

INTRODUCTION

Hermann von Helmholtz's is celebrated—or notorious—for originating a hybrid form of the Kantian philosophy in which contingent facts serve simultaneously as transcendental conditions on knowledge. According to such a naturalised Kantian epistemology, the make-up of our perceptual organs, of our brains, or even certain regularities in the natural world condition the form of our knowledge. Even as theoretical science extends to the point of permitting us to recognise these contingencies as such, still we cannot truly step outside their confines. This epistemological conception is evident in much of what Helmholtz says about geometry; however, this is, as I shall argue in my conclusion, only a late example of an earlier development. Helmholtz started from a philosophical position representing a quite orthodox form of Kant's epistemology, but he was progressively radicalised in the course of his scientific career. By examining his early memoir, the *Erhaltung der Kraft*, as well as his later comments and correctives to this text, I will show how

1

V.F. Hendricks, K.F. Jørgensen, J. Lützen and S.A. Pedersen (eds.), Interactions: Mathematics, Physics and Philosophy, 1860-1930, pp. 1–44.
© 2006 Springer.

Helmholtz's proof of the equivalence of central forces to the conservation of *vis viva* rests on two hypotheses or principles which Helmholtz derives from the regulative demand that the science of nature be determinate. These principles are the following: the principle of the decompositionality of actions, as Olivier Darrigol has called it, and the principle of positional determinacy. It is just these hypotheses that Helmholtz reinterpreted over the course of his scientific career. Early on, he seeks to justify them by means of transcendental arguments borrowed in part from Kant's *Metaphysical Foundations of Natural Science* (hereafter, *MFNS*); however, in his later work, he adopts a position akin to what Martin Carrier has called "metrogenic apriorism".[1] According to Helmholtz's mature philosophy, such principles are inductive truths; however, they are singled out among physical principles by virtue of their making possible what he calls the "physical application" of scientific laws.

My discussion is divided into four sections: First, I shall examine the mathematical proof for the centrality of force given in the first, philosophical section of the *Erhaltung*, which depends essentially on the assumption that forces can be geometrically decomposed. From there I will turn to a longer discussion of Kant's criticisms of Newton's proof of the additivity of forces in the *Metaphysical Foundations of Natural Science*. I then argue that Helmholtz's replies to the early criticisms of his centrality proof raised by the physicist Rudolf Clausius reiterate Kant's demand that the fundamental concepts of physical theory must be "constructable" in intuition, and thus that they cannot involve appeals to absolute space. For Kant, all quantitative relations must be founded in relations among spatio-temporal magnitudes— for the latter are the only "extensive" magnitudes, thus the only structures in our experience that can ground mathematical principles. Every magnitudinal concept involved in physical theory, including those of motion and of force, must as a result be schematised in space and time if mathematics is to be applied to nature in a *determinate* manner. This schematisation of mathematical concepts is a specific instance of the more general schematisation of concepts demanded in the first *Critique*.

When I return to Helmholtz in the third section, I will show how his proof and his replies to his critics represent an extension of Kant's arguments in two respects: the one concerns the principle of decomposition, that is to say Helmholtz's interpretation of Newton's parallelogram of forces; the other concerns the conservation of energy. In his replies to Clausius's objections to his memoir, Helmholtz separates these two assumptions in his proof, and connects the positionality principle to a further assumption he had not explicated previously, namely the possibility of determining the congruence of pairs of points. In the last section, I suggest that this principle of positional determinacy forms the background for his first two papers on geometry. Thus I shall say a few words about that work, and about the more general philosophical position that I ascribe to him in his mature phase. On this later view, the determinacy of physical theory is seen as depending on *contingent* conditions. The principles of decomposition and congruence measurement, despite their contingency, are still conditions on the "physical application" of other laws, and thus conditions on the latter's having a determinate *meaning*. They are regulative conditions on science that are transcendental even though they have an empirical content.

1. HELMHOLTZ'S PROOF OF THE CENTRALITY OF FORCES

In his 1847 *Über die Erhaltung der Kraft*, Helmholtz had argued that the principle of *vis viva* conservation was equivalent to the hypothesis that all forces of nature were central forces holding between mass-points. But, he claimed, this latter hypothesis was, in a certain sense, a necessary truth. For science aims at providing a complete description of the natural world, and this requires that all phenomena it treats of must be empirically *determinate*. Helmholtz thought that he could derive both of the two defining characteristics of central forces from epistemological requirements placed on the possible form of laws and phenomena: first, the forces had to be directed along the line connecting the mass-points, for this was the only spatial magnitude determined by the two points; second, in order that science be maximally unified, their intensity also had to be a function of this one magnitude. Both of these demands supposedly followed from the essential *in*determinacy of spatial relations, that is to say from their relativity, which will be discussed in detail in the following. Thus the characteristics of central forces derive from conditions on the *determinate representability* of motions and forces.

When the *Erhaltung* was republished in the first volume of his collected works in 1882, Helmholtz appended a series of supplementary, and on some points critical remarks on his earlier positions. Here, he modified both of the above claims. He conceded that his aprioristic argument for the necessity of central forces was incorrect, and thus that it represented at most an empirical generalisation. Furthermore, he admitted that this empirical generalisation had itself been called into question by current theories of electromagnetism. This state of affairs pointed, in his view, to a critical tension in the state of physical theory. The epistemological privilege he had ascribed to central forces remained unchanged, for the philosophical argument from determinacy was not wholly mistaken. But the new theories of Weber and others had met with undeniable success. These new theories thereby threatened to force physics to abandon to the goal of providing a completely determinate theory of nature, and therefore also the hope of "completing" natural science. In order to clarify this dilemma, I shall briefly contrast his early account with the analysis he gave in 1882.

1.1. The philosophical argument

In his original conception of 1847, Helmholtz had maintained that forces had to be central because this postulate followed from the regulative demand that nature be completely comprehensible [*vollständig begreiflich*]. This demand would be fully satisfied just in case we had articulated laws allowing us to predict the (future) behaviour of each natural system whose material properties were fully known to us in the present. In such a complete science, our knowledge of the physical world would cleave neatly into two parts: (1) temporally invariant general laws and (2) descriptions of particular physical systems. In a pure mathematical theory of the sort Helmholtz required, the properties allowed to these systems are restricted to the mass and the position of their material points, along with the velocities of, and the forces holding between these particles. The exact nature of these forces will depend on the kinds of matter we are dealing with. Indeed, the differences among kinds of matter will, in a fully mathematical theory, be represented solely as differences in their masses and in the forces the various matters generate.[2] According to Helmholtz, we identify forces in general with the causes of changes in the motions of mass-points. But the intensity of such forces can evidently vary with time. Thus a final reduction of physical phenomena to invariant laws will require our identifying the fundamental *in*variant forces characterising the various species of matter. These fundamental forces will have to be described by functions that depend on empirically determinate properties only, and which, in particular, do not depend on time. But since the mass of material particles is assumed to be constant, the empirically determinate properties spoken of here can consist only in the positional relations among the points.

This result implied, according to Helmholtz, the dual connection between force and position that I outlined above: on the one hand, forces are the ultimate causes of observed changes in position; on the other hand, if these forces themselves are subject to change, the latter changes must depend on these same positions. Helmholtz then applies a principle deriving from spatial relativity, which I shall call the "principle of positional determinacy", in order to prove centrality. This amounts to requiring that any magnitudes used to characterise the motion of a system be defined only in terms of the relative positions of its mass-points. Such magnitudes, which include directions and distances, we will call the *internal properties* of the system. For, while directions and distances in absolute space may enter into our mathematical description of the system, they cannot be objects of experience.

The principle of positional determinacy restricts the range of motive concepts that can be applied to systems. A single point in space determines no spatial magnitudes at all, thus it cannot be said to undergo motion, let alone accelerated or forced motion. When two points are given there can indeed be relative motion, but only one direction and only one magnitude are determinate, *etc.* Helmholtz, almost without comment, restricts the results concerning the positional dependence of forces to the elements of a two-point system, and thereby derives the two elements of centrality from positional determinacy: (1) Force intensities must be functions of the distance determined by the two mass-points. For if this were not the case, then changes in the forces acting among the points (changes in the causes of change), would depend on non-observable, and thus experientially indeterminate properties of the system. And this would mean in turn that nature was not completely comprehensible in the desired sense. Furthermore, (2) the only observable effect of the force acting between two mass-points can be to alter their distance. Here as well, to suppose other effects would be to posit differences between systems which were identical with respect to their internal properties. Finally, by (1) and (2) it follows that all forces in nature must be central forces, and the demand that nature be completely comprehensible entails that all forces are central.

This transcendental deduction of central forces is followed by a mathematical demonstration of the equivalence of the postulate of force centrality to the principle of the conservation of *vis viva*. The latter principle, in Helmholtz's version, states that whenever a system is in the same state—that is to say whenever all the internal properties of the system are the same—the kinetic energy of the system is the same as well, whatever the path followed by the system in the intervening time. Helmholtz's proof, which we shall examine in greater detail below, contained a number of flaws, one of which was shared with the philosophical deduction just outlined. For this deduction assumes that we can draw conclusions about the nature of fundamental forces by restricting ourselves to two-point systems and then applying the principle of positional determinacy. Helmholtz did the same in his mathematical demonstration of the equivalence of centrality to *vis viva* conservation. In addition, he overlooked the possibility that forces could depend on the velocities and accelerations of masses. But let us, for the moment, ignore these difficulties, so that we can get a clear view of the purpose of these two proofs.

Suppose that both the philosophical deduction of centrality from the comprehensibility of nature *and* the mathematical demonstration of the equivalence of centrality to *vis viva* conservation had been valid. What exactly would Helmholtz have shown? We have,

I. comprehensibility of nature $\Rightarrow$ force centrality

II. force centrality $\Leftrightarrow$ conservation of *vis viva*

From this it follows that,

III. comprehensibility of nature $\Rightarrow$ conservation of *vis viva*

Our proposition III. expresses a Leibnizian intuition, namely that if it were possible to construct a perpetuum mobile (if the same system, in the same internal state, had a different kinetic energy) this would violate the principle of sufficient reason. Perpetual motion is not merely empirically false, but it is indeed logically impossible.[3] But the implication that expresses this supposed truth, however persuasive it may be intuitively, is hardly rigorous enough to do physics. Whereas Helmholtz's deduction of force centrality from the comprehensibility of nature is not only philosophically persuasive, but it also yields a proposition with a precise physical content. Furthermore, Helmholtz has also refined the antecedent to the implication. His definition of the comprehensibility of nature makes specific demands on the forms of laws and of phenomena, and he is able to show that these specific demands entail a result that is logically equivalent to energy conservation. Helmholtz's analysis is therefore not merely of philosophical interest—it can be, and it was used by Helmholtz to argue against physical theories that were not in agreement with the requirement of force centrality. But for this very reason, the coherence of the proofs and the modal status of the premises are of decisive importance. As we have seen, the proofs of both I and II contained additional premisses, and these must have at least as much of a claim to *a priori* validity as the principle of the comprehensibility of nature if the philosophical argument is to do any physical work.

In moving from the claim that the forces within a system must depend on the positions of the points alone to the conclusion that they must be central, Helmholtz appealed implicitly to two such conditions: (1) what Olivier Darrigol has called "the principle of decomposition",[4] and (2) the principle of positional determinacy. According to (1), the force acting on a single point in a system is the (geometric) sum of the forces deriving from the other points in the system. Thus in order to characterise any one of these forces, we may ignore the positions and motions of the other masses in the system, and confine our attention to just these two points. Only by invoking this principle, can Helmholtz prove that the intensity of a force holding between two arbitrary points varies with *their* positions only. According to (2), neither directions nor distances are determinate unless they are referable to empirically given points. Thus only spatial properties defined with reference to the points involved in the system under consideration can be employed to characterise its motion. By invoking (2), Helmholtz is able to argue that "dependence on position" in the case of two points can mean only "dependence on distance".

As we shall see in the second section of this paper, Helmholtz's arguments throughout these introductory sections draw on those used by Kant in his *Metaphysical Foundations of Natural Science*. But it is not necessary to know the details in order to appreciate the sense in which the principles just mentioned resemble Kant's transcendental principles of natural science. The introductory section of the *Erhaltung* argues for the logical connection between a systemic principle—the complete comprehensibility of nature— and what would appear to be a contingent law of natural science, namely the conservation of energy. But one should not be misled. It is not a proof that force centrality must be, or probably is to be found in nature. It does not demonstrate the necessity of central forces, if one means by this either their logical or mathematical necessity, or indeed a metaphysical necessity deriving from rational first principles. There is no guarantee that nature *is* completely comprehensible, nor indeed should one conclude from such a proof that one has good reasons for *believing* that forces are, in fact, central. The point is purely methodological: *if* we approach nature with the intention of producing a complete set of invariant laws, *then* it follows that only certain kinds of laws are going to work, in the sense of being adequate to the task we have set ourselves.

Helmholtz purports to have shown that these laws will involve central forces, because any other sort of description will fail to realise our aims. But even if this is a methodological, and not a metaphysical thesis, it still involves a strong negative claim, namely that certain kinds of laws are non-starters. Furthermore, the proof involves more premisses than just the regulative principle. Lastly, the negative claim is just as little an empirical claim as the positive conclusion of centrality. And so the premisses involved cannot be merely empirical, for otherwise we should have drawn a conclusion concerning the possible form of scientific theories from propositions that are up for grabs. Helmholtz therefore attempts to argue for these principles by suggesting that they derive from conditions on the possibility of representing nature *at all*, and in this sense they can be called transcendental in Kant's sense of the term. For instance, according to Helmholtz, forces can depend only on the relative distances holding between points because (1) regulatively, we must seek to simplify forces by regarding them as characterised by regular functions, (2) constitutively, the distances that suggest themselves as candidates for the parameters of these functions must be possible experiences, i.e. spatially determinate magnitudes.

Helmholtz does not explicitly state the principle of decomposition in the *Erhaltung* (he does so only in his later reply of 1854), but he does, in effect, apply it when making the transition from the actions of forces in complex systems to the actions of the forces holding between pairs of their components. He sees this step as unproblematic because it apparently follows from his definition of a force: a force is the cause of a change in position, and such a change cannot be observed unless there is a second point relative to which the change takes place. Thus motive force is "to be defined as the striving of two masses to change their relative positions."[5] But the reasoning is fallacious: because a change in position can be observed only when there are "at least two points", Helmholtz concludes that all complex systems must be resolved into sums of two point systems, i.e. into cases in which there are *at most* two points. As we shall see in a moment, Helmholtz was forced to retrench drastically from this part of his analysis in 1882. But he did not step down from the principle of positional determinacy, so much as use it as the starting point for his later work on geometry.

By purportedly showing that the supplementary assumptions in his proofs derive from conditions on the intuitive representation of motions and forces, Helmholtz secures a connection between the comprehensibility of nature and the central force hypothesis that he derives from it. The latter hypothesis thereby inherits the status of a regulative principle: it could be frustrated by experience, but in such an event, no alternative, comprehensive theory of nature would be possible. Thus we must seek to formulate theories by means of central force laws. Given the structure of this argument, one might conclude that the errors in his deduction would have led Helmholtz to abandon his commitment to the central force hypothesis. But even in 1882, after he had acknowledged the formal errors and well after he had reinterpreted both of the transcendental principles as empirical propositions, he continues to argue that the central force hypothesis is epistemologically privileged. This fact is at first glance puzzling—however, Helmholtz's position is just as coherent, or just as contradictory, as his later views on geometry. For there as well, he argued for an interpretation of physical theory in which certain principles (the axioms of geometry) were *necessary* for the description of physical phenomena, even though the content of these principles was of inductive origin. The distinction between the regulative and constitutive premisses that I have used to describe the various premisses of the *Erhaltung*'s transcendental arguments suggests the direction in which Helmholtz retrenched: regulative principles point at methods of representation that are systemically preferable, we might say; whereas in saying that a principle is constitutive, I make a stronger claim. Constitutive principles are not *preferable* to their alternatives; rather, the alternatives are incoherent, they are not alternatives at all. But if one demotes a principle from constitutive to regulative status, one can continue to employ it as a premise in a transcendental argument. Even at that date in his career when Helmholtz presented himself as an ardent empiricist, he continues to ascribe these two principles a privileged role—they continue to be necessary for the construction of comprehensive physical theories. But they are no longer, for all that, necessary truths in the strict, constitutive sense of that term.

When the *Erhaltung* was reprinted in his 1882 collection of physical papers,[6] Helmholtz reinterpreted his early arguments along just these lines. He gave a detailed comment on the passage where he had argued that forces had to be resolved into the forces acting between point-masses if science was to be comprehensive and determinate. He now distinguished between two senses in which this claim might be true. It might mean that (1) the motion of every point in the system be determinable once the forces present in the system were known. And it could also mean that (2) the force acting on each point could be decomposed into the forces emanating from all the other points in the system. The first interpretation was indeed epistemologically necessary. For if there were individual points in a system whose motion was not determined by the forces acting within the system, then certain properties of the system would indeed fail to fall under general physical laws. The second interpretation is our principle of decomposition, whose importance Helmholtz had first isolated in his 1854 "Reply to Clausius", but which he had then viewed as unproblematic. This principle, Helmholtz now admitted, was empirical in origin and it represented "the real content of Newton's second axiom."[7] According to Helmholtz, it states that the acceleration undergone by a particle when several causes act at once is equal to the geometric sum of the accelerations it undergoes when these same causes act severally. In other words, that Newton's parallelogram of forces can be applied in reverse in order to resolve the accelerations onto components directed towards the other points in the system.

This way of stating the principle of decomposition makes clear its importance to the proof of central forces. For if one assumes—as both Helmholtz and Kant before him had done—that force is *by definition* the capacity of one body to alter its position relative to another, then this principle can easily seem apodictically true. In a three-point system, for example, the motion of a point *can* only be resolved onto those directions determined by the other two, for no other motions are observable. They cannot, as Kant would put it, be constructed in intuition. And so it may seem as though the principle of decomposition follows logically from the principle of positional determinacy and the definition of force. The error in this reasoning lies in assuming that the change in the position of the two points referred to in the definition of force *must* be a change in the distance along the straight line connecting them, as is the case in such a simple system. But so long as the system is complex enough to define other directional and metrical relations, this conclusion no longer follows. Thus, as Helmholtz is now forced to admit, the principle states an empirical truth. Nevertheless, he contends that all theories that do not assume it have also been found to contradict the principle of energy conservation, and indeed the law of equal action and reaction. And the same holds true of attempts to do without the assumption that forces are determinate "once the positions of the masses are completely given".[8] Finally, to suggest that forces might be "made dependent on an absolute motion, that is to say on an alteration of the relation of a mass to something that could never the object of a possible perception" is to abandon all hope of "a complete solution of natural scientific problems".[9]

In other words, Helmholtz does not conclude that his original analysis was wholly mistaken. On the contrary, he suggests that the conflict between his earlier philosophical views and the state of electrodynamic theory points to a fundamental tension. Implicitly referring to Weber's electrodynamic theory, he maintains that (1) theories that allow forces that are not central threaten both the *Eindeutigkeit* and the definiteness of electrical theory. And to assume that, (2) the intensity of a force depends on its velocity relative to absolute space is to abdicate the fundamental responsibility of physics, which is to comprehend nature completely. Although it is not logically incoherent to deny the necessity of central forces, to do so is to deny the validity of hypotheses which, although they are now admitted to be empirical, nevertheless remain conditions on the determinacy and completeness of natural science. And this step can only be taken, Helmholtz maintained, once it is clear that all other alternatives have been exhausted.

Both of Helmholtz's retractions in 1882 derived from criticisms of his book that had been raised early on by Clausius and Lipschitz.[10] Lipschitz's objection centred on Helmholtz's mathematical proof of the equivalence of force centrality with energy conservation. He showed that velocity- and acceleration-dependent forces were consistent with Helmholtz's definition of conservation, and thus that positional dependence did not follow without further assumptions. Clausius raised a number of objections to the text, of which only one is of immediate concern here, namely his contention that the forces acting on material points did not have to be central merely because they were position-dependent. Helmholtz had made this inference from positional dependence to centrality twice: first in the philosophical deduction outlined above, and again in his mathematical demonstration of the equivalence of centrality and conservation. In both proofs, he had employed both the principle of decomposition and the principle of positional determinacy. We have already examined Helmholtz's reasoning in the philosophical proof in some detail; however, in order to understand Clausius's objection and Helmholtz's response, it is worth examining the reasoning of the mathematical demonstration in greater detail.

1.2. The mathematical argument

This demonstration contained two parts, each of which proved one of the two implications involved in the equivalence of force centrality to *vis viva* conservation. Obviously it was not difficult to demonstrate that centrality implies conservation (that central forces are conservative), so we shall focus on the problematic implication, namely that centrality follows from *vis viva* conservation. Helmholtz's proof ran as follows. He first defines *vis viva* conservation as the proposition that "when an arbitrary number of masspoints move under the influence of only those forces that they exert on each other ... the sum of the living force of all of these is the same at all points in time in which they adopt the same relative position to one another."[11] This definition implies the existence of a potential,

$$(1) \qquad d(q^2) = \frac{d(q^2)}{dx}dx + \frac{d(q^2)}{dy}dy + \frac{d(q^2)}{dz}dz$$

where q is the tangential velocity of the mass-point relative to a system A, and where the coordinates are defined relative to that system. If u, v, and w are then the components of the motion along the axes x, y, and z, we obtain the following expressions for the component forces on the point:

$$X = m\frac{du}{dt}, \quad Y = m\frac{dv}{dt}, \quad Z = m\frac{dw}{dt}$$

from which it follows immediately that,

$$du = \frac{Xdt}{m}, \text{ etc.}$$

Now, since $q^2 = u^2 + v^2 + w^2$, thus $d(q^2) = 2udu + 2vdv + 2wdw$, and since,

$$u = \frac{dx}{dt}, \text{ etc.}$$

We can substitute for each term $2u\,du$ one of the form

$$\frac{2X}{m}dx, \text{ etc.}$$

in order to derive

(2) $$d(q^2) = \frac{2X}{m}dx + \frac{2Y}{m}dy + \frac{2X}{m}dx$$

Helmholtz thought that he could immediate derive from (1) and (2) that,

(3) $$\frac{d(q^2)}{dx} = \frac{2X}{m}, \quad \frac{d(q^2)}{y} = \frac{2Y}{m}, \quad \text{and} \quad \frac{d(q^2)}{z} = \frac{2Z}{m}$$

And then, on the assumption that q^2 is a function only of the coordinates x, y, and z, it follows that the component forces X, Y, and Z are also functions only of these coordinates, and thus by definition they are functions of the relative position of the mass-point to the system A. Nevertheless, in deriving (3), Helmholtz overlooked the possibility identified by Lipschitz, namely that the force along each axis could depend on the velocity and acceleration of the particle. But if we grant him that simplification, Helmholtz has shown that the force acting on each individual point is a function of the position of the point *relative to the system as a whole*.[12]

There then remains one last step to the proof, for the positional dependence of the force is not the same as centrality. To complete his proof, Helmholtz needed to get from this sense of positional dependence to the narrower, two-point case. Tacitly invoking the principle of decomposition, he restricted this result to the case of two material points. Since by hypothesis, the coordinate system being used is determined by the mass-points involved in the system, in the two-point case this coordinate system can be reduced to the single dimension determined by the two points. Thus, (1) the energetic state of the one particle is a function only of its distance from the second. It then follows trivially that (2) the positions in which the first point is in the same energetic state form concentric shells about the second point, and thus that the force acting on the first is directed toward the second. By (1) and (2) the force is a central force.

In his critical report on the *Erhaltung*, Clausius objected that this part of the proof simply begged the question. For Helmholtz had merely assumed that which needed to be demonstrated, namely that the intensity of the force holding between two points was a function of their distance. In a long reply to this and other objections of Clausius, Helmholtz defended his procedure on the following grounds.[13] There were two fundamental assumptions involved in his earlier proof. The first of these was that the forces acting on a given point in a system could be decomposed by means of geometrical addition, that is to say by employing Newton's parallelogram of forces. But this principle, he contended, was independent of Clausius's objection. Their main difference of opinion, Helmholtz went on, concerned the latter's supposition that we can coherently imagine forces holding between masses that are not functions of their relative position, that is to say of their distance. In the second part of his proof, Helmholtz argued, he had merely confined the more general result, namely that force was a function position alone, a result which Clausius accepted, to its minimal case. Under the assumption that it was only meaningful to speak of *relative* positions, in other words to make reference to coordinate systems that were empirically given, it followed that in the case of a two-point system, dependence on position was synonymous with dependence on distance. From this one could derive the requisite direction of the force, and therefore its centrality. Clausius's non-central forces, by contrast, could be defined only in terms of an absolute coordinate system. But this was absurd, because it meant supposing that they were determined by something which could never be an object of possible experience, namely absolute space.

In order to understand this emphasis on possible experience, one must recall Helmholtz's insistence on the completeness and determinacy of natural science. As I indicated previously, it was an essential tenet of the *Erhaltung der Kraft* that the ultimate causes in nature are unchanging, even though the forces we find in nature do quite obviously change in intensity. Nevertheless, this demand can still be satisfied if these changes in intensity are describable as law-like functions of spatial magnitudes. For in this case, the intensity of the force depends wholly on properties of a system that are independent of the time. Now, since space is undifferentiated, we can only speak of determinate spatial magnitudes when there are empirical points demarcating intervals in space. Thus, in the case of an isolated system, the forces will have to vary in accordance with the magnitudes determined by the relative positions of the points in the system. In other words, changes in the relative positions of mass-points in space are not only the experiential *consequences* of the actions of forces. They are also the only possible *determinants* of changes in the forces' intensities.[14] The problem with the deviant forces that Clausius imagines is that they would have to depend on features of reality that are not possible experiences. They are, in consequence, experientially indeterminate: if the intensities of forces so defined change with time, their changes depend on factors that cannot be empirically identified. Claiming that they are functionally dependent on changes relative to an absolute co-ordinate system is thus an empty gesture. For in admitting non-experiential determinations of physical values, we would in effect be admitting transcendent elements in our theory.

Nevertheless, as I have already suggested, Helmholtz's reply to Clausius cleaves the problem at the wrong joint. Suppose we allow him proposition (3) from his proof, which states that the energetic state of a particle depends only on its position. And suppose we agree that references to absolute coordinate systems are to be disallowed because they are transcendent. Does the proof then go through? No, because the application of the principle of decomposition effectively eliminates determinate spatial relations which *could* be used to define non-central forces. That Helmholtz apparently failed to see this, while devoting several pages of his "Reply" to a deduction of the mathematical consequences of his fallacious philosophical argument provides ample evidence of what he admitted in 1882, namely that the philosophical portions of the *Erhaltung* are "influenced by epistemological views of Kant that are stronger that what I would be prepared to accept today."[15]

Helmholtz expands on this statement by explaining how his views on causality and matter have changed since his early years; however, as I shall show in the following section, the connections to Kant's philosophy of physics run deeper than this, and they include the ongoing confusion concerning the epistemological status of the principle of decomposition. Helmholtz, like Kant, acts as if the latter principle follows directly from the principle of positional determinacy, for he repeatedly suggests that centrality is entailed by the requirement that all basic magnitudes be empirically determinate. Now, as we have just seen, *if* the principle of decomposition holds, positional determinacy entails centrality. And decomposition itself might appear to follow from Kant's definition of force as a relation between two mass-points and, once again, positional determinacy. For it might seem that, *if* all the properties of a force must be characterisable with reference only to the properties of a *pair* of points, then the force can only change as a function of their distance. This means supposing that the positions of other points in a system *a fortiori* do not contribute to the properties of the point-pair. And that might in turn appear to be nothing more than a special case of the relativity of spatial relations. That is to say, since it is illegitimate to refer to absolute space in characterising the position of two points, it might also seem to be illegitimate to refer to any other points but the two under consideration, for these are irrelevant to characterising their position *relative to one another*. But this reasoning, while seductive, is circular. For the position of a pair of points relative to some third point can very well be seen as a property of that pair, and thus as something that might determine a force that nevertheless acts only on the pair. If one denies this, then one has simply postulated that their distance is the only property they have. In sum, if decomposition holds, then forces determine, and are determined by, the distances of pairs of points; whereas assuming that only these distances are relevant to characterising the actions and variation of forces is in effect to assume decompositionality.

2. THE KÖNIGSBERG CONNECTION

I have suggested that Helmholtz's reply to Clausius, and indeed his insistence on the epistemological priority of positional relations, should be read as a transcendental argument. Clausius's "arbitrary" forces are ruled out because they can be given only a hypothetical, "mathematical" definition, whereas Helmholtz will allow only forces agreeing to an epistemological demand: Not only must their effects be determinate, but they themselves should be *determined*, meaning that their intensities must depend only on spatial relations which are themselves, in turn, determinate. Helmholtz's arguments therefore contain two distinct appeals to the notion of logical determination that are easily confused. The first is the more straightforward, "downward" demand that the extension of a concept be determinate. It should be defined precisely enough that we may say of each object whether or not the concept applies to it. For instance, one might argue that a precise definition of the concept of force will also have to contain some reference to an inertial frame, because without this specification, the concept of an accelerated motion, and thus the concept of the cause of an acceleration is indeterminate. But Helmholtz also appeals to logical determination in a second, "upward" sense: each concept must be *determined*, in that it should be seen as a specific case of a higher-order concept. The specific forces that causes changes in motion should be seen as positional determinations of a single force; furthermore, this single force should derive from a material point, and it should itself be a member of the family of "basic forces" that define the various kinds of matter.

In my previous discussion, I have distinguished loosely between these two aspects of Helmholtz's arguments by using the Kantian distinction between "regulative" and "constitutive" principles. In this section, I will explain in somewhat greater detail how this distinction applies within Kant's philosophy of natural science. Furthermore, I shall show how Kant, too, attempts to derive consequences concerning the centrality of force and Newton's parallelogram law by appealing to such constitutive and regulative principles. I take it that Helmholtz was borrowing these arguments more or less directly from Kant, and that this also explains why he inherits a number of the specific flaws in Kant's reasoning, most notably the circular justification of the principle of decomposition just mentioned. But I shall not seek to establish historiographic links between Kant's text and that of Helmholtz, for the systematic congruence of the arguments is strong enough that an analysis of Kant's reasoning is of immediate utility in interpreting Helmholtz's.

2.1. Regulative and constitutive principles

According to the first *Critique*, both regulative and constitutive principles derive from the categories, that is to say from pure logical operations of the understanding. The categories form the building blocks of logical reasoning not because they correspond to basic metaphysical properties (which was Aristotle's view), but simply because they are the root concepts employed by the understanding to subsume intuitions. But this presents a well-known problem. How can one claim that the categories necessarily apply to all possible intuitions while maintaining that they are distinct of these intuitions? The categories could apply to objects other than those given in human intuition. And the mere fact that all human intuitions are spatio-temporal magnitudes does not logically entail anything about, say, their causal interrelation. Kant's answer, which we shall not analyse further here, is contained in the transcendental deduction, which purports to show that the manifold of intuition can be unified in conscious experience only by subsumption under the categories. Conversely, the categories must have what Kant terms "schemata" if they are to have any empirical application at all. They must be, as it were, projected onto the structure of pure intuition if they are to have an empirical extension. The schematisation of the categories yields what Kant calls the "pure principles of the understanding": (1) the axioms of intuition (principles concerning the magnitudinal structure of space and time), (2) the anticipations of perception (which concern the magnitudinal structure of possible sensations), (3) the analogies of experience (basic laws concerning the substantial and causal structure of reality), and (4) the postulates of empirical thought (modal principles concerning possibility, actuality, and necessity).

Kant terms the axioms of intuition and the anticipations of perception "mathematical" principles, for they reflect the possibility of applying mathematics to experience. The analogies of experience are "dynamical", because they concern the substantial and causal nature of possible objects. The "mathematical" principles are *constitutive* of experience, in that the truths they express are strictly and necessarily true of every appearance. The "dynamical" principles, on the other hand, express *rules* for organising experience, for instance, that one must correlate individual appearances by means of necessary connections. Thus the latter are "regulative" and not constitutive. Nevertheless, both the mathematical and dynamical principles reflect the conditions under which experience can be unified in consciousness, and as such they also express "general laws of nature".[16] The critical epistemology thus contains the germ of natural science in these principles. And it is they that Kant extends in the *Metaphysical Foundations of Natural Science* to produce a "pure empirical" metaphysics of nature.

As is so often the case in Kant, the distinction between regulative and constitutive principles drawn *within* the principles of the understanding is also applied on a larger scale. Kant calls the "transcendental ideas" regulative principles in order to distinguish them from other "constitutive" principles "from which, strictly speaking, the truth of the general rule ... follows" (B675).[17] Generally, when one speaks of a "regulative" demand, one means it in this larger sense. These transcendental, or "cosmological" ideas are also derived from the categories, but they are the results of the operation of Reason, as opposed to the Understanding. For instance, the faculty of reason produces the idea of an infinite causal chain, or the idea of the totality of existence, by recursively invoking principles that are in themselves unproblematic: from "Each event has a cause", Reason derives the undecidable proposition "Reality consists of an infinity of causes and effects".

In their illegitimate, non-regulative employment, such transcendental ideas correspond to rules which are not "strictly speaking" true. Treating them as if they were true leads us into the various "antinomies" of reason, for instance into asserting both that the world must have a first cause, and that there can be no first cause. Taken regulatively, however, they perform a valuable role, for they direct the understanding to systematise its rules hierarchically, and thus to construct a unified totality of natural laws. Such a system of laws remains an ideal we aim at, for it cannot actually be completed. But if it were completed, it would amount to a complete *determination* of nature, for every phenomenon and every concept would be regarded as a determinate instance of a higher-level concept. Regulative principles can thus be seen from two points of view: as illegitimate statements concerning the totality of the natural world, or as methodological, meta-theoretical principles concerning the organisation of theories. Only in the latter sense can they be taken to be valid rules for thought.

Since our purpose here is to understand Kant's philosophy of *science*, I will not pursue these aspects of his epistemology further. We need to retain only two essential points: (1) the distinction between *a priori* principles that are constitutive (because they express an essential cognitive link between the categories and the structure of intuition) and those which are regulative (because they mandate the hierarchical organisation of empirical concepts and laws). And, more straightforwardly, (2) the claim that because these principles derive from conditions on our possible thought about the natural world, they are simultaneously the most fundamental principles of natural science. In the *Metaphysical Foundations of Natural Science* Kant makes good on the promise of (2) by extending the principles of (1) to cover the domain of physical science. In order to achieve this, Kant needs to supplement the pure principles of experience with an empirical content. For the principles enumerated in the *Critique* are *a priori* schemata of the categories, and they do not involve reference to anything beyond the *structure* of possible experience. Thus they form what Kant calls the transcendental part of the metaphysics of nature.[18] Once one specifies the species of natural object with which one is concerned, and thus the species of empirical concept, one gets either "pure empirical" physics or pure empirical psychology. Kant, and we, shall be concerned only with the former. Here, the empirical concept in question is that of *matter*.

Because the transcendental metaphysics of nature is conditioned by both regulative and constitutive principles, the concepts involved in the pure physics that Kant develops in the *Metaphysical Foundations* are similarly constrained. The constitutive, mathematical part of the task consists in providing empirical schemata to a series of material concepts. In the four main sections of the work (the Phoronomy, Dynamics, Mechanics, and Phenomenology) the concept of matter is successively "determined" with respect to the table of categories. This four-fold division cleaves closely to the analysis of the pure principles of experience. Matter is successively defined as follows:

(1) in Phoronomy, as the movable in space (corresponding to the axioms of intuition, i.e. the theory of extensive magnitudes),

(2) in Dynamics, as something which fills a space and resists motion (corresponding to the anticipations of perception, i.e. the theory of sensations as intensive magnitudes),

(3) in Mechanics, as something that causes motion in another substance (corresponding to the analogies of experience, i.e. the "dynamical" principles of deriving from causality and community), and

(4) in Phenomenology, as an object of possible experience (corresponding to the postulates of empirical thought, i.e. the modal determination of empirical propositions).

This entire enterprise takes place under additional systemic constraints, which Kant outlines in the introduction to the *MFNS*. For instance, the metaphysics of nature that results should not only extend the pure principles of experience, but it should also produce a *unified system of nature*. It is, in other words, directed by regulative considerations deriving from the transcendental or cosmological ideas. In order to exemplify this interaction between the regulative and constitutive parts of the project, we may take as our stalking horse the concept of force. This concept is indeed the very one that Kant himself chooses in the *Critique* to illustrate what he means by the regulative application of a cosmological idea; furthermore, in choosing it as our example, we will connect our discussion of Kant directly to the arguments of Helmholtz that we are considering.

In this discussion ("On the Regulative Use of the Ideas of Pure Reason"), Kant defines a force as "the causality of a substance" (B677), pointing out that this definition puts no restriction on the number of forces that we might find in nature. Its extension is as wide as the number of sets of appearances that exhibit a regularity, and which we therefore subsume under the categorical relation of cause and effect. But the regulative idea that there is an absolute and complete dependence among changes in appearance (to paraphrase slightly the fourth of Kant's cosmological ideas) leads us to try to organise these various forces as species of more basic forces, and ultimately of what Kant calls a "*Grundkraft*". This is the "upward" determinacy requirement: not only must the concept of a specific (*e.g.* chemical) force determine the phenomena it subsumes, but it should also be seen as the determination of a more basic force. In the case of Helmholtz's arguments, this principle is invoked to justify the claim that all forces observed in nature must be seen as determinations of a set of basic forces that characterise the various species of matter. It also finds specific mathematical employment in the argument that force intensity must depend on position.

Nevertheless, a body of knowledge that satisfies the regulative demand to systematise need not qualify as a science in the strict sense. Biological taxonomies, for instance, organise concepts in determinate hierarchies, but according to Kant, they lack a *constitutive* core. And without such a core, they can never establish apodictic relations among their principles. In general, natural motion (the object of physics, in Aristotle's wide sense of the term) can only be described by strict apodictic laws if these motions are quantifiable. Thus Kant argues in the introduction to the *MFNS* that a natural science is "proper science" [*eigentliche Wissenschaft*] only to the extent that it is mathematical. The concepts of a proper science must accordingly be given a precise extension by schematising them on the magnitudinal structure of intuition. But even in mathematical physics, this task is not easily accomplished. Key physical concepts, above all that of force, are not among the strictly constitutive principles of the understanding. The concept of force belongs to those which are "dynamical" and "regulative" in that they enjoin us to correlate experiences by means of necessary connections. Whereas properly mathematical concepts are distinguished by our ability to "construct" their corresponding intuitions *a priori*.

This demand can in fact be met in "Phoronomy", that part of the pure metaphysics of corporeal nature that is concerned with purely kinematic properties of matter. Constructing the motion of an idealised material point differs from a geometric construction in only two respects: we suppose an absolutely general "something" which is moved, and we imagine the motion taking place in a single interval of time. This empirical extension of the axioms of intuition can indeed be carried out with mathematical precision. But it is not immediately evident how the other principles of the understanding (the "general laws of nature" referred to in the *Prolegomena*) are to be mathematised in this fashion, for the notion of a cause is not inherently mathematical, as Kant himself insists. Thus a major goal of the *MFNS* is to explain how the pure empirical schemata of the so-called "dynamical" categories can be constructed in intuition. This is achieved, as we shall see in greater detail in a moment, by schematising these categories on kinematic appearances, whose concepts are themselves schematised on geometric intuitions. This corresponds to the "downward" determination of these concepts: they must be given an extension composed of possible intuitions, each of which is also fully determinate. In the case of Helmholtz's arguments, this demand is reflected in the principle of positional determinacy, which requires that the ultimate spatial referents of motive concepts be determinate.

Thus on Kant's analysis, the concept of motive force is squeezed between the upward and downward, that is to say the regulative and constitutive determinacy requirements. Forces, as causes of change, must be related determinately to lower-level kinematic concepts, such as the speed and the path of a material particle. At the same time, the forces thus defined are subject to regulative pressure from above. Because the science of nature must be a unified system of laws, all forces must be seen as special cases—determinations—of higher-level laws. Both Kant and Helmholtz believe that one can draw specific consequences concerning the kinds of forces that are possible in nature from these dual requirements. Because the ultimate referents of the motive concepts must be determinate spatial magnitudes, the motions that are caused by forces must all be *relative* motions. For every motion of a particle, in order to be determinate, must be relative to some other, empirically given particle. And because the differentia of high-level, basic forces [*Grundkräfte*] must also be properties of the system in which they act, these forces can only change as functions of the same, determinate spatial magnitudes.

As a result of these considerations, both the philosopher and the physicist believe that the centrality of the basic forces is transcendentally required. Their reasoning is clearest in the one case where it is also valid, namely the two-point system: here motion can only take place along a single dimension, and the same holds true *a fortiori* of acceleration. Thus the force can only be "constructed" as acting along the line connecting the points. That is the downward determinacy requirement. Furthermore, changes in the magnitude of the force can only be a function of the distance separating the points (the upward requirement). Thus the force must be central. Now, as we have seen already, in order to generalise this argument to complex systems, one must assume that the principle of decomposition is *a priori* true. That would mean showing that the motions of the points in a complex system *can only be described* with reference to the spatial magnitudes determined by the various pairs of points making up the system. And such a claim can indeed be defended for systems with up to four points (in three dimensions). But it is evidently not true in general.

The mathematical construction of motions and forces that Kant undertakes in the *MFNS* is flawed in just this last respect: Kant thinks that the centrality of force is logically entailed by his various determinacy requirements in part because he draws too general conclusions from the relativity of motion. These arguments are essentially repeated by Helmholtz in the *Erhaltung*. Before examining them in greater detail, however, I should emphasise that my analysis is skewed towards that aspect of the *MFNS* which concerns us, namely the link between central forces and the principle of decomposition. Of course, Kant's principal aim is not to establish such a link, nor does he accord either of these notions an extended treatment. On the contrary, he argues for them in passing while pursuing bigger game. He purports to show that Newton's parallelogram law of force composition is an apodictic principle. To imagine a composite force is to imagine, by definition, (at least) two point-sources acting on a third. The law of equal action and reaction, which Kant derives from the pure principles, requires that we interpret this interaction by means of a centre of mass construction. That the composite force equals the sum of the component forces is then a necessary consequence of the method of construction, and not an empirical proposition. In consequence, the mathematical schematisation of the concept of causality will be a principle of pure science, and not an inductive generalisation.

2.2. *The parallelogram law*

In order to understand the central importance of the parallelogram law in Kant's project, one must keep in mind his aim of providing an apodictic core

to physical science. The central difficulty is that the "dynamical" categories—
force among them—are not inherently quantitative. Providing *a priori* definitions of the additive relations between such dynamical concepts is, as a
result, of critical importance. For without such definitions, the apodictic
mathematical core of physics will not have been secured, and it will therefore fail to meet the requirements on "proper science" that Kant laid down in
the introduction to the book. In the following, I will briefly describe Kant's
analysis of the parallelogram law through the four sections main sections of
the *MFNS*, contrasting his approach to that of Newton. Although my treatment is selective, I follow Kant's lead, for he begins his book by calling
Newton's proof into question, and he returns to it in conclusion in order to
illustrate the importance of his mechanical laws in the metaphysics of nature.

The first section of the *MFNS*, entitled "Phoronomy", is concerned with
the construction of what we would call kinematic concepts. The concept
of matter to be constructed is that of a movable point in space, and Kant
sets himself the task of defining the additive relations that hold between motions. Since, for Kant, all mathematics derives its synthetic *a priori* necessity from the structure of space and time, the two essential properties of
phoronomic motion, namely speed and direction, must be defined in terms
of spatio-temporal magnitudes. He approaches the problem using the theory
of magnitudes developed in the *Critique* in the "Axioms of Intuition" and
the "Anticipations of Perception", where he explains that all appearances
are simultaneously *intensive* and *extensive* magnitudes. Because they are
spatio-temporal, they have a magnitudinal structure deriving from the pure
intuitions: they are *extended* in space and in time. And because they have
a specific sensory content (a colour, a degree of hardness, etc.) they also
have a particular *intensity*. Extensive and intensive magnitudes differ with
regard their additive properties. Because the representation of an extensive
magnitude entails the representation of its parts, it is an analytic truth that the
whole contains its parts. Conversely, the addition of the parts produces the
whole with synthetic *a priori* necessity. Thus an additive proposition whose
terms refer to extensive magnitudes is a synthetic *a priori* proposition. The
additivity of intensive magnitudes, by contrast, requires a further specification of the addition operation, for an intensive magnitude does not literally
contain lesser intensive magnitudes as its parts. For instance, colours are
intensive magnitudes, in that they can be ordered in a sequence of intensity.
But in representing two shades of a given colour, we do not to thereby produce the colour which is their sum. That two colours add to form a third
is not (yet) a synthetic *a priori* truth. Once an additive procedure has been
defined, however, it may be one.[19]

In the "Phoronomy" Kant characterises speed and direction as intensive magnitudes not because he considers them to be sensations like colour, but because they do not literally contain lesser speeds and directions as their proper parts. In order to mathematise these concepts adequately, we need to provide them with definitions that secure determinate additivity relations. And here we are aided by a peculiarity of the phoronomic magnitudes. In contrast to sensations, they have an implicit connection to the extensive magnitudes of space and time: a speed can be represented in intuition by means of the distance that a material point covers in a unit of time; different directions can be represented by means of distinct paths. But these extensive constructions of the intensive concepts of speed also reveal a difficulty. To say that speeds and directions can be added is to say that the same material point, at the same time, has multiple speeds and directions. And if these are to be constructed as distinct line segments, we are immediately faced with a contradiction: two line segments, precisely because they are *distinct* extensive magnitudes, are not contained either in one another, or in a third which would correspond to their sum. How can we meaningfully say of a single particle that its motion is the sum of two distinct (and non-collinear) motions?[20] On the one hand, if the propositions of kinematics are to be apodictically true, they must express geometric truths. But the strict construction of individual motions as line segments produces geometrical magnitudes that do not embody the requisite additive characteristics. Thus we are at an impasse:

> Geometrical *construction* requires that one magnitude, or two
> magnitudes in their conjunction be *identical* with another, and
> not that they produce the third as causes, which would be a
> mechanical construction. Complete similarity and equality,
> insofar as it can be cognised in intuition, is *congruence*. All
> geometrical construction of complete identity rests on congru-
> ence. This congruence of two conjoined motions with a third
> ... can never take place if each of them is imagined in the
> same space, e.g. [the same] relative space.[21]

Kant overcomes this dilemma by appealing to the relativity of space: we can make sense of the idea that two distinct motions are parts of a third if we imagine each of the two motions as relative to a distinct frame of reference. This "construction" of the addition of motions has far-reaching consequences. The addition of two velocities requires two reference-frames for its construction. Since these must be empirically given, a purely kinematic description of velocity addition requires at least three empirical points: the first is conceived as moving relative to the second, and this pair is then represented as moving with respect to the third.

Standing on its own, both the problem and its solution may well strike the reader as tendentious. Thus it may be useful to recall the general direction of Kant's arguments. The complete determinacy of natural science requires that all the concepts employed in physics have a determinate content. This means that we must provide pure empirical "constructions" of these concepts, which will specify precisely the possible intuitions to which they apply. Higher-level "dynamical" physical concepts, such as that of force, must therefore be tied to the lower-level "mathematical" concepts, such as motion and distance, whose changes are supposedly determined by the forces. But this requires that we specify the additive relations holding among these lower-level concepts. The *definienda* must be apodictically determined if their *definiens* is to be so as well. For instance, the additivity of force presupposes the additivity of velocities. And the latter presupposes the additivity of distances. But, Kant is arguing in the Phoronomy, these *cannot* be added in the strict sense of part-whole containment, because two distinct distances are, by definition, not coextensive. Kant's conclusion is that the addition of distances, and thus by extension, the addition of motions, instantaneous changes in motions, and, finally, forces, presupposes the specification of frames of reference relative to which the motions can be "constructed". Only in this case will we have successively tied the dynamic concepts to the extensive magnitudes that ground mathematics.

Furthermore, Kant has a specific target in mind here. Both in the "Phoronomy" and in the third section of the *MFNS*, the "Mechanics" (our dynamics) Kant contrasts his construction to alternative proofs of the composition of motions. He evidently has Newton's proof of the parallelogram law in the *Principia* in mind. Such demonstrations, Kant claims, generate the empirical content of the concept of composite forces "mechanically", but they do not *construct* it mathematically.[22] In fact, Kant is simply mistaken in this, for Newton did indeed formulate a proof for the composition of velocities that would correspond to Kant's phoronomic construction in his "Tract of October 1666",[23] and he is a good bit clearer than Kant in distinguishing between the phoronomic (kinematic) and mechanical (dynamical) parallelogram laws. Newton's early kinematic proof uses his method of fluxions (the differential calculus) to show that every rectilinear motion is simultaneously a motion in any other arbitrary direction, where the magnitude of this second motion is the product of the magnitude of the first with the cosine of the angle they contain.[24] From this relation, he can derive the result that the two sides of a parallelogram correspond to the component motions of the diagonal motion. This result is assumed without mention in the *Principia* proof of force composition, which is what leads Kant to think that Newton has overlooked the need for the kinematic proof.

For Newton just as well as for Kant, the dynamic parallelogram law pre-supposes a proof of the kinematic one. The latter can be provided either by differentiation within a single frame, so that one follows Newton in allowing that the same point can have multiple (instantaneous) motions at the same time. Or one denies this possibility, arguing with Kant that the motions must be relative to distinct frames, each of which must be determined by at least one empirical point. Since Kant doesn't know of Newton's kinematic proof, he falsely thinks they disagree on the very need for one. But the crux of the matter concerns the empirical determinacy of the motions being added. Newton does not *require* that motions be directed towards empirical points (though of course his demonstration is compatible with this interpretation), whereas Kant does. In so doing, Kant lays the ground for the claim that forces cannot have absolute directions.

Indeed this phoronomic analysis is intended to feed directly into the def-inition of force that Kant provides in the next section of the *MFNS*, the "Dy-namics". Here, a force is defined as the capacity of one body to "resist the approach", or "to cause others to move away from it".[25] Finally, in the "Mechanics", force is characterised as the capacity of a body to change the motion of another through its own motion.[26] Kant introduces a law of equal action and reaction, which in turn permits us to transform the kinematic def-inition of the composition of motions into a dynamical one (in Kant's ter-minology, a phoronomic into a mechanical one). According to this a priori "Law of Mechanics", the dynamical interaction of two bodies entails the motion of *both* of these with respect to that reference frame in which mo-mentum is conserved, namely that determined by the centre of mass of the bodies. This frame of reference, Kant explains in the fourth section of the *MFNS*, the "Phenomenology", can effectively stand in for Newton's abso-lute space: it provides an empirically determinate space relative to which the motions of bodies are themselves fully determinate. So the successive introduction of constitutive and regulative laws, each of which results from applying one of the principles of pure experience to the concept of matter, determines the concepts of matter, motion and force completely.[27]

That forces are central is assumed by Kant almost *en passant*. A force obtains when one body affects the motion of another. Since in Kant's pure metaphysics of nature "we regard each of these only as a point" it follows that the motion the one body causes in the other, must be seen as taking place along the line connecting them. But there are only two possible motions along this straight line: one in which these points *move apart*, and one in which they *approach* each other. The force which is the cause of the first [sort of] motion is called a *repulsive force*, and that of the second is caused an *attractive force*. Thus we can conceive only of these two sorts of forces as those to which all motive forces in nature must be reduced.[28]

This reasoning is the same argument from the principle of positional determinacy that we saw in Helmholtz: because two points determine only one spatial magnitude, they can change their relation only as this single magnitude changes. The argument can in fact be extended to systems of several points, so long as their number does not exceed the dimensionality of the space in question by more than one. But it does not hold generally.[29] The moment we consider more complex systems, the possibility of reducing the interactions among the points to interactions of pairs is no longer apodictically given. And this means quite simply that the principle of decomposition is not *necessarily* true, in that it does not follow from the possibility of constructing, in Kant's sense, the motions of mass-points in space. But since Kant relies on this principle just as much as Helmholtz in order to argue for the centrality of force, it follows that non-central forces are conceivable. The centrality of force is neither regulatively nor a constitutively required, thus it can only be an empirical postulate.

Nevertheless, so long as one restricts oneself to the simple cases Kant discusses in the *MFNS*, it can indeed appear that forces *must* be conceived as acting along the lines connecting pairs of masses. From here, it is a short step to the conclusion that in order to "construct" the *dynamic* parallelogram law, we must imagine at least three points: that which is subject to the two forces, and the two which determine the directions along which the forces are taken to be acting. For, according to Kant, we are supposing that without these supplementary points, it would be meaningless to speak of the component forces at all. Finally, we can apply the results of the Phoronomy, where Kant demanded that resultant motions be strictly "congruent" to their components. This must hold in the dynamic case as well, meaning that the motions of the particle relative to the two points that determine the forced motion sum to form a single motion that is *identical* to these. That is to say, all three motions (the component motions, which are relative motions of the pairs of points, and the resultant motion) must be conceived as relative to a further space. This demand is satisfied just in the case of a centre of mass construction, for here the motions of a point relative to the centre is indeed the geometric sum of its motions relative to the other points in the strict sense demanded by Kant in the Phoronomy: they are one and the same motion described with regard to distinct frames of reference.

Kant therefore takes his analysis to show that the truth of the parallelogram law follows from the very possibility of constructing accelerated motions in intuition. We tend to see things the other way round. In a conservative system where all forces are central, the action of a particle P relative to the centre of mass can be arbitrarily decomposed into the geometric sum of its action relative the centre of mass of any arbitrary subsets of masses in the system. All the actions sum, and can be decomposed, geometrically. But this says only that the parallelogram law of forces, when supplemented with the assumption that forces are central, will refer the total force acting on a point to the individual forces centred on the other points. And then, since momentum conservation is assumed, both the individual and the composite forces will determine actions relative to the respective centres of mass which sum as do the forces—in accordance with the parallelogram law. Kant turns this deduction on its head. He too assumes conservation. He claims that in order to "construct" a composite motion, we must assume the existence of point-masses relative to which each component is determined. The phoronomic parallelogram law then demands that the total motion be referred to a frame of reference *in which it is true* that the composite motion is the geometric sum of the components. Thus actions must sum geometrically, and forces, which are the causes of the relative displacement of pairs, do so as well.

This all assumes, however, that the frame of reference whose existence is postulated in the last step must exist. Furthermore, as I have already pointed out, it is not in general the case that the action of a point can always be decomposed into independent linear actions relative to the other points in the system. And these two demands, taken together, give a clear indication of what Kant wants, but cannot have. One way of defining an inertial frame is as a frame in which every acceleration corresponds to an impressed force.[30] If it were the case that each motion of a point, in order to be described at all, *had* to be described as the sum of its motions relative to the other individual points in the system, then the decomposition of actions would be apodictically true. Every observable acceleration would be in the direction of an empirical point. And if it were also true that this decomposition *could* always be carried out in such a way that the actions were independent of one another, then the required inertial frame could always be constructed—it would be one whose motion was equal and opposite to the constructed sum (if, on the other hand, they were not completely independent, it could be the case that the required frame could not be constructed). If these propositions were, taken together, always satisfiable, we would need neither a principle of inertia, nor would we need Newton's second law—and indeed both of these are conspicuously missing from Kant's laws of motion. Furthermore, it would follow that the constructable predicates of a system (if we include mass among these, which Kant does seem to allow) were also sufficient to permit a complete determination of its motive properties, in that determinate motions relative to the inertial frame, and the additivity of forces and motions in the strict sense demanded by Kant, would be satisfied.

In fact, to get all this one needs to introduce just those principles that Kant seeks to circumvent as explicit hypotheses: that all forces are central, that the parallelogram law of forces (or, more generally, Newton's second law) is valid, and that the law of inertia holds. Only then does one get the sort of empirical determinacy that Helmholtz, as late as 1882, requires from a regulatively complete system of physical laws, namely the "principle that the forces that two masses exert on one another are necessarily determinate when the positions of the masses are completely given."[31]

3. HELMHOLTZ'S LATER CRITICISMS OF HIS DETERMINACY ARGUMENT

As my principle concern in this paper is Helmholtz's application of these Kantian arguments, I will not pursue this interpretation of Kant's text further here. But before we pick up the thread of Helmholtz's arguments, we would do well to summarise the results of our treatment of the *Metaphysical Foundations*. We might characterise the line of argumentation that I have extracted here as aiming at that single difficulty in Newton's theory which Kant first raises in the Phoronomy. Newton must introduce the notion of independently existing forces in absolute space in order to prove certain propositions concerning their interaction. Whereas Kant wishes to eliminate both the notion of independently existing forces and that of absolute space. Because he holds the notion of absolute space to be experientially transcendent, he denies that there can be forces whose properties are determined with respect to it. But there can be no doubt but that we require the notion of directed causes in order to organise experience by means of universal laws. Thus the class of admissible forces is "squeezed" between regulative and constitutive demands. The magnitudes and directions of the forces must be dependent on constitutive mathematical relations among empirical givens. For if they were not, the regulative requirement that nature be completely determinate would draw a blank: it would end up asserting that the forces were determined by magnitudes that were not constructable in intuition.

These constructable intuitions are just those magnitudes that are determined by ideal material points in space. Their relative positions must provide us with the basic resources required to reconstruct the notion of directed forces, as well as the laws of composition that apply to such forces. This attempt, with all its twists and turns, results in Kant's claiming both that forces must be central, and that motion is empirically determinate relative to the centre of mass of a system. The motions caused by the forces then satisfy the requirement of strict congruence imposed by Kant's parallelogram construction of motion in the Phoronomy. If Kant's analysis had been valid, it would have meant that Newton's second law was not a mere inductive generalisation, but a pure empirical law. The parallelogram law of force composition would not express an observed regularity among independent forces in absolute space, namely that they produce the same total effect when acting simultaneously as they do when acting in succession. It would say something quite different: Every system of bodies determines a frame of reference relative to which the acceleration of each individual body is the geometric sum of its accelerations relative to the reference frames determined pairwise by the other points in the system. This proposition is a regulative demand: if it were not true, then we would have admitted the existence of accelerations, and thus of forces, that could not be determinately described by means of rules. That does not mean that their effects could not be described *at all* (thus the demand is only regulative), but that if there were such effects, their causes could not be subsumed under higher laws.

Now, as I suggested in the first section, Helmholtz's reasoning is close to Kant's not only in its general aims, but also in the details. Common to both men is the belief that *a priori* constraints on the intuitive construction of force disqualify "absolute" forces—forces whose magnitudes and directions are, as it were, anchored in absolute space. Helmholtz thought that he could employ such arguments in order to invalidate competing electrodynamic theories: if a theory made appeal to forces that were not experientially determinate, then it could be rejected. To be experientially determinate, a force would have to be central, for all other options amounted to reifying or absolutising forces. In Kantian terminology, both Kant and Helmholtz object to definitions (empirical schemata) of force that involve reference to absolute space because they invoke features of reality that are *in principle* unobservable.

Helmholtz, as we saw, employed two principles in his derivation of centrality from positional dependence: the principle of decomposition, and the principle of positional determinacy. Only the second of these is explicitly

stated in the *Erhaltung* itself. The principle of decomposition is first identified in the 1854 reply to Clausius, and Helmholtz only admits that it is an
empirical proposition in his 1882 reappraisal of the monograph. Nevertheless, in both cases, Helmholtz makes it clear that both principles play a special epistemological role in the physical sciences. For, (1) The principle of
positional determinacy stipulates that the terminal magnitudes employed in
physical theory must be defined in terms of materially determinate, and not
merely mathematical quantities. (2) The principle of decomposition, even
if imperfectly articulated by Helmholtz, says that complex forced motions
must be represented as the sum of simpler, independent ones. Thus it argues that in order to for such complex motions to be fully determinate, they
must be resolved onto motions that are determinate in the sense demanded
by (1). In fact, as has already emerged in our discussion of Kant, this claim
cannot be sustained in its strongest form. For it holds necessarily only in
those few cases where the number of points in a system does not determine
more magnitudes than are required to span the space. Once we are dealing with a complex system, we have enough empirically determinate spatial
relations available to describe motions without it following that they are de-
composable in the required sense. But Helmholtz needed this principle in
order to get centrality out of positional determinacy, for without it, he could
not reduce such complex systems to component pairs.

   This difficulty was precisely that which Clausius homed in on, even
though Helmholtz did not at first grasp the full import of his opponent's attack. Clausius's critique, we may recall, was directed at Helmholtz's mathematical derivation of the equivalence of force centrality to his principle of *vis
viva* conservation. The tricky bit was to prove one of the two implications
making up the equivalence, namely,

   conservation of *vis viva* $\Rightarrow$ postulate of central forces,

   To prove this proposition in the *Erhaltung*, Helmholtz applied the principle of decomposition (without explicitly flagging this step) to reduce the
case of a complex system to a conjunction of two-point systems. Invoking
the principle of positional determinacy, he claimed that the magnitude of the
force could depend only on the distance between the two points. He then
proved that the force also had to be directed along the line connecting the
points. Clausius then objected that this was a *petitio*: Helmholtz assumed
one half of centrality (dependence on distance) in order to prove the other
(directionality). Now, as we saw earlier, Helmholtz assumes the principles of
decomposition in the original text of the *Erhaltung* without special mention
or justification. However, in his "Reply" to Clausius, he does acknowledges
his use of it, and reformulates his arguments in order to distinguish clearly

between the two assumptions he had tacitly made before: (1) the kinetic energy of a system is the same *whenever the system is in the same (relative) state* (positional determinacy), and (2) the force holding between any two points of the system is independent of the other points in the system (the principle of decomposition). In other words:

(positional determinacy & principle of decomposition) $\Rightarrow$ (conservation of *vis viva* $\Rightarrow$ postulate of central forces)

He then goes on to explain the significance of the positional determinacy requirement. This concept of "same relative position", "has not been applied by all mechanists who have made use of this principle [the conservation of energy], but it is obviously necessary to its physical application". And he goes on to define it as follows: "Movable points have the same relative position to one another whenever a coordinate system can be constructed in which all their coordinates have the same corresponding values."[32] To assume positional dependence is to assume the possibility of establishing congruence relations, so that our implication from above can be rewritten as:

(principle of decomposition & determinability of congruence relations) $\Rightarrow$ (constant *vis viva* $\Rightarrow$ postulate of central forces)

Unfortunately, Helmholtz continues to assume, as did Kant before him, that the conditions on the positional determination of two points remain the same in more complex systems. That is, he assumes that the principle of decomposition goes without saying. For Kant, as we have just seen, the reasoning is similar, meaning that both philosopher and physicist try to extract more work from the positionality principle than it can possibly do for them. In consequence, although Helmholtz goes on to expend great mathematical effort in showing that, given these assumption, he can secure the truth of his implication, he overlooks an essential weakness in his arguments. Nevertheless, in separating the two assumptions, he identifies for the first time the importance of our being able to empirically characterise an inertial frame. I will address this point more fully in the final section.

Helmholtz concluded his 1882 comments on the opening sections of the *Erhaltung* by observing that *both* the principle of decomposition, *and* the result that forces depend on position only had been called into question in electrodynamics. As he had himself maintained in 1872, "Weber's hypothesis concerning electrical forces is the first, at least partially successful attempt to base an explanation of a class of phenomena ... on the assumption of forces that depend not only on the position of mass-points, but also on their motion."[33] Furthermore, as I mentioned briefly above, the consistency of such forces with his conservation principle was an objection raised early on by Lipschitz, whom Helmholtz had been unable to refute as he had Clausius. Finally, as he himself admitted, he had been wrong to assume that these principles could be *proven* by means of *a priori* arguments. Nevertheless, despite his recognition of the contingency of the principles he had used in his early work, Helmholtz continued to maintain that they were conditions for the "determinacy and univocity" [*Eindeutigkeit und Bestimmtheit*] of physical theory. In other words, Helmholtz retained the transcendentalist argumentation he had learned from Kant—certain physical principles are singled out by virtue of their making possible a determinate description of reality— all while relativising the status of these principles. As such, he was the first to make a move that has become characteristic of modern forms of Kantianism, which is to assign certain contingent propositions the role of transcendental conditions of experience. This is best exemplified by Helmholtz's subsequent treatment of the third assumption listed above, which was that one could construct *empirically* the coordinate systems required to verify the congruence of systems of points.[34]

4. EMPIRICAL DETERMINACY AND GEOMETRY

Even though Helmholtz's notebooks in the period before the *Erhaltung*[35] reveal that he had been concerned with the empirical conditions on spatial measurement at an earlier phase, he first puts these reflections to epistemological work in his reply to Clausius's objections to the *Erhaltung*. He does this by explicating the meaning of the term "relative position" that he had used in the *Erhaltung* to single out those states of a system that qualify as identical from the point of view of the conservation law. The original formulation of the principle in the *Erhaltung* was the following:

> When an arbitrary number of mass-points move under the influence of only those forces that they exert on each other, **or that are directed at fixed centres,** then the sum of the living force of all of these is the same at all points in time in which they adopt the same relative position to one another **and to the possibly given fixed centres**, whatever their paths and speeds in the intervening time may have been.[36]

In this first statement of the principle, Helmholtz had not explained what the phrase "same relative position" meant. He had also allowed that a system might be considered to be in the same state with reference to a fixed centre; however, in the reply to Clausius, Helmholtz eliminates the passages in boldface without comment. And he adds a further specification of what is meant by "being in the same relative position":

Moving points have the same position relative to one another whenever a coordinate-system can be constructed in which all of their coordinates receive the same respective values.[37]

The reasons for the retraction and expansion of the definition of *vis viva* conservation are evident enough: Clausius's objection to Helmholtz was that there was no reason *a priori* why the force acting between two points might not be "an arbitrary function of the coordinates".[38] According to Clausius, Helmholtz had illegitimately assumed that the magnitude of the force could vary only as a function of the distance between the two points, and thus he had assumed half of the centrality he claimed to be proving.

Helmholtz responded that the forces envisaged by Clausius would depend on magnitudes that were not determined by the mass-points of the system alone. Thus his previous admission of "fixed" centres had to be eliminated, for such a notion permits directional and positional relations that are not internal to the system. Furthermore, the scope of the term "same relative position" had to be more precisely defined, in order to block all appeals to properties of absolute space. Once these refinements had been made, he could argue that Clausius's assumption that the potential about a single point in space might vary with direction rests on a confusion between mathematically and epistemologically legitimate properties. The rebuttal of Clausius hinges, Helmholtz emphasises, on the demand that we must "seek the grounds of real effects only in the relations of real things to one another."[39]

The theoretical import of this change is two-fold: first, the energy principle is now formulated exclusively in terms of the "internal" properties of a system; second, the question of what it means to construct an *empirical* coordinate-system is raised, if only implicitly. Regarding the first point, Helmholtz now requires that the energetic state of a closed system must be definable without reference to any external frame. Forces that violate this requirement, in that they would characterise systems for which Helmholtz's principle does not hold, are to be dismissed as mathematically possible, but physically meaningless. The assumption of such forces cannot "be applied [*übertragen*] to physical reality."[40] Even in his 1882 comments on the *Erhaltung*, in which he admits that the principle of decomposition is an empirical proposition, Helmholtz continues to insist on this point when discussing the state of electromagnetism. Those theories that would make forces depend on absolute space, although they cannot be ruled out on purely logical grounds, are still to be seen as the option of last resort.

The second point to observe is the modified status of geometrical relations entailed by this analysis. In order to rebut Clausius, Helmholtz is led to deny that directions and magnitudes which are not determined by

"the real relations of things to one another" may be introduced into theoretical definitions. But his own definitions continue to employ the notion of
a "constructible coordinate system" relative to which we can say of a system that it is in the same state at two points in time. This puts the ball back
in Helmholtz's court. If it is true, as he maintains, that a single point does
not determine *directions* in its vicinity, in what sense can it be said that two
points determine a distance? It may be the case that they determine a single spatial magnitude, that is to say a single line element (assuming we are
not dealing with a spherical geometry). But that sense of determination is
not sufficient to do the work required by Helmholtz's definitions: the magnitude they determine must be congruent with another magnitude determined
by those points *at a second point in time*. And Helmholtz, in contrast to
Kant, is quite aware that there is a problem lurking here—that congruence is
not the "complete similarity and identity" of two spatial magnitudes, to re-
peat Kant's definition from the Phoronomy, but that every claim concerning
the congruence of two spatial magnitudes contains an implicit reference to
motion.

Helmholtz, the physicist, sees a difficulty that the philosopher overlooks.
In a sense, physics places less stringent demands on geometry than philosophy does. The sense of congruence that one needs in physics is only that of
metrical equality: two different distances must be the same in the sense that
they are ascribed the same measure. This need not involve any appeal to the
possible coincidence of the distances, for we might decide to call two distances equal when they satisfy some arbitrary operational definition. Establishing metrical equality is, on the face of it, a merely practical problem. But
this physical sense of congruence has its geometrical counterpart: in geometry as well, we speak of two *distinct* line segments as being congruent, and
Kant's explanation of the grounds of geometry does not adequately explain
this possibility. According to him, geometrical axioms rest on the results
of pure operations of the productive imagination: the successive synthesis
of the spatial manifold ensures that spatial magnitudes are *quanta*, for each
spatial magnitude is "drawn" (*gezogen*) in intuition by successively adding
its parts. But the resultant quanta are not, as Kant himself emphasises, *quantities*, which he identifies as the subjects of arithmetical propositions. And
physics needs magnitudes that are arithmetised. In other words, Kant does
not adequately appreciate the *metrical* aspect of the notion of congruence,
which demands not only that each of two magnitudes consist of parts, but
that we know the measure of these parts. This means that the parts of distinct spatial magnitudes must be comparable to one another if we are to speak
of congruence in a manner that is useful to physics. When Kant insists in the

Phoronomy that the parallelogram law be formulated so as to make component and resultant motions *identical*, he is in essence circumventing a central problem in his philosophy of science: on the one hand, he assumes the validity of certain geometrical axioms, but he is unwilling to accept metrical equality as a sufficient definition of congruence. So he insists that all metrical equality must be reduced to what he calls "complete similarity and equality, insofar as it can be cognised in intuition".[41] But physics cannot hope to satisfy this demand. If certain physical principles, such as Helmholtz's *vis viva* principle, makes claims about the relations between the energetic states of a system when it is in distinct, but metrically identical configurations, then these claims must be empirically determinate.

In the early memoir reproduced by Königsberger, Helmholtz had already identified this point explicitly, even if he too failed to respond to it adequately. Here, Helmholtz distinguishes between what he calls "mathematical bodies" and "material bodies", suggesting that the first are like rigid bodies surrounding or containing the latter. Spatial magnitudes may therefore be conceived as "continuous rigid systems" in which the relative determinations of the system (by which Helmholtz means the relative distances and orientations of its points) are unchanged. Helmholtz then proceeds to define congruence as possible superimposition:

> Rigid systems are congruent when the one can be moved onto the other in such a way that each point of the one coincides with a point of the other. Pairs of equally distant points are congruent. ... Motion must belong to matter quite aside from its special forces; but then the only remaining characteristic of a determinate piece of matter is the space in which it is enclosed; but since it is robbed of this characteristic as well by motion, we can only speak of its identity if we can intuit the transition from the one space to the other, i.e. motion must be continuous in space.[42]

In other words, Helmholtz explicitly problematises the notion of geometrical determination before he begins work on the *Erhaltung*, thus well before he appeals to the notion of a possible coordinate system in the reply to Clausius. In this manuscript, as in the much later papers on geometry, he considers the comparison of spatially distinct parts to be an ineliminable part of the *concept* of congruence.[43] Systems, whether "mathematical" or "material", are congruent if and only if they can be superimposed. Kant would so far be in agreement. But Helmholtz discerns a further difficulty: the motion of either kind of body involves a change in place, and this calls into question the identity of the system after the motion with that before the motion. He does not at this point provide a satisfactory analysis of what is involved in the concept of comparison itself, reverting instead to the supposition that we can intuit the formal identity of an object in motion, and thus that we can somehow assure ourselves of invariance of the system by appealing to continuity.

These early writings demonstrate that Helmholtz is already aware by the time of his reply to Clausius that the assertion that a single material system is in the same state at two distinct points in time cannot mean that the system occupies "equal" portions of *undifferentiated* space. Furthermore, if the system in question has undergone changes in the relative positions of its masses in the intervening period, then we cannot appeal to its continuous identity in order to intuit that they are now in the initial configuration. Since Helmholtz insists that such a statement must involve only "real" relations, we can infer that the possibility of constructing a coordinate system to which Helmholtz refers must at the very least be the possibility of determining congruence relations by comparison with rigid "mathematical bodies" of the sort he referred to in the earlier manuscripts. In sum, since Helmholtz denies that directions can be employed in physical theories as if they were independent properties of space, he cannot himself assume that congruence relations can be so employed. This injunction against absolute directions is not, as Helmholtz emphasises, a "logical" one; rather it derives from a distinction between those coordinate systems which one "draws on paper"[44] and those which are determined by "real things". Now, the meaning of the principle of energy conservation is that the energetic state of a system is the same whenever it is in the same position. But if we are to avoid covertly appealing to coordinate systems drawn on paper, we must admit that this statement assumes the existence of material bodies independent of the system in question that we can use to compare the two states of the system.

5. CONCLUSION

Thus Helmholtz's position in his reply to Clausius is highly unstable, and this for several reasons. First of all, the principle of decomposition does not have the *a priori* status that Helmholtz implicitly accorded it, as he admits the moment he identifies its role. But since he needs it in order to reduce complex systems to two-point systems, and thus to apply the principle of positional determinacy, he must also admit that non-central forces are not impossible on constitutive grounds. Second, the appeal to non-arbitrary coordinate systems does more work than Helmholtz could have wanted. According to Helmholtz, I cannot appeal to properties of space itself in saying that a system is in the same configuration at two different times. But according to his own analysis in the early memoir, I also cannot appeal to the spatial determinations provided by the system itself, for these have, by hypothesis, changed in the intervening period. Thus Helmholtz is already committed to the existence of an empirically given coordinate system used to define those sets of a system's states that qualify as congruent. If he doesn't assume such a coordinate system, he has no empirically given magnitudes to ground his definition. Unfortunately, if he does admit one, he runs the risk that his opponent Clausius can employ it as well. For if I have enough measuring instruments at my disposal to determine the distances and angles that characterise the state of a complex system, why can't I use them to define asymmetries of the sort invoked by Clausius? The state of the system is no longer characterisable *solely* by means of purely "internal" relations. But Helmholtz cannot afford to relinquish that position, for to do so would mean to threaten the entire basis of his transcendental argumentation.

What Helmholtz needs is an argument proving the following: (1) statements concerning the relative positions of points in a material system are always statements concerning the (possible) coincidence of these points with other systems of points (the "mathematical bodies" he had discussed in the memoir); (2) the congruence relations determined by these mathematical bodies satisfy, or indeed entail, the axioms of Euclidean geometry; furthermore, (3) these bodies must not be so numerous as to permit the definition of arbitrary directions in space. If (1) is true, then one is justified in rejecting any physical theory that appeals to intrinsic magnitudinal and directional properties of space on the grounds of empirical indeterminacy. In other words (1) amounts to saying that such relations are *necessary* to the empirical sciences. Conversely, (2) is needed in order to ensure that the system of measurement that results satisfies all demands that physics places on its elementary magnitudes. In other words, (2) amounts to saying that such relations are *sufficient* to the requirements of the empirical sciences. Lastly, (3) is required to ensure that the supplementary empirical relations do not invalidate the arguments from the relativity of space that Helmholtz was directing against his opponents. Without (3), we might just as well reintroduce the idea of an absolute space.

This is a strange list of requirements. For it amounts to demanding that one prove the validity of Euclidean geometry from conditions on the comparison of arbitrary spatial magnitudes. But this is just what Helmholtz tried to do in first paper on geometry from 1868, "Über die Tatsachen, die der Geometrie zum Grunde Liegen".[45] Although it contains the bulk of his mathematical arguments, and although the philosophical argument is unmistakably a transcendental one, this paper is generally read as an abortive attempt at the arguments for the empirical status of geometry presented in the three papers following it. In this paper, Helmholtz argues there that since physics presupposes our ability to equate spatial magnitudes, it also assumes that we make comparisons of the distances determined by pairs of points by transporting the same measuring instrument between them. Helmholtz then uses analytical methods to deduce the metrical characteristics of those manifolds in which the requisite measurement operations can be carried out. If such comparisons can be made regardless of the location and orientation of the points, this entails that there be rigid bodies that can be freely transported and rotated, and therefore, he concludes, that the possible manifolds thus singled out have a constant curvature. Adding to these assumptions the demand that space be unbounded, he concludes that only a space with a Euclidean geometry is compatible with the demands of measurement.[46]

Had he been right to conclude that Euclidean geometry was entailed by his measurement postulates, Helmholtz would have established that the basic magnitudes referred to in physical theories were empirical in the sense that Kant had called the propositions of the metaphysics of nature "pure empirical" propositions. They would refer to relations of coincidence among pairs of ideal material points. The supposition that geometry was true would then be equivalent to supposing that certain sets of relations among such ideal points in fact obtained. This would not be a constitutive *a priori* truth, but a regulative one: if it didn't hold, then certain kinds of physical descriptions would not be possible. Furthermore, the negative arguments that appealed to the indeterminacy of absolute spatial relationships would remain untouched. Helmholtz could still have maintained, against Clausius and others, that a theory which required reference to purely mathematical magnitudes was invalid. Here again, it would be so not because such theories were logically impossible, in that they would describe states of affairs which were unimaginable. Rather, they too would violate the regulative demand of the complete comprehensibility of nature.

Because Helmholtz formulated his arguments as "empiricist" arguments directed against dogmatic Kantians, and because his later readers have considered these writings in isolation from his research programme in physics itself, they have also overlooked the strong transcendentalist programme that motivated the first two papers on geometry. On my reading, this aspect of the papers is not a philosophical atavism, but is directly related to the line of transcendental argument that we find in the *Erhaltung*. Indeed, since Helmholtz wrote these first two papers on geometry at the time when he turned his attention back to electrodynamics, in order to reopen the quarrel with Weber's and Neumann's theories, it is quite possible that his intent was to pick up his defence against Clausius and Lipschitz at the point he had left off in 1854.

NOTES

[0]My thanks to Richard Arthur, who responded generously to my inquiries regarding Newton's early parallelogram proof, and to Olivier Darrigol, whose (Darrigol, 1994) was invaluable to my understanding of Helmholtz's early work. Audiences at the Copenhagen conference, at the Institut d'Histoire et Philosophie des Sciences et des Techniques in Paris, and at the Universität Marburg provided useful criticisms. In particular, Jacques Dubucs and Peter Janich pointed out a number of ambiguities in my earlier presentations. Konstantin Pollok's commentary on Kant's *Metaphysical Foundations of Natural Sceince* (Pollok, 2001) was a constant help. My reading of the *Metaphysical Foundations* goes back to a seminar Michael Friedman offered at the University of Indiana at Bloomington in 1998.

[1]Carrier (1994) argues that this position typifies the arguments of the early Bertrand Russell and of Hugo Dingler.

[2]Cf. Helmholtz's manuscript from the period before the *Erhaltung* reproduced in (Königsberger, 1903, 126-138,131).

[3]Indeed, Leibniz contended that the possibility of deriving infinite work from the same system was analogous to being able to derive any proposition from contradictory premisses.

[4](Darrigol, 1994, 217). Helmholtz's principle is "a particular case of what I shall call the *principle of decomposition*, according to which all actions in nature must be resolved into actions involving only two elements of volume."

[5](Helmholtz, 1996, 6)

[6]This was the first volume of his *Wissenschaftliche Abhandlungen*, (Helmholtz, 1883).

[7](Helmholtz, 1996, 54)

[8](Helmholtz, 1996, 54)

[9](Helmholtz, 1996, 55)

[10]See (Bevilacqua, 1993, 314), (Darrigol, 1994, 221), (Bevilacqua, 1994).

[11](Helmholtz, 1996, 9)

[12]In the 1882 comments, Hemholtz suggests adding the requirements of equal action and reaction, and of the reducibility of force into point-masses (i.e. the principle of decomposition) in order to rule out forces of the sort admitted by Lipschitz.

[13](Helmholtz, 1854)

[14](Heidelberger, 1993, 470)

[15](Helmholtz, 1996, 53)

[16](Kant, 1911b, 306)

[17]It is not clear whether the dichotomy of the principles of pure experience understanding to which constitutive/regulative ≡ mathematical/dynamical is intended to coincide with that drawn between the transcendental ideas and the (unidentified) principles "from which, strictly speaking, the truth of the general rule ... follows".

To settle that, we would have to decide between two interpretations: (1) Kant takes the proposition that "each event has a cause" to be true "strictly speaking" and thus to be constitutive (in which case there would be two separate meanings of constitutive); or, (2) he holds that only the mathematical principles are strictly speaking true (in which case there would be only one sense of constitutive). But we do not have to settle this question in order to pursue the analysis at hand.

[18] MFNS, (Kant, 1911a, 469–470).

[19] If, for instance, rules are given for mixing particular quantities of paints, or quantities of spectral light, the statement that two colours mix to form a third will have a determinate value, since it may or may not be true. Whether or not it is true will depend on three factors: the physical properties of the substances mixed, the mixture rules, and the psycho-physiological make-up of the perceiving subject. Whether or not such a statement should be interpreted as *a priori* true (or false) is a subtle question, which cannot be treated here in greater detail. I discuss Helmholtz's attempts to define this additive operation and thereby the structure of the colour-space in (Hyder, 2001). This research was, on my reading, of fundamental importance for Helmholtz's understanding of metrical relations on manifolds.

[20] Kant anticipate the obvious objection that the sum of two collinear motions can be represented by laying the two line segments they determine end to end. In such a case, he argues, we would have represented two successive, as opposed to simultaneous motions.

[21] (Kant, 1911a, 493)

[22] (Kant, 1911a, 494)

[23] Printed in (Cohen and Westfall, 1995, 377–385). Richard Arthur provides a thorough treatment of this early parallelogram proof in (Arthur, 2006).

[24] In both cases, the motions in question are considered to be instantaneous. Had Kant known of this text, he might still have objected that Newton's kinematic proof, because it deals with the motion of a single point, must still make reference to the geometrical properties of an empty, and thus indeterminate background space.

[25] (Kant, 1911a, 498)

[26] The "dynamical" and "mechanical" definitions of force are not truly distinct, even though Kant maintains that they differ in that the dynamical definition "could regard the matter [i.e. the body causing the motion of the other] as at rest" (Kant, 1911a, 536). In fact, the strict relativity of motion postulated in the "Phoronomy" makes this distinction spurious. Kant's reasons for distinguishing between his dynamical and mechanical concepts of force derive from his desire to ensure a strict correspondence between the four principles of the understanding and the four sections of the *MFNS*.

[27] As Michael Friedman (1986) has argued, Kant effectively inverts the relation between force and absolute space suggested by Newton, and argues that true motions, thus the concept of absolute space, are definable only within systems conforming to Newton's third law.

[28] (Kant, 1911a, 498)

[29]Suppose I know how three non-collinear points in the plane have changed their relative positions. Then, from my knowledge of the motion of some fourth point relative to two of these, I can infer how it has changed its position relative to the third. I am no longer free to regard its motion as the sum of three independent variables that depend only on its distance to the other three points, because the relations among the first three points—whether expressed in angular or linear coordinates—span the space in question.

[30]See: (DiSalle, Summer 2002, Section 1.6) "The Emergence of the Concept of Inertial Frame". Kant assumes not only that such a frame is apodictically required, but also that the accelerations defined in such a frame all resolve onto independent linear displacements relative to the other bodies in the system.

[31](Helmholtz, 1996, 54)

[32](Helmholtz, 1854, 83)

[33](Helmholtz, 1872, 645)

[34](Helmholtz, 1868a)

[35]Quoted in (Königsberger, 1903, 126–138). Königsberger does not give an exact source or date for the manuscript he reproduces.

[36]"Wenn sich eine beliebige Zahl beweglicher Massenpunkte nur unter dem Einfluss solcher Kräfte bewegt, welche sie selbst gegen einander ausüben, **oder welche gegen feste Centren gerichtet sind:** so ist die Summe der lebendigen Kräfte aller zusammen genommen zu allen Zeitpunkten dieselbe, in welchen alle Punkte dieselben relativen Lagen gegen einander **und gegen die etwa vorhandenen festen Centren** einnehmen, wie auch ihre Bahnen und Geschwindigkeiten in der Zwischenzeit gewesen sein mögen." (Helmholtz, 1996, 9) The exact wording of the new definition in (Helmholtz, 1854, 82–83) is: "Wenn in beliebiger Zahl bewegliche Massenpunkte sich nur unter dem Einflusse solcher Kräfte bewegen, die sie selbst gegeneinander ausüben, so ist die Summe der lebendigen Kräfte aller zusammengenommen zu allen Zeitpunkten dieselbe, in welchen alle Punkte dieselben relativen Lagen gegeneinander einnehmen, wie auch ihre Bahnen und Geschwindigkeiten in der Zwischenzeit gewesen sein mögen."

[37]"Gleiche relative Lage zu einander haben bewegliche Punkte, so oft ein Coordinatensystem zu construiren ist, in welchem alle ihre Coordinaten beziehungsweise dieselben Werthe wiederbekommen." (Helmholtz, 1854, 83)

[38](Helmholtz, 1854, 84)

[39](Helmholtz, 1854, 84)

[40](Helmholtz, 1854, 84)

[41](Kant, 1911a, 493)

[42](Königsberger, 1903, 185)

[43]Again, compare Kant on the importance of coincidence [*Deckung*] in *Prolegomena* §12: "All proofs of the complete equality of two given figures amount to this: that they coincide with each other [*daß sie einander decken*]; which is obviously nothing other than a proposition resting on immediate intuition." Kant's immediate point is that the proposition does not depend on purely conceptual relations; however, he does not problematise here or elsewhere the sense in which

two spatially *distinct* figures could be said to coincide. And, as we know from the Phoronomy, when he is confronted with physical laws that require the equality of distinct motions, he responds by insisting that they must be made strictly identical if geometrical propositions are to apply to them.

[44](Helmholtz, 1854, 84)

[45](Helmholtz, 1868a)

[46]This claim was, however, quickly corrected in his second paper on the subject, in which he admitted that pseudo-spherical spaces are also compatible with his demands. (Helmholtz, 1868b)

REFERENCES

Arthur, R. (2006). Newton's Proof of the Vector Addition of Motive Forces, *in* W. Harper and W. Myrvold (eds), *Infinitesimals*, Springer, Dordrecht.

Bevilacqua, F. (1993). Helmholtz's über die Erhaltung der Kraft: The Emergence of a Theoretical Physicist, *in* D. Cahan (ed.), *Hermann von Helmholtz and the foundations of nineteenth-century science*, Univ. of California Press, Berkeley, pp. 291–333.

Bevilacqua, F. (1994). Theoretical and Mathematical Interpretations of Energy Conservation: The Helmholtz-Clausius Debate on Central Forces 1852–54, *in* L. Krüger (ed.), *Universalgenie Helmholtz*, Akademie Verlag, Berlin, pp. 89–106.

Carrier, M. (1994). Geometric Facts and Geometric Theory: Helmholtz and the 20th Century Philosophy of Physical Geometry, *in* L. Krüger (ed.), *Universalgenie Helmholtz*, Akademie Verlag, Berlin, pp. 276–291.

Cohen, B. and Westfall, K. (eds) (1995). *Newton: Texts, Backgrounds, Commentaries*, Norton, New York.

Darrigol, O. (1994). Helmholtz's Electrodynamics and the Comprehensibility of Nature, *in* L. Krüger (ed.), *Universalgenie Helmholtz*, Akademie Verlag, Berlin, pp. 216–242.

DiSalle, R. (Summer 2002). *Space and Time: Inertial Frames*, The Stanford Encyclopedia of Philosophy. Accessed August 2003.
URL: *http://plato.stanford.edu/archives/sum2002/entries/spacetime-iframes/*

Friedman, M. (1986). The Metaphysical Foundations of Newtonian Science, *in* R. Butts (ed.), *Kant's Philosophy of Physical Science*, Reidel, Dordrecht, pp. 25–60.

Heidelberger, M. (1993). Force, Law and Experiment: The Evolution of Helmholtz's Philosophy of Science, *in* D. Cahan (ed.), *Hermann von Helmholtz and the foundations of nineteenth-century science*, Univ. of California Press, Berkeley, pp. 461–497.

Helmholtz, H. (1854). Erwiderung auf die Bemerkungen von Hrn. Clausius, *Wissenschaftliche Abhandlungen*, Vol. 1., Johann Ambrosius Barth, Leipzig, pp. 76–96.

Helmholtz, H. (1868a). Über die Tatsachen, die der Geometrie zum Grunde liegen, *Wissenschaftliche Abhandlungen*, Vol. 2., Johann Ambrosius Barth, Leipzig, pp. 618–639.

Helmholtz, H. (1868b). Über die tatsächlichen Grundlagen der Geometrie, *Wissenschaftliche Abhandlungen*, Vol. 2., Johann Ambrosius Barth, Leipzig, pp. 610–617.

Helmholtz, H. (1872). Über die Theorie der Elektrodynamik: Vorläufiger Bericht, *Wissenschaftliche Abhandlungen*, Vol. 1., Johann Ambrosius Barth, Leipzig, pp. 636–646.

Helmholtz, H. (1883). *Wissenschaftliche Abhandlungen*, Johann Ambrosius Barth, Leipzig.

Helmholtz, H. (1996). *Über die Erhaltung der Kraft [u.a.]*, Harri Deutsch, Frankfurt a.m.

Hyder, D. (2001). Physiological Optics and Physical Geometry, *Science in Context* **14**(3): 419–456.

Kant, I. (1911a). Metaphysische Anfangsgründe der Naturwissenschaften, *Kants Werke*, Vol. 4., Preußiche Akademie der Wissenschaften, Berlin, pp. 465–565.

Kant, I. (1911b). Prolegomena zu einer jeden künftigen Metaphysik, *Kants Werke*, Vol. 4., Preußiche Akademie der Wissenschaften, Berlin, pp. 253–383.

Königsberger, L. (1903). *Hermann von Helmholtz*, Vieweg, Braunschweig.

Pollok, K. (2001). *Kants Metaphysische Anfangsgründe der Naturwissenschaft. Ein Kritischer Kommentar*, Vol. 13 of *Kant-Forschungen*, Meiner, Hamburg.

JESPER LÜTZEN

A MECHANICAL IMAGE: HEINRICH HERTZ'S PRINCIPLES OF MECHANICS

1. INTRODUCTION

Heinrich Hertz's book *The Principles of Mechanics Presented in a New Form* was innovative in all of the three areas discussed in this volume: In physics it presented the first foundation of mechanics avoiding force as a basic concept; in mathematics it presented the first use by a physicist of the new Riemannian geometry to geometrize configuration space; and in philosophy it presented a radically new theory of the (mental) images we can make of nature.

In this paper I shall analyze how Hertz's interest in and work on mechanics and images grew naturally out of his earlier work in physics and his epistemological reflections on this work. In that way I shall illustrate the fruitful interaction between the physical and the philosophical aspects of Hertz's *Mechanics*. There are also close links between these two aspects and the mathematical aspects of Hertz's work. However, these links will only be discussed briefly in a postscript to the present paper. For a more detailed discussion see (Lützen, 2005).

2. A STRUGGLE WITH MECHANICS

In 1891 Heinrich Hertz (1857–1894) could look back on five very successful years of research on electro-magnetism. From 1886 to 1888 he had produced electro-magnetic waves (radio waves) in his laboratory and had shown that they behave like light. These experiments made him world famous. The following years he had written two theoretical papers on Maxwell's theory in which he presented Maxwell's equations in the form they are known today. After these great achievements Hertz turned to new experiments, but when they only resulted in one small observation regarding cathode rays, he felt "worn out" and became "fed up with physics."[1] He then turned to an entirely new field of research: mechanics. In March 1891 he informed Felix Klein that he had begun a work on physical mechanics, in particular on the concept of energy, and he expected that the work on "these difficult things" would last one half to one year (Fölsing, 1997, 474). As it turned out Hertz's research was soon extended into an investigation of the foundations of mechanics which was not completed until one month before his untimely death on January 1, 1894. The resulting book *Prinzipien der Mechanik in neuem Zusammenhange dargestellt* was seen through press by Hertz's last assistant Philipp Lenard.

45

V.F. Hendricks, K.F. Jørgensen, J. Lützen and S.A. Pedersen (eds.), Interactions: Mathematics, Physics and Philosophy, 1860-1930, pp. 45–64.
© *2006 Springer.*

Hertz struggled hard with his research on mechanics and often regretted that he had begun it.[2] As in his previous research his moods often swung between utter depression and great joy, but the depressive periods were more dominant in his research on mechanics than in his previous researches. There seem to be at least three reasons why Hertz's work on mechanics was a predominantly depressive experience: First, this kind of theoretical work did not provide the splendid breakthroughs that his experimental work had often done. Second, as Hertz himself admitted to his parents, he had been spoiled by his earlier success, and found it hard to accept when his work did not lead to immediate results. Third, during the last one and a half years he was seriously ill from an infection of his jaw and nose, an infection that in the end led to blood poisoning and his death.

But why did it take Hertz so much longer to finish his research on mechanics than he had estimated in 1891? First, he was impeded by his illness that kept him from serious work for extended periods of time. Second, while he had earlier worked to find new effects or new theoretical explanations that could be published quickly, Hertz now strove for perfection. The foundational nature of the subject demanded it and he knew that with his newly won status as a physics celebrity his colleagues expected perfection from him and would severely criticize any imperfection. The great pains he went through to secure a perfect result can be seen from the many drafts of the book that he left behind: five drafts of the mathematical formalism, four drafts of the mechanics of free systems, three drafts of the mechanics of unfree systems and two drafts of the preface and the philosophical introduction. Third, Hertz found the project more difficult than he had expected, and fourth, he extended the project from a paper on energy to a book on the foundations of mechanics.

3. WHY THEORETICAL RESEARCH ON MECHANICS?

3.1. Not because of illness

It may seem surprising that a physicist like Hertz would abandon his successful research on the new and lively area of electro-magnetism for a work on a classical subject as mechanics. First, let me point out that this change of subject had nothing to do with his fatal illness. He did not become ill until one year after embarking on the new research project and when he had acquired the infection he did not work on theoretical matters because of it but rather in spite of it. Indeed, as he explained to his parents on October 10, 1893 after the book was finished, he felt much better when he did experiments than when he sat at his desk:

I am glad, because it was a great burden for me and I blame a large part of my last year's infirmity on it: first because it may be so, second because it gives me a sort of consolation to think so. In any case I feel infinitely better when I am up and about and keeping busy with my hands than when I sit at my desk or squat in my room, lost in thought. So I promise myself that I shall master my suffering more easily now. Today I even went to the laboratory and started to make some preparations for working there. Long ago I made a solemn vow not to enter on theoretical work for a long time to come. But this one had to be finished. (Hertz, 1977, 341–343)

3.2. Mechanistic philosophy

The principal reason why Hertz found research in mechanics important was that he adhered to a mechanistic philosophy of nature. He appealed to this philosophical standpoint in the opening words of the preface to his book: "All physicists agree that the problem of physics consists in tracing the phenomena of nature back to the simple laws of mechanics". This reductionist program had enjoyed great triumphs over the previous century: Optics had been reduced to the study of waves in the luminiferous ether, the kinetic theory of heat had reduced heat to a kinetic phenomenon, and even in electromagnetism Maxwell had suggested that one ought to consider the electromagnetic field as a result of matter in motion. Thus for Hertz mechanics was not just another branch of physics, it was the most fundamental physical discipline to which every other discipline should ultimately be reduced.

3.3. Foundational problems in mechanics

"But", Hertz continued his preface "there is not the same agreement as to what these simple laws are". In fact, despite its long history the field of mechanics was amazingly active during the last third of the 19th century. More than 50 expositions of the basics of mechanics and almost as many critical works on the foundations of mechanics were published during this period. Hertz himself was exposed to 6 different introductions to mechanics during his student days: In 1875 in Dresden he began to follow Königsberger's lectures but found them too difficult. In 1877 while in München he read Lagrange's *Mécanique analytique* and Laplace's *Mécanique celeste*. After he had moved to Berlin he followed Borchardt's lectures in 1878 and Kirchhoff's and Kummer's lectures in 1879. Later in life he also read Thomson and Tait's *Treatise on Natural Philosophy*. These expositions and other works that he consulted revealed to him that there was no general agreement about the foundations of mechanics, and consequently about the foundation of all of physics. In particular Mach's penetrating analysis (Mach, 1883) of these foundational issues was a great source of inspiration for Hertz.

By 1891 Hertz had already shown aptitude for foundational research in the area of electro-magnetism. Dissatisfied with Maxwell's own presentation of his field theory Hertz succeeded in presenting it in a stark almost axiomatic form:

> The structure of the system [of Maxwellian electromagnetic theory] ought to allow a clear apprehension of its logical foundation. All non-essential concepts ought to be removed from the system and the connections between the essential concepts ought to be reduced to their simplest form. In this respect, Maxwell's own exposition does not represent the attainable goal. It often swings back and forth between the views that Maxwell inherited (vorfand) and those to which he was guided. Maxwell begins by assuming unmediated distance forces and he investigates the laws, according to which the hypothetical polarization of the dialectic ether will change under the action of such distance forces, and he ends with the claim that the polarizations really changes in this way, even though the changes are in fact not caused by distance forces. This procedure leaves one with an unsatisfactory feeling that either the final results or the way in which they were obtained must be wrong. Moreover this procedure leaves, in the formulas, a number of superfluous rather rudimentary concepts behind that only had a proper meaning in the old theory of immediate action at a distance. (Hertz, 1890, 208–209)

Such an attempt to remove all superfluous concepts from a theory in order to attain absolute clarity is also characteristic of Hertz's mechanics.

3.4. Forces: a special problem

One concept struck Hertz as particularly problematic: The concept of force. Both Mach and Kirchhoff had tried to introduce this concept in a consistent way, but Hertz was not satisfied with their definitions. Moreover, the existence and nature of actions at a distance was a problem that had haunted mechanics since Newton's days. According to Hertz it had not found a satisfactory solution and it was responsible for the most blatant problems in the foundations of mechanics. Hertz had met the discussion about distance forces in connection with his work on electro-magnetism. Around 1885 there were two promising theories of this branch of physics: Weber's theory that explained electro-magnetic interactions in terms of actions at a distance between moving electrically charged particles, and Maxwell's theory that explained such interactions as a result of contiguous actions in the electro-magnetic field, that was thought of as a mechanical state in the ether. In between these two extremes were other theories such as Riemann's and Carl Neumann's potential theories as well as Helmholtz's version of Maxwell's theory that operated with distance actions as well as field actions. Hertz, who was a student of Helmholtz, naturally started out as an adherent of the latter's view, but gradually developed into a pure Maxwellian. To a large degree this shift of allegiance was due to his discovery of electro-magnetic waves:

> A considerable part of this approval [of Hertz's own experimental results] was due to reasons of a philosophic nature. The old question as to the possibility and nature of forces acting at a distance was again raised. The preponderance of such forces in theory has long been sanctioned by science, but has always been accepted with reluctance by ordinary common sense; in the domain of electricity these forces now appeared to be dethroned from their position by simple and striking experiments. (Hertz, 1892, 19–20/18)

At the Naturforscherversamlung in Heidelberg in 1899, Hertz even suggested that there might not exist actions at a distance in nature at all. In his talk he was rather prudent:

> We are at once confronted with the question of direct actions-at-a-distance. Are there such? Of the many in which we once believed there now remains but one – gravitation. Is it too a deception? The law according to which it acts makes us suspicious. (Hertz, 1889, 353/326)

After his talk he was invited to Königsberger's home together with other of the leading participants in the meeting and they asked him "why he had not in his talk openly declared that he also wanted to eliminate gravity as an action at a distance". "I am still too much of a coward for that" Hertz answered (Koenigsberger, 1903, vol 3, 26). However, he seems to have explored various methods for experimentally verifying the field theoretic nature of gravitation. In 1889 he corresponded with Lehmann-Filhles and the astronomer George Darwin about the possibility of observing a finite velocity of gravitational action, and in his diary he made an entry: "Made experiments on polarization through gravitational effect" (Hertz, 1977, 313). The meaning of this note is not immediately obvious, but it may refer to an attempt to detect a kind of gravitational Faraday effect, the existence of which would clearly suggest a field theoretic nature of gravitation. This entry in

Hertz's notebook is dated January 5. 1891, which indicates a connection to his subsequent interest in the foundations of mechanics.

3.5. Ether

Hertz was also aware of the problems involved in a mechanical description of the ether. He had already argued forcefully for the existence of such a medium in 1884 in a popular series of lectures at the University in Kiel where he discussed its possible constitution and its somewhat contradictory properties (Hertz, 1999). At the Naturforscherversamlung in Heidelberg in 1889 in the talk from which I have already quoted his reflections regarding distance forces, his rhetoric reached a high point when he pointed to the problem of the ether as the ultimate problem in physics:

> Directly connected with these is the great problem of the nature and properties of the ether which fills space, of its structure, of its rest or motion, of its finite or infinite extent. More and more we feel that this is the all-important problem, and that the solution of it will not only reveal to us the nature of what used to be called imponderables, but also the nature of matter itself and of its most essential properties – weight and inertia.[3] The quintessence of ancient systems of physical science is preserved for us in the assertion that all things have been fashioned out of fire and water. Just at present physics is more inclined to ask whether all things have not been fashioned out of the ether? These are the ultimate problems of physical science, the icy summits of its loftiest range. Shall we ever be permitted to set foot upon one of these summits? Will it be soon? Or have we long to wait? We know not: but we have found a starting-point for further attempts which is a stage higher than any used before. Here the path does not end abruptly in a rocky wall; the first steps that we can see form a gentle ascent, and amongst the rocks there are tracks leading upwards. There is no lack of eager and practiced explorers: how can we feel otherwise than hopeful of the success of future attempts? (Hertz, 1889, 354/326–7)

At the Naturforscherversamlung in 1891 in Halle it was rumored that Hertz was working on the mechanics of the ether but Hertz himself rejected the idea:

> What you have heard about my works via Halle is unfortunately without any foundation and I do not know how this opinion has been formed. I have not at all worked with the

mechanics of the electric field, and I have not obtained anything concerning the motion of the ether. This summer I have thought a great deal about the usual mechanics, but I do not think I spoke about that in Halle at all. In this area I would like to put something straight and arrange the concepts in such a way that one can see more clearly what are the definitions and what are the facts of experience, such as, for example, concepts of force and inertia. I am also already convinced that it is possible to obtain great simplifications; for example, I have only recently clarified for myself in a satisfactory manner what a mechanical force is. However, I have neither written the thing down nor do I know if others will afterwards find it satisfactory. In any case, It is something that can only ripen slowly. [Hertz to Emil Cohn November 29, 1891. Deutsches Museum]

This does not preclude that the ether was a source of inspiration for Hertz's work on mechanics. In fact in the *Principles of Mechanics* Hertz described his new foundation of mechanics as a necessary step on the road to understanding the ether:

It is in the treatment of new problems that we recognize the existence of such open questions as a real bar to progress. So, for example, it is premature to attempt to base the equations of motion of the ether upon the laws of mechanics until we have obtained a perfect agreement as to what is understood by this name. (Hertz, 1894, xxv)

3.6. An energetic beginning

By 1891 an alternative to the usual Newtonian theory of mechanics was gradually forming. This theory took energy to be a basic concept instead of forces. This possibility had been opened up by Helmholtz's work on energy conservation from 1847. Thomson's and Tait's *Treatise on Natural Philosophy* was to a certain degree written in this style, but only in the period 1887 to 1893 was a purely energetic program developed by the chemists Ostwald and Helm. The proponents of the energetic program claimed that their approach to nature was phenomenalistic in the sense that it only dealt with directly perceivable objects or even with the perceptions themselves and the laws that relate them. It did not seek mechanical explanations for the laws. In particular the proponents of the energetic program were proud to point out that they avoided any reference to hypothetic atoms or other microscopic unobservables. They only needed macroscopic observables.

In 1895, more than a year after Hertz had died, Ostwald and Helm clashed frontally with the proponents of the traditional mechanistic world view, in particular Boltzmann, who often referred to Hertz mechanics as a possible foundation of a mechanistic cosmology. However it is interesting to notice that Hertz initially planned to develop an energetic foundation of mechanics. This is evident form his previously quoted letter to Felix Klein and in particular from the following passage in the introduction to *The Principles of Mechanics*:

> I have discussed the second mode of representation [i.e. the energetic mode] at some length, not in order to urge its adoption, but rather to show why, *after due trial, I have felt obliged to abandon it.* [My italics] (Hertz, 1894, 29/24)

It is even possible to reconstruct a possible route that might have led Hertz from the energetic approach to the one he ended up following in his book. In the energetic approach to mechanics the principle of least action is often taken as a point of departure. In order to get to grips with this principle Hertz naturally turned to a paper by Helmholtz on this matter. In one of the very few letters in which he mentioned his mechanical work he explained to his former teacher:

> Recently I have been confined to theoretical work on topics suggested by a study of your papers on the principle of least action. I asked myself what form mechanics should be given right from the outset if the principle of least action is to appear at the point of departure and if its various forms are to show up not as the results of complicated derivations but as obvious truths of simple significance, and to present themselves clearly and distinctly as various forms of one and the same theorem. I am to a degree satisfied with my results, but I still have six months' or a year's work to go on this matter, ... (Hertz to Helmholtz, Dec. 25, 1892 (Hertz, 1977, 332/33))

Helmholtz's paper might have suggested to Hertz that potential energy can be considered as resulting from cyclic motion of a system of hidden masses. Such an idea is also more or less explicitly present in Maxwell's description of the electro-magnetic field. But if potential energy can be so conceived there is no need for neither force nor energy as basic notions in mechanics. The only things that are needed are hidden masses and rigid connections. Neither is as objectionable as the concepts of force and energy: hidden mass because it is just like ordinary mass, except it is not immediately perceptible, and connections because they are purely geometric in nature.

4. HERTZ'S MECHANICAL SYSTEMS

The theory of mechanics presented in Hertz's *Principles of Mechanics* operates with three basic concepts: Time, space and mass (ordinary as well as hidden). This is not as radical as the image suggested by William Thomson in which even ordinary masses are not basic concepts but only an epiphenomenona resulting from the vortex motion of the ether. Moreover Hertz's theory does not really avoid distance actions, since the connections can in principle act over arbitrary distances. However it does away with forces or energy as basic concepts.

In Hertz's mechanics there is only one law of motion:

> Fundamental Law: Every free system persists in its state of
> rest or of uniform motion along a straightest path. (Hertz,
> 1894, §309)

Here, unifom motion and straightest path have to be understood in terms of the "geometry of systems of points", a differential geometric formalism (a Riemannian geometry) that Hertz introduced in configuration space in order to simplify the presentation of his mechanics. According to Hertz, an isolated mechanical system therefore consists of a system of ordinary mass points connected to each other and to a system of hidden masses. The task of the physicist is to describe the motion of the system of ordinary masses without making any direct appeal to the motion of the hidden system that is unknown to us. Hertz showed that if we define force exerted on the ordinary system as a Lagrangean multiplier that is the result of the connection with the hidden system, and if we define potential energy as the kinetic energy of the hidden system, the ordinary system will approximately obey the usual laws of mechanics.

How, then, did Hertz argue for his version of mechanics? He did not argue that his theory is true about nature in the sense that nature really consists of such ordinary and hidden masses connected the way he described and moving according to the fundamental law. He emphasized that we have no way of knowing if that is the case or if there are in fact forces acting at a distance in the real world. Instead he argued in terms of a philosophical theory of images to which we shall turn now.

5. HERTZ'S IMAGE THEORY

In the introduction to his *Pinciples of Mechanics* Hertz first gave a three page introduction to his theory of images and then over the next 45 pages used it as a framework for comparing three images of mechanics: The usual

Newtonian image, the energetic image and Hertz's own image. Here I shall focus on the general ideas of images as described in the first three pages.

According to Hertz our theories of nature are mental images of external reality:

> We form ourselves images or symbols of external objects; and the form which we give them is such that *the necessary consequents of the images in thought are always the images of the necessary consequents in nature of the things pictured.* [my italics] (Hertz, 1894, 1/4)

This means in ordinary language that the predictions we can make of observable phenomena on the basis of the image turn out to be correct. This italicized requirement is the *basic requirement* to an image of external nature. Hertz formulated three requirements of an image:

1. It must be (logically) *permissible*. That means that it must be consistent with our laws of thought and our a priori intuitions.
2. It must be *correct* in the sense that it obeys the basic requirement stated above.

These two are absolute requirements of an image. However, there will in general be many images of nature or a section of nature that satisfy these two requirements. In order to chose between them Hertz invoked a third (relative) requirement:

3. It must be as *appropriate* as possible, which means that
   (a) It is as *distinct* as possible, i.e depicts more essential relations than competing images and
   (b) It must be as *simple* as possible, i.e. include fewer inessential relations (idle wheels) than any equally distinct competitor.

Hertz stressed that the requirement of correctness is the only empirical requirement and that we cannot require any other similarity between our image and external nature:

> The images which we here speak of are our conception [Vorstellungen] of things. With the things themselves they are in conformity in one important respect, namely in satisfying the above-mentioned requirement. For our purpose, it is not necessary that they should be in conformity with the things in any other respect, whatever. As a matter of fact, we do not know, nor have we any means of knowing, whether our conceptions of things are in conformity with them in any other than this one fundamental respect. (Hertz, 1894, 2/1–2)

This rather weak claim about conformity between image and nature has made G. Schiemann speak of "the loss of world in the image" [Schiemann 1998].

Hertz had to abstain from any further claim to truth of his images. Indeed, he was convinced that any image of nature had to contain hidden or unobservable entities or empty relations:

> Empty relations cannot be altogether avoided: they enter into the images because they are simply images. (Hertz, 1894, 3/2)

> We have felt sure from the beginning that unessential relations could not be altogether avoided in our images. (Hertz, 1894, 15/12)

However we obviously do not know what kind of unobservables exist in nature. The requirement of simplicity is a way to guide the choice of inessential elements in our images, but unlike many of his predecessors Hertz did not present any philosophical or theological argument for the simplicity of nature:

> We cannot *a priori* demand from nature simplicity nor can we judge what, in her opinion, is simple... hence our requirement of simplicity does not apply to nature, but to the images thereof which we fashion. (Hertz, 1894, 28/33–34)

Therefore images may be very dissimilar from nature except for the accordance required by correctness.

Hertz's image theory influenced philosophers as Wittgenstein, physicists as Boltzmann and Heisenberg and even mathematicians as Hilbert.[4] It owed elements to Kant, to the British physicists' ideas of analogies and to Helmholtz's ideas about images and symbols.[5] In this paper, however I shall only discuss the relations between Hertz's mature image theory and his earlier epistemological ideas.

6. IMAGES, COLORLESS THEORIES, AND THE GAY GARMENT

6.1. *Image theory and mechanics*

What is the relation between Hertz's image theory and the specific image of mechanics that he presented in his *Principles of Mechanics*? One could imagine two clearcut alternatives:

1. The image theory could have been the starting point and the basis for Hertz's development of his image of mechanics.
2. The particular choice of mechanical theory could have been the starting point and the basis for Hertz's development of his theory of images.

The first alternative is the one that at first seems to fit the presentation in the book best. In the introduction the image theory is presented first and then it is apparently made the basis of the discussion of the various images, a discussion that ends with the adoption of Hertz's image as the best. However there are problems with this alternative. First, a closer reading of the introduction will reveal that the comparison of the three images of mechanics do not really follow the pattern suggested by the image theory. Second, as we have seen above, Hertz's own route to his theory of mechanics seem to indicate that the comparison made in the introduction is a rational reconstruction rather than a reproduction of the reflections that guided him in the first place. Third, Hertz's manuscripts of the book reveal that the first draft of the image theory was written *after* the physical part of the book had reached an almost final stage. An early outline of the contents does not even stipulate a section on the image theory, but mentions in its place a historical introduction including a comparison of the different mechanical principles. So it does not look as if Hertz developed the image theory first and then used it as a guide in his search for a satisfactory foundation of mechanics.

So, did Hertz construct the image theory after he had already decided his own approach to mechanics as a philosophy that would help him argue for this choice of image? No, not really. In fact one can find early versions of the image theory in his lectures from 1884 on the constitution of matter, as well as in the introduction to his book on electric waves from 1892.

6.2. Hertz's 1884 ideas on images

It is ironic that the image theory, which has made Hertz famous as a "modern philosopher" (Baird et al., 1998) first appeared as a defense *against* philosophers. In his popular lectures in Kiel on the constitution of matter, Hertz described a fictitious dialogue between a philosopher and a physicist (himself). He did not have any particular philosopher in mind, but only the embodiment of the philosophical reflections of every human being, including physicists. In the dialogue Hertz did not want to denounce philosophical investigation, but he wanted to emphasize that "we [Hertz and his philosopher opponent], not only can, but must, pursue our goals independently of each other". Where the physicist's goal is to find an empirically correct description of the world, the philosopher's goal is to show its logical consistency. Where the physicist investigates the facts of nature experimentally, the philosopher investigates the difficulties that the human mind encounters while trying to understand facts of nature. In particular the philosopher may argue that the physicist's investigations of (microscopic) nature is impossible. For example a physicist may claim that matter is made up from atoms which are like small balls of diameter around 10^{-6} in an ether that conducts light waves. In fact much of Hertz's lecture course is devoted to arguing just that. However, the philosopher could justifiably argue that such a claim makes no sense: First we have not solved the enigma of the constitution of matter just by saying that matter consists of little balls; for we can always imagine such a ball cut into two half balls, so we must ask what make up the ball or the half balls. Second, we cannot imagine the balls or the constituents of the ether without ascribing to them properties that they cannot possibly have. For example we cannot imagine them without any color but since color is manifested by the length of a wave in the ether, it makes no sense to attribute color to the ether.

However, Hertz maintained that we cannot avoid such a situation. Even in the most perfect of sciences mathematics, one operates with straight lines which are by definition without width but cannot be imagined without a very small width.

> It is a general and necessary property of the human mind that we can neither intuitively represent nor conceptually define, the things without attributing properties to them that do not at all exist in them. (Hertz, 1999, 35)

Such addition of inessential properties "is not false intuitions, but the condition for imagining at all" (Hertz, 1999, 36) and according to Hertz it poses no problem as long as one keeps in mind which of the properties are essential and which are inessential.

> Thus let us guard ourselves from believing that we can investigate the nature of the things themselves by considering the atoms; let us also guard ourselves from confusing the unnecessary properties, that we must necessarily ascribe to them with the essential properties, that are merely time and space relations. However, let them [the philosophers] not make us believe that we have worked in vain when we have made ourselves images [Bilder] of the things that are real but do not enter into our mind, images that correspond to those things in some respects, while in other respects they bear the imprint of our imagination. We have then, in our field, followed the general course of the human mind. (Hertz, 1999, 36)

In this way the idea of an image helped Hertz reject the philosopher's objections. Twice in his 1884 lectures he mentioned which properties images of the real world must have. First (Hertz, 1999, 35) he required that its essential elements must be 1. logically possible and appropriate and the entire image should be as probable as possible. Later (Hertz, 1999, 62) he asked that they be 1. possible mathematically (corresponding loosely to permissibility in the *Mechanics*), 2. possible physically (corresponding to correctness in the *Mechanics*), and 3. advantageous which means that it will facilitate our understanding. It must be emphasized that advantageousness does not correspond to appropriateness as it is understood in the *Mechanics*. Indeed the addition of inessential elements or properties will per definition decrease simplicity and therefore appropriateness but it may well increase advantageousness. For example Hertz explained how a sufficiently vivid image of matter might lead us to immediately intuit consequences that one could otherwise only find out through lengthy calculations. In this way a good image can work as a computer.

Already in 1884 Hertz explicitly rejected the question "is our intuition correct" as meaningless. Of course correct here does not mean correct in the 1894 sense but rather it means true of the world.

In his 1884 lectures Hertz also told a parable of paper money: Just as paper money symbolizes real values, so our mental images symbolize the external world. Some things are shared by the symbol and the signified and other things are not. It is important to distinguish between the two types of things. Otherwise we will behave like a man who puts bills into the melting pot, because he has heard that one can change money into silver.

Hertz acknowledged that some physicists were in favor of a purely phenomenalistic description of the world, but in 1884 he does not seem to believe that such a lawful theory is feasible, and he definitely favored his images.

6.3. *1892 ideas on the simple and homely figure and the gay garment*

When Hertz in 1892 collected his papers on electromagnetic waves in a book he added an introduction in which he explained how his theoretical papers on Maxwell's equations related to Maxwell's own presentation of his theory. Hertz's presentation was very axiomatic in spirit: he postulated that in space there were vector fields (he did not use that word) and he postulated differential equations between them (Maxwell's equations) from which he deduced experimentally verifiable consequences. He admitted that this approach lacks intuitive appeal.

> Nevertheless I believe that we cannot, without deceiving ourselves, extract much more from experience than is asserted in the papers referred to. If we wish to lend more color to the *theory*, there is nothing to prevent us from supplementing all this and aiding our powers of imagination by *concrete representations* of the various conceptions as to the nature of electric

polarization, the electric current, etc. But scientific accuracy requires of us that we should in no wise confuse the *simple and homely figure*, as it is presented to us by nature, with the *gay garment* which we use to clothe it. Of our free will we can make no change whatever in the form of the one, but the cut and color of the other we can chose as we please. [My italics](Hertz, 1892, 30–31/28)

In order to exemplify what Hertz had in mind we can think of a representation of polarization by a host of similarly oriented dipoles. This is a colorful description, but the homely figure is nothing but a vector field.

This distinction between a phenomenalistic theory (a simple and homely figure) and a representation (a gay garment) corresponds quite well to the 1884 distinction between a phenomenalistic theory and an image. But where he came down in favor of the image in 1884 he now favored the theory. In particular he now praised and emphasized the simplicity of the theory:

I have further endeavored in the exposition [of the theory] to limit as far as possible the number of those concepts which are arbitrarily introduced by us, and only to admit such elements as cannot be removed or altered without at the same time altering possible experimental results. (Hertz, 1892, 30/28)

6.4. 1884, 1892, and 1894 compared

Thus we can conclude that Hertz was not in possession of his mature image theory when he began to write his *Principles of Mechanics*, but he had developed similar ideas in connection with his earlier reflections about the constitution of matter and electro-magnetism. More specifically his mature image theory seems to be a result of incorporating the 1892 requirement of simplicity into the 1884 idea of an image, which had not contained such a requirement. Why did he not stick to the rather phenomenalistic 1892 theories? Apparently because he had come to the conclusion that we cannot strip an image of all inessential relations or concepts:

If we try to understand the motions of bodies around us, and to refer them to simple and clear rules, paying attention only to what can be directly observed, our attempt will in general fail. We soon become aware that the totality of things visible and tangible do not form an universe conformable to law, in which the same results always follow from the same conditions. We become convinced that the manifold of the actual universe must be greater than the manifold of the universe which is directly revealed to us by our senses. If we wish to obtain an image of the universe (Weltbild), which shall be well-rounded, complete, and conformable to law, we have to presuppose, behind the things which we see, other, invisible things – to search for confederates concealed beyond the limits of our senses. (Hertz, 1894, 30/25)

Thus Hertz was convinced that unobservables or inessential relations are needed in order to make a law-like description of nature. That means that the phenomenalistic theory is impossible; we need both essential and inessential elements, for example hidden mass and connections in mechanics.

But in that case, why did Hertz not stick with the not so simple colorful images of his 1884 lectures? I shall mention three reasons:

1. A strategic reason: By requiring simplicity of an image Hertz had a great tool for arguing for the superiority of his own image of mechanics as compared with the Newtonian and the energetic image.
2. Moreover in his mechanics Hertz was first of all interested in creating a logically permissible image. He emphasized that one cannot remove a logical contradiction by adding more to an image although this was often what other authors had tried. The only way one can get rid of a contradiction is to remove something from the image. Thus a requirement about simplicity is indirectly also a requirement that tend to increase permissibility.
3. The difference of the subject matter may also give an explanation of the addition of the requirement of simplicity. Indeed, in 1884 Hertz suggested an image of an atom as a number of smaller balls that are connected by elastic bands that will allow them to vibrate relative to each other. Thus the image is phrased in terms of even more fundamental things (mechanical entities). However in the *Mechanics* Hertz was at the bedrock. Mechanics was supposed to be the mother image in terms of which one should be able to construct all of physics. The only concepts that are more fundamental are the concepts of time and space relations. Indeed it is interesting to notice that Hertz did in fact define his concept of mass in terms of time and space relations. He defined a *Massenteilchen* as what we would call a function of time with values in space. A point mass is a collection of such *Massenteilchen*. In this way his *Mechanics* presented a stark and colorless image, which is not very different from his 1892 theory, except for the fact that it contains one type of unobservable things, the hidden masses (or two if we add connections).

7. CONCLUSION

Hertz's work on mechanics did not represent a break with his earlier work on electro magnetism but rather a natural continuation. It was an attempt to provide a mechanical foundation of all of physic including electro magnetism. It was an attempt to provide mechanics with a logically sound minimalistic axiomatic foundation of the same kind as the one he had provided for Maxwell's theory. In particular it was an attempt to rid mechanics of the concept of force which would be superfluous if distance forces can be avoided as his experiments with electro magnetic waves suggested. And finally, it provided a theoretical basis for the study of the ether which according to Hertz and many of his colleagues was the most urgent subject in fundamental physics.

Similarly, Hertz's epistemological theory of images was a reworking of ideas he had developed in connection with his earlier lectures on the constitution of matter. In these lectures the idea of an image enabled him to speak about such unobservables as atoms, molecules and the ether without fearing objections from positivist phenomenologists. These ideas may have been at the back of his mind when he decided to leave his initial energetic approach to mechanics in favor of an approach involving hidden masses. And at the same time his concept of an image changed from a rather colorful type to a starker colorless type. This happened when he incorporated the requirement of simplicity that he had originally formulated for his rather phenomenalistic Maxwellian theory.

8. POSTSCRIPT ABOUT PHILOSOPHY, PHYSICS AND MATHEMATICS

The conclusion illustrates the connections found in Hertz's *Mechanics* between philosophy and physics. The relation between physics and mathematics is even clearer. Although Hertz emphasized that the physical content and the mathematical form of his image of mechanics were logically independent, he also stressed that they fit each other very well. Indeed a mechanics without forces but with connections will naturally focus on systems rather than single point masses. Hertz's geometry of systems of points was created precisely in order to be able to treat systems in a way similar to the treatment of single particles in ordinary mechanics.

Also in the details of Hertz's mechanics there are many ways in which the mathematics influenced the physics or conversely. Let me just mention that Hertz seems to have introduced the concept of Massenteilchen in order to give a derivation of the line element in his Riemannian geometry (Lützen, 1999). Conversely, the introduction of the concept of reduced components along a coordinate in his differential geometry (corresponding to the concept of covariant components of a vector) seems to have been suggested to Hertz by the Lagrange and the Hamilton formalisms.

The relation between mathematics and philosophy is less obvious in Hertz's Mechanics. However, this last side of the triangle was supplied by Hilbert. He explicitly viewed his *Grundlagen der Geometrie* as an image of space in Hertz's sense. He even followed Hertz and removed as many empty relations as possible in order to obtain consistency. He also saw his own requirements to an axiomatic system: consistency, completeness and independence as parallels to Hertz's requirements of permissibility, correctness and simplicity. (Corry, 1997).

University of Copenhagen
Denmark

NOTES

[1] Quote from Hertz's diary (Hertz, 1977, 313).
[2] Hertz to Sarassin May 19 1893 (Fölsing, 1997, 500).
[3] This is probably a reference to William Thomson's attempts to describe matter as vortices in the ether.
[4] (Corry, 1997), (Majer, 1998).
[5] (Friedman, 1997).

REFERENCES

Baird, D., Hughes, R.I.G. and Nordmann, A. (1998). *Heinrich Hertz: Classical Physicist, Modern Philosopher*, Kluwer, Dordrecht.

Corry, L. (1997). David Hilbert and the axiomatization of physics (1894–1905), *Archive for History of Exact Science* **51(2)**: 83–198.

Fölsing, A. (1997). *Heinrich Hertz. Eine Biographie*, Hoffmann und Campe, Hamburg.

Friedman, M. (1997). Helmholtz's Zeichentheorie and Schlick's allgemeine Erkenntnislehre: Early empiricism and ist nineteenth century background, *Philosophical Topics* **25**: 19–50.

Hertz, H. (1889). Über die Beziehung zwischen Licht und Elektricität, *Vorträge gehalten bei der 62. Versammlung deutscher Naturforscher und Ärztezu Heidelberg am 20. September 1889, Gesammelte Werke von Heinrich Hertz* **1**: 339–354. Emil Strauss, Bonn, 1895.

Hertz, H. (1890). Ueber die Grundgleichungen der Electrodynamik für ruhende Körper, *Nachrichten von der Königl. Gesellschaft der Wissenschaften zu Göttingen* pp. 106–149.

Hertz, H. (1892). *Untersuchungen ueber die Ausbreitung der elektrischen Kraft*, Barth, Leipzig. English translation: *Electric Waves*. Macmillan, London, 1900.

Hertz, H. (1894). Die Prinzipien der Mechanik in neuem Zusammenhange dargestellt, *Gesammelte Werke*, Vol. 3, 1910, Barth, Leipzig. English translation: *The Principles of Mechanics Presented in a New Form*. Macmillan, 1900.

Hertz, H. (1999). *Die Constitution der Materie*, Springer Verlag, Berlin. Edited by A. Fölsing.

Hertz, J. (1977). *Heinrich Hertz. Erinnerungen, Briefe, Tagebücher / Memoirs, Letters, Diaries, , 2. ed.*, Physik Verlag / San Fransisco Press, Weinheim / San Fransisco.

Koenigsberger, L. (1903). *Hermann von Helmholtz, vol. 3*, Vieveg, Braunschweig.

Lützen, J. (1999). A Matter of Matter or a Matter of Space, *Archives Internationales d'Histoire des Sciences* **49**: 103–121.

Lützen, J. (2005). *Mechanistic Images in Geometric Form: Heinrich Hertz's Principles of Mechanics*, Oxford University Press, Oxford.

Mach, E. (1883). *Die Mechanik in Ihrer Entwickelung. Historisch-kritisch dargestellt*, Brockhaus, Leipzig.

Majer, U. (1998). Heinrich Hertz's Picture-Conception of Theories: Its Elaboration by Hilbert, Weyl, and Ramsey, pp. 225–242. In Baird et al. (1998).

MICHEL JANSSEN AND MATTHEW MECKLENBURG

FROM CLASSICAL TO RELATIVISTIC MECHANICS:
ELECTROMAGNETIC MODELS OF THE ELECTRON

1. INTRODUCTION

"Special relativity killed the classical dream of using the energy-momentum-velocity relations as a means of probing the dynamical origins of [the mass of the electron]. The relations are purely kinematical" (Pais, 1982, 159). This perceptive comment comes from a section on the pre-relativistic notion of electromagnetic mass in '*Subtle is the Lord* ... ', Abraham Pais' highly acclaimed biography of Albert Einstein. 'Kinematical' in this context means 'independent of the details of the dynamics'. In this paper we examine the classical dream referred to by Pais from the vantage point of relativistic continuum mechanics.

There were actually two such dreams in the years surrounding the advent of special relativity. Like Einstein's theory, both dreams originated in the electrodynamics of moving bodies developed in the 1890s by the Dutch physicist Hendrik Antoon Lorentz. Both took the form of concrete models of the electron. Even these models were similar. Yet they were part of fundamentally different programs competing with one another in the years around 1905. One model, due to the German theoretician Max Abraham (1902a), was part of a revolutionary effort to substitute the laws of electrodynamics for those of Newtonian mechanics as the fundamental laws of physics. The other model, adapted from Abraham's by Lorentz (1904b) and fixed up by the French mathematician Henri Poincaré (1906), was part of the attempt to provide a general explanation for the absence of any sign of the earth's motion through the ether, the elusive 19th-century medium thought to carry light waves and electromagnetic fields. A choice had to be made between the objectives of Lorentz and Abraham. One could not eliminate all signs of ether drift and reduce all physics to electrodynamics at the same time. Special relativity was initially conflated with Lorentz's theory because it too seemed to focus on the undetectability of motion at the expense of electromagnetic purity. The theories of Lorentz and Einstein agreed in all their empirical predictions, including those for the velocity-dependence of electron mass, even though special relativity was not wedded to any particular model of the electron. For a while there was a third electron model, a variant on Lorentz's proposed independently by Alfred Bucherer (1904, 57–60; 1905) and Paul Langevin (1905). At the time, the acknowledged arbiter between these models and the broader theories (perceived to be) attached to them was a series of experiments by Walter Kaufmann and others on the deflection of high-speed

65

V.F. Hendricks, K.F. Jørgensen, J. Lützen and S.A. Pedersen (eds.), Interactions: Mathematics, Physics and Philosophy, 1860-1930, pp. 65–134.
© *2006 Springer.*

electrons in β-radiation and cathode rays by electric and magnetic fields for the purpose of determining the velocity-dependence of their mass.[1]

As appropriate for reveries, neither Lorentz's nor Abraham's dream about the nature and structure of the electron lasted long. They started to fade a few years after Einstein's formulation of special relativity, even though the visions that inspired them lingered on for quite a while. Lorentz went to his grave in 1928 clinging to the notion of an ether hidden from view by the Lorentz-invariant laws governing the phenomena. Abraham's electromagnetic vision was pursued well into the 1920s by kindred spirits such as Gustav Mie (1912a, 1912b, 1913). By then mainstream physics had long moved on. The two dreams, however, did not evaporate without a trace. They played a decisive role in the development of relativistic mechanics.[2] It is no coincidence therefore that relativistic (continuum) mechanics will be central to our analysis in this paper. The development of the new mechanics effectively began with the non-Newtonian transformation laws for force and mass introduced by Lorentz (1895, 1899). It continued with the introduction of electromagnetic momentum and electromagnetic mass by Abraham (1902a, 1902b, 1903, 1904a, 1905, 1909) in the wake of the proclamation of the electromagnetic view of nature by Willy Wien (1900). Einstein (1907b), Max Planck (1906a, 1908), Hermann Minkowski (1908), Arnold Sommerfeld (1910a, 1910b), and Gustav Herglotz (1910, 1911)—the last three champions of the electromagnetic program[3]—all contributed to its further development in a proper relativistic setting. These efforts culminated in a seminal paper by Max Laue (1911a) and were enshrined in the first textbook on relativity published later that year (Laue, 1911b).

There already exists a voluminous literature on the various aspects of this story.[4] We shall freely draw and build on that literature. One of us has written extensively on the development of Lorentz's research program in the electrodynamics of moving bodies (Janssen, 1995, 2002b; Janssen and Stachel, 2004).[5] The canonical source for the electromagnetic view of nature is still (McCormmach, 1970), despite its focus on Lorentz whose attitude toward the electromagnetic program was ambivalent (cf. Lorentz, 1900; 1905, 93–101; 1915, secs. 178–186). His work formed its starting point and he was sympathetic to the program, but never a strong advocate of it. (Goldberg, 1970) puts the spotlight on the program's undisputed leader, Max Abraham. (Pauli, 1921, Ch. 5) is a good source for the degenerative phase of the electromagnetic program in the 1910s.[6] For a concise overview of the rise and fall of the electromagnetic program, see Ch. 8 of (Kragh, 1999), aptly titled "A Revolution that Failed."

Another important source for the electromagnetic program is Ch. 5 in (Pyenson, 1985), which discusses a seminar on electron theory held in Göttingen in the summer semester of 1905. Minkowski was one of four instructors of this course. The other three were Herglotz, David Hilbert, and Emil Wiechert. Max Laue audited the seminar as a postdoc. Among the students was Max Born, whose later work on the problem of rigid bodies in special relativity (Born, 1909a, 1909b, 1910) was inspired by the seminar.[7] The syllabus for the seminar lists papers by Lorentz (1904a, 1904b), Abraham (1903), Karl Schwarzschild (1903a, 1903b, 1903c), and Sommerfeld (1904a, 1904b, 1905a). This seminar gives a good indication of how active and cutting edge this research area was at the time. Further evidence of this vitality is provided by debates in the literature of the day over various points concerning these electron models such as those between Wien (1904a, 1904b, 1904c, 1904d) and Abraham (1904b, 1904c),[8] Bucherer (1907, 1908a, 1908b) and Ebenezer Cunningham (1907, 1908),[9] and Einstein (1907a) and Paul Ehrenfest (1906, 1907). The roll call of researchers active in this area also included the Italian mathematician Tullio Levi-Civita (1907, 1909).[10] One may even get the impression that in the early 1900s the journals were flooded with papers on electron models. We wonder, for instance, whether the book by Bucherer (1904) was not originally written as a long journal article, which was rejected, given its similarity to earlier articles by Abraham, Lorentz, Schwarzschild, and Sommerfeld.

The saga of the Abraham, Lorentz, and Bucherer-Langevin electron models and their changing fortunes in the laboratories of Kaufmann, Bucherer, and others has been told admirably by Arthur I. Miller (1981, secs. 1.8–1.14, 7.4, and 12.4). Miller (1973) is also responsible for a detailed analysis of the classic paper by Poincaré (1906) that introduced what came to be known as "Poincaré pressure" to stabilize Lorentz's purely electromagnetic electron.[11] The model has been discussed extensively in the physics literature, by Fritz Rohrlich and by such luminaries as Enrico Fermi, Paul Dirac, and Julian Schwinger.[12] It is also covered elegantly in volume two of the Feynman lectures (Feynman et al., 1964, Ch. 28). (Pais, 1972) and (Rohrlich, 1973) combine discussions of physics and history in an informative way.

Given how extensively this episode has been discussed in the historical literature, the number of sources covering its denouement with the formulation of Laue's relativistic continuum mechanics is surprisingly low. Max Jammer does not discuss relativistic continuum mechanics at all in his classic monograph on the development of the concept of mass (cf. Jammer, 1997, Chs. 11–13). Miller prominently discusses Laue's work, both in (Miller, 1973, sec. 7.5) and in the concluding section of his book (Miller, 1981, sec.

12.5.8), but does not give it the central place that in our opinion it deserves. To bring out the importance of Laue's work, we show right from the start how the kind of spatially extended systems studied by Abraham, Lorentz, and Poincaré can be dealt with in special relativity. We use modern notation and modern units throughout and give self-contained derivations of almost all results. Our treatment of these electron models follows the analysis of the experiments of Trouton and Noble in (Janssen, 1995, 2002b, 2003), which was inspired in part by the discussion in (Norton, 1992) of the importance of Laue's relativistic mechanics for the development of Gunnar Nordström's special-relativistic theory of gravity. The focus on the conceptual changes in mechanics that accompanied the transition from classical to relativistic kinematics was inspired in part by the work of Jürgen Renn and his collaborators on pre-classical mechanics (Damerow et al., 2004). Ultimately, our story is part of a larger tale about shifts in such concepts as mass, energy, momentum, and stresses and the relations between them in the transition from Newtonian mechanics and the electrodynamics of Maxwell and Lorentz to special relativity.

2. ENERGY-MOMENTUM-MASS-VELOCITY RELATIONS

2.1. Special relativity

In special relativity, the relations between energy, momentum, mass, and velocity of a system are encoded in the transformation properties of its four-momentum. This quantity combines the energy U and the three components of the ordinary momentum $\mathbf{P}$:[13]

$$(4) \qquad P^\mu = \left(\frac{U}{c}, \mathbf{P} \right)$$

(where c is the velocity of light). In the system's rest frame, with coordinates $x_0^\mu = (ct_0, x_0, y_0, z_0)$, the four-momentum reduces to:

$$(5) \qquad P_0^\mu = \left(\frac{U_0}{c}, 0, 0, 0 \right),$$

i.e., $\mathbf{P}_0 = 0$. The system's rest mass is defined as $m_0 \equiv U_0/c^2$.

We transform P_0^μ from the x_0^μ-frame to some new x^μ-frame, assuming for the moment that P_0^μ always transforms as a four-vector under Lorentz transformations. Let the two frames be related by the Lorentz transformation $x^\mu = \Lambda^\mu_{\ \nu} x_0^\nu$, where the transformation matrices $\Lambda^\mu_{\ \nu}$ satisfy $\Lambda^\mu_{\ \rho} \Lambda^\nu_{\ \sigma} \eta^{\rho\sigma} = \eta^{\mu\nu}$, the defining equation for Lorentz transformations, with $\eta^{\mu\nu} \equiv \mathrm{diag}(1, -1, -1, -1)$ the standard diagonal Minkowski metric. Here and in the rest of the paper summation over repeated indices is implied. We follow the convention that

Greek indices run from 0 to 3 and Latin ones from 1 to 3. Since, in general, the four-momentum does *not* transform as a four-vector, the Lorentz transform of P_0^μ will, in general, not be the four-momentum in the x^μ-frame. We therefore cautiously write the result of the transformation with an asterisk:

$$(6) \qquad P^{*\mu} = \Lambda^\mu_{\ \nu} P_0^\nu.$$

Without loss of generality we can focus on the special case in which the motion of the x^μ-frame with respect to the x_0^μ-frame is with velocity v in the x-direction. The matrix for this transformation is:

$$(7) \qquad \Lambda^\mu_{\ \nu} = \begin{pmatrix} \gamma & \gamma\beta & 0 & 0 \\ \gamma\beta & \gamma & 0 & 0 \\ 0 & 0 & 1 & 0 \\ 0 & 0 & 0 & 1 \end{pmatrix}$$

with $\gamma \equiv 1/\sqrt{1 - \beta^2}$ and $\beta \equiv v/c$. In that case,

$$(8) \qquad P^{*\mu} = \left(\gamma\frac{U_0}{c}, \gamma\beta\frac{U_0}{c}, 0, 0 \right) = (\gamma m_0 c, \gamma m_0 \mathbf{v}).$$

If the four-momentum of the system *does* transform as a four-vector, $P^{*\mu}$ in eq. 8 is equal to P^μ in eq. 4 and we can read off the following relations between energy, momentum, mass, and velocity from these two equations:

$$(9) \qquad U = \gamma U_0 = \gamma m_0 c^2, \quad \mathbf{P} = \gamma m_0 \mathbf{v}.$$

Eqs. 9 hold for a relativistic point particle with rest mass m_0. Its four-momentum is given by

$$(10) \qquad P^\mu = m_0 u^\mu = m_0 \frac{dx^\mu}{d\tau} = \gamma m_0 \frac{dx^\mu}{dt}.$$

Since $u^\mu \equiv dx^\mu/d\tau$ is the four-velocity, this is clearly a four-vector. The relation between proper time τ, arc length s, and coordinate time t is given by $d\tau = ds/c = dt/\gamma$.[14] If the particle is moving with velocity $\mathbf{v}$, $dx^\mu/dt = (c, \mathbf{v})$ and eq. 10 becomes:

$$(11) \qquad P^\mu = (\gamma m_0 c, \gamma m_0 \mathbf{v}).$$

Eqs. 9 also hold for spatially extended *closed* systems, i.e., systems described by an energy-momentum tensor $T^{\mu\nu}$ with a vanishing four-divergence, i.e., $\partial_\nu T^{\mu\nu} = 0$ (where ∂_ν stands for $\partial/\partial x^\nu$). The energy-momentum tensor brings together the following quantities. The component T^{00} gives the energy density; T^{i0}/c the components of the momentum density; cT^{0i} the components of the energy flow density;[15] and T^{ij} the components of the momentum flow density, or, equivalently, the stresses.[16] The standard definition of the four-momentum of a spatially extended (not necessarily closed) system described by the (not necessarily divergence-free) energy-momentum tensor $T^{\mu\nu}$ is:

$$(12) \qquad\qquad P^\mu \equiv \frac{1}{c} \int T^{\mu 0} d^3x.$$

Before the advent of relativity, this equation was written as a pair of separate equations:

$$(13) \qquad\qquad U = \int u d^3x, \quad \mathbf{P} = \int \mathbf{p} d^3x,$$

where u and $\mathbf{p}$ are the energy density and the momentum density, respectively. Definition 12 is clearly not manifestly Lorentz invariant. The space integrals of $T^{\mu 0}$ in the x^μ-frame are integrals in space-time over a three-dimensional hyperplane of simultaneity in that frame. A Lorentz transformation does not change the hyperplane over which the integration is to be carried out. A hyperplane of simultaneity in the x^μ-frame is *not* a hyperplane of simultaneity in any frame moving with respect to it. From these last three observations, it follows that the Lorentz transforms of the space integrals in eq. 12 will not be space integrals in the new frame. But then how can these Lorentz transforms ever be the four-momentum in the new frame? The answer to this question is that *if the system is closed* (i.e., if $\partial_\nu T^{\mu\nu} = 0$), it does not matter over which hyperplane the integration is done. The integrals of the relevant components of $T^{\mu\nu}$ over any hyperplane extending to infinity will all give the same values. So for closed systems a Lorentz transformation does map the four-momentum in one frame to a quantity that is equal to the four-momentum in the new frame even though these two quantities are defined as integrals over different hyperplanes.[17]

The standard definition of four-momentum can be replaced by a manifestly Lorentz-invariant one. First note that the space integrals of $T^{\mu 0}$ in the x^μ-frame can be written in a manifestly covariant form as[18]

$$(14) \qquad\qquad P^\mu = \frac{1}{c} \int \delta \left(\eta_{\rho\sigma} x^\rho n^\sigma \right) T^{\mu\nu} n_\nu d^4x,$$

where $\delta(x)$ is the Dirac delta function, defined through $\int f(x)\delta(x-a)dx = f(a)$, and n^μ is a unit vector in the time direction in the x^μ-frame. In that frame n^μ has components $(1,0,0,0)$. The delta function picks out hyperplanes of simultaneity in the x^μ-frame. The standard definition 12 of four-momentum can, of course, be written in the form of eq. 14 in any frame, but that requires a different choice of n^μ in each one. This is just a different way of saying what we said before: under the standard definition 12, the result of transforming P^μ in the x^μ-frame to some new frame will *not* be the four-momentum in the new frame unless the system is closed. If, however, we take the unit vector n^μ in eq. 14 to be some *fixed* timelike vector—typically the unit vector in the time direction in the system's rest frame[19]—and take eq. 14 with that fixed vector n^μ as our new definition of four-momentum, the problem disappears.

Eq. 14 with a fixed timelike unit vector n^μ provides an alternative manifestly Lorentz-invariant definition of four-momentum. Under this new definition—which was proposed by, among others, Fermi (1922)[20] and Fritz Rohrlich (1960, 1965)[21]—the four-momentum of a spatially extended system transforms as a four-vector under Lorentz transformations no matter whether the system is open or closed. The definitions 12 and 14 are equivalent to one another for closed systems, but only coincide for open systems in the frame of reference in which n^μ has components $(1,0,0,0)$. In this paper, we shall use the admittedly less elegant definition 12, simply because either it or its decomposition into eqs. 13 were the definitions used in the period of interest. Part of the problem encountered by our protagonists simply disappears by switching to the alternative definition 14. With this definition energy and momentum always obey the familiar relativistic transformation rules, regardless of whether we are dealing with closed systems or with their open components. As one would expect, however, a mere change of definition does not take care of the main problem that troubled the likes of Lorentz, Poincaré, and Abraham. That is the problem of the stability of a spatially extended electromagnetic electron.

2.2. Pre-relativistic theory

The analogues of relations 9 between energy, mass, momentum, and velocity in Newtonian mechanics are the basic formulae for kinetic energy and momentum:

$$(15) \qquad U_{\text{kin}} = \frac{1}{2}mv^2, \quad \mathbf{p} = m\mathbf{v}$$

In the years before the advent of special relativity, electrodynamics was a hybrid theory in which Galilean-invariant Newtonian mechanics was supposed to govern matter while Maxwell's equations, which are inherently Lorentz invariant, governed the electromagnetic fields. This hybrid theory already harbored the relativistic energy-momentum-velocity relations.

Initially, the starting point of physicists working in this area had unquestionably been Newton's second law, $\mathbf{F} = m\mathbf{a}$, force equals mass times acceleration. Electrodynamics merely supplied the Lorentz force for the left-hand side of this equation. Eventually, however, some of the leading practitioners were leaning toward the view that matter does not have any Newtonian mass at all and that its inertia is just a manifestation of the interaction of electric charge distributions with their self-fields. Lorentz was reluctantly driven to this conclusion because, as we shall see in secs. 3 and 4, it would help explain the absence of any signs of ether drift. Abraham enthusiastically embraced it because it opened up the prospect of a purely electromagnetic basis for all of physics. With $\mathbf{F} = m\mathbf{a}$ reduced to $\mathbf{F} = 0$, Newton's second law only nominally retained its lofty position as the fundamental equation of motion. All real work was done by electrodynamics. Writing $\mathbf{F} = 0$ as $d\mathbf{P}_{tot}/dt = 0$, one can read it as expressing momentum conservation. Momentum does not need to be mechanical. Abraham introduced the concept of electromagnetic momentum.[22] Lorentz was happy to leave Newtonian royalty its ceremonial role. Abraham, of a more regicidal temperament, sought to replace $\mathbf{F} = m\mathbf{a}$ by a new purely electrodynamic equation that would explain why Newton's law had appeared to be the rule of the land for so long.

Despite their different motivations, Lorentz and Abraham agreed that the effective equation of motion for an electron in some external field is[23]

$$(16) \qquad \mathbf{F}_{ext} + \mathbf{F}_{self} = 0,$$

with $\mathbf{F}_{ext}$ the Lorentz force coming from the external field and $\mathbf{F}_{self}$ the Lorentz force coming from the self-field of the electron. The key experiments to which eq. 16 was applied were the experiments of Kaufmann and others on the deflection of fast electrons by electric and magnetic fields. Both Lorentz and Abraham conceived of the electron as a spherical surface charge distribution. They disagreed about whether the electron's shape would depend on its velocity with respect to the ether, more specifically about whether it would be subject to a microscopic version of the Lorentz-FitzGerald contraction. Lorentz believed it would, Abraham believed it would not.

The Lorentz force that an electron moving through the ether at velocity $\mathbf{v}$ experiences from its self-field can be written as minus the time derivative of the quantity that Abraham proposed to call the electromagnetic momentum:

$$(17) \qquad \mathbf{F}_{self} = \int \rho(\mathbf{E} + \mathbf{v} \times \mathbf{B})d^3x = -\frac{d\mathbf{P}_{EM}}{dt}.$$

In this expression ρ is the density of the electron's charge distribution, and $\mathbf{E}$ and $\mathbf{B}$ are the electric and magnetic field produced by this charge distribution. The electromagnetic momentum of these fields is defined as

$$(18) \qquad \mathbf{P}_{EM} \equiv \int \varepsilon_0 \mathbf{E} \times \mathbf{B}d^3x,$$

and doubles as the electromagnetic momentum of the electron itself. In general there will be an extra term on the right-hand side of eq. 17. In general, the components of $\mathbf{F}_{\text{self}}$ are given by:

$$(19) \qquad F^i_{\text{self}} = -\frac{dP^i_{\text{EM}}}{dt} + \int \partial_j T^{ij}_{\text{Maxwell}} d^3x,$$

where

$$(20) \qquad T^{ij}_{\text{Maxwell}} \equiv \varepsilon_0 \left(E^i E^j - \frac{1}{2}\delta^{ij} E^2 \right) + \mu_0^{-1} \left(B^i B^j - \frac{1}{2}\delta^{ij} B^2 \right)$$

is the Maxwell stress tensor (the Kronecker delta δ^{ij} is defined as follows: $\delta^{ij} = 1$ for $i = j$ and 0 otherwise). Gauss's theorem tells us that this additional term vanishes as long as T^{ij}_{Maxwell} drops off faster than $1/r^2$ as $\mathbf{x}$ goes to infinity. Simple derivations of these results can be found in many sources, old and new.[24] With the help of eq. 17 the electromagnetic equation of motion 16 can be written in the form of the Newtonian equation $\mathbf{F} = d\mathbf{p}/dt$ with Abraham's electromagnetic momentum replacing ordinary momentum:

$$(21) \qquad \mathbf{F}_{\text{ext}} = \frac{d\mathbf{P}_{\text{EM}}}{dt}.$$

Like Newton's second law, which can be written either as $\mathbf{F} = m\mathbf{a}$ or as $\mathbf{F} = d\mathbf{p}/dt$, this new law can, under special circumstances, be written as the product of mass and acceleration. Assume that the momentum is in the direction of motion,[25] i.e., that $\mathbf{P}_{\text{EM}} = (P_{\text{EM}}/v)\mathbf{v}$. We then have

$$(22) \qquad \frac{d\mathbf{P}_{\text{EM}}}{dt} = \frac{dP_{\text{EM}}}{dt}\frac{\mathbf{v}}{v} + P_{\text{EM}}\frac{d}{dt}\left(\frac{\mathbf{v}}{v}\right).$$

The first term on the right-hand side can be written as

$$(23) \qquad \frac{dP_{\text{EM}}}{dt}\frac{\mathbf{v}}{v} = \frac{dP_{\text{EM}}}{dv}\frac{dv}{dt}\frac{\mathbf{v}}{v} = \frac{dP_{\text{EM}}}{dv}\mathbf{a}_{//},$$

where $\mathbf{a}_{//}$ is the longitudinal acceleration, i.e., the acceleration in the direction of motion. The second term can be written as

$$(24) \qquad P_{\text{EM}}\frac{d}{dt}\left(\frac{\mathbf{v}}{v}\right) = \frac{P_{\text{EM}}}{v}\mathbf{a}_{\perp},$$

where $\mathbf{a}_{\perp}$ is the transverse acceleration, i.e., the acceleration perpendicular to the direction of motion. The factors multiplying these two components of the acceleration are called the *longitudinal (electromagnetic) mass*, $m_{//}$, and the *transverse (electromagnetic) mass*, $m_{\perp}$, respectively. This terminology is due to Abraham (1903, 150–151). Eq. 22 can thus be written as

$$(25) \qquad \frac{d\mathbf{P}_{\text{EM}}}{dt} = m_{//}\mathbf{a}_{//} + m_{\perp}\mathbf{a}_{\perp},$$

with[26]

(26)
$$m_{//} = \frac{dP_{\text{EM}}}{dv}, \quad m_{\perp} = \frac{P_{\text{EM}}}{v}.$$

The effective equation of motion 21 becomes:

(27)
$$\mathbf{F}_{\text{ext}} = m_{//}\mathbf{a}_{//} + m_{\perp}\mathbf{a}_{\perp}.$$

We shall see that, for $v = 0$ (in which case the electron models of Abraham and Lorentz coincide), $m_{//} = m_{\perp} = m_0$, and that, for $v \neq 0$, $m_{//}$ and $m_{\perp}$ differ from m_0 only by terms of order v^2/c^2. For velocities $v \ll c$, eq. 27 thus reduces to:

(28)
$$\mathbf{F}_{\text{ext}} \approx m_0(\mathbf{a}_{//} + \mathbf{a}_{\perp}) = m_0\mathbf{a}.$$

Proponents of the electromagnetic view of nature took eq. 21 to be the fundamental equation of motion and derived Newton's law from it by identifying the ordinary Newtonian mass with the electromagnetic mass m_0 of the relevant system at rest in the ether.

Eq. 26 defines the longitudinal mass $m_{//}$ of the electron in terms of its electromagnetic momentum. It can also be defined in terms of the electron's electromagnetic energy. Consider the work done as an electron is moving in the x-direction in the absence of an external field. The work expended goes into the internal energy of the electron, $dU = -dW$. According to eq. 16, the work is done by $\mathbf{F}_{\text{self}}$.[27] The internal energy is identified with the electromagnetic energy U_{EM}:

(29)
$$dU_{\text{EM}} = -dW = -\mathbf{F}_{\text{self}} \cdot d\mathbf{x}.$$

Using eqs. 17 and 25, we can write this as

(30)
$$dU_{\text{EM}} = \frac{d\mathbf{P}_{\text{EM}}}{dt} \cdot d\mathbf{x} = m_{//}\mathbf{a}_{//} \cdot d\mathbf{x} = m_{//}\frac{dv}{dt}dx = m_{//}vdv.$$

It follows that[28]

(31)
$$m_{//} = \frac{1}{v}\frac{dU_{\text{EM}}}{dv}.$$

As we shall see in sec. 4, given the standard definitions 13 of electromagnetic energy and momentum, the neglect of non-electromagnetic stabilizing forces in the derivation of eqs. 26 and 31 leads to an ambiguity in the expression for the longitudinal mass of Lorentz's electron.

If the combination of the energy U (divided by c), and the momentum $\mathbf{P}$ for any system, electromagnetic or otherwise, transforms as a four-vector under Lorentz transformations, then $m_{//}$ calculated from eq. 26 (with P substituted for P_{EM}) is equal to $m_{//}$ calculated from eq. 31 (with U substituted for U_{EM}).[29] Consider the transformation from a rest frame with coordinates x_0^μ to the x^μ-frame. In that case (see eqs. 4–8):

$$(32) \qquad P^\mu = \left(\frac{U}{c}, \mathbf{P}\right) = (\gamma m_0 c, \gamma m_0 \mathbf{v}).$$

The energy U gives the longitudinal mass (see eq. 31)

$$(33) \qquad m_{//} = \frac{1}{v}\frac{dU}{dv} = \frac{1}{v}\frac{d}{dv}(\gamma m_0 c^2) = \frac{m_0 c^2}{v}\frac{d\gamma}{dv}.$$

The momentum $\mathbf{P}$ gives the longitudinal mass (eq. 26):

$$(34) \qquad m_{//} = \frac{dP}{dv} = \frac{d}{dv}(\gamma m_0 v) = m_0 \frac{d(\gamma v)}{dv}.$$

Noting that[30]

$$(35) \qquad \frac{d\gamma}{dv} = \gamma^3 \frac{v}{c^2}, \quad \frac{d(\gamma v)}{dv} = \gamma^3,$$

we find that eqs. 33 and 34 do indeed give the same result:

$$(36) \qquad m_{//} = \frac{1}{v}\frac{dU}{dv} = \frac{dP}{dv} = \gamma^3 m_0.$$

The momentum $\mathbf{P}$ in eq. 32 gives the transverse mass (eq. 26):

$$(37) \qquad m_\perp = \frac{P}{v} = \gamma m_0.$$

Eqs. 36 and 37 give mass-velocity relations that hold for any relativistic particle. These equations thus have much broader applicability than their origin in electrodynamics suggests. This is exactly what killed the dreams of Abraham and Lorentz of using these relations to draw conclusions about the nature and shape of the electron.

3. LORENTZ'S THEOREM OF CORRESPONDING STATES, THE GENERALIZED CONTRACTION HYPOTHESIS, AND THE VELOCITY DEPENDENCE OF ELECTRON MASS

Lorentz had already published the relativistic eqs. 36 and 37 for longitudinal and transverse mass, up to an undetermined factor l, in 1899. To understand how Lorentz originally arrived at these equations we need to take a look at his general approach to problems in the electrodynamics of moving bodies.[31] The basic problem that Lorentz was facing was that Maxwell's equations are not invariant under Galilean transformations, which relate frames in relative motion to one another in Lorentz's classical Newtonian space-time. Lorentz thus labored under the impression that Maxwell's equations only hold in frames at rest in the ether and not in the terrestrial lab frames in which all our experiments are done.

Consider an ether frame with space-time coordinates $(t_0, \mathbf{x}_0)$ and a lab frame with space-time coordinates $(t, \mathbf{x})$ related to one another via the Galilean transformation

$$t = t_0, \quad x = x_0 - vt_0, \quad y = y_0, \quad z = z_0,$$

(38)

$$\mathbf{E} = \mathbf{E}_0, \quad \mathbf{B} = \mathbf{B}_0, \quad \rho = \rho_0.$$

The second line of this equation expresses that the electric field, the magnetic field, and the charge density remain the same even though after the transformation they are thought of as functions of $(t, \mathbf{x})$ rather than as functions of $(t_0, \mathbf{x}_0)$.

The equations for the fields produced by a charge distribution static in the lab frame as functions of the space-time coordinates $(t, \mathbf{x})$ are obtained by writing down Maxwell's equations for the relevant quantities in the lab frame, adding the current $\mu_0 \rho \mathbf{v}$[32] and replacing time derivatives by the operator $\partial/\partial t - v\partial/\partial x$.[33] We thus arrive at:

$$\operatorname{div} \mathbf{E} = \rho/\varepsilon_0, \quad \operatorname{curl} \mathbf{B} = \mu_0 \rho \mathbf{v} + \frac{1}{c^2}\left(\frac{\partial \mathbf{E}}{\partial t} - v\frac{\partial \mathbf{E}}{\partial x}\right),$$

(39)

$$\operatorname{div} \mathbf{B} = 0, \quad \operatorname{curl} \mathbf{E} = -\frac{\partial \mathbf{B}}{\partial t} + v\frac{\partial \mathbf{B}}{\partial x}.$$

Lorentz now replaced the space-time coordinates $(t, \mathbf{x})$, the fields $\mathbf{E}$ and $\mathbf{B}$, and the charge density ρ by auxiliary variables defined as:

$$\mathbf{x}' = l\operatorname{diag}(\gamma, 1, 1)\mathbf{x}, \quad t' = l\left(\frac{t}{\gamma} - \gamma\left(\frac{v}{c^2}\right)x\right),$$

$$\rho' = \frac{\rho}{\gamma l^3},$$

(40)

$$\mathbf{E}' = \frac{1}{l^2}\operatorname{diag}(1, \gamma, \gamma)(\mathbf{E} + \mathbf{v} \times \mathbf{B}),$$

$$\mathbf{B}' = \frac{1}{l^2}\operatorname{diag}(1, \gamma, \gamma)\left(\mathbf{B} - \frac{1}{c^2}\mathbf{v} \times \mathbf{E}\right),$$

where l is an undetermined factor that is assumed to be equal to one to first order in v/c. Since the auxiliary time variable depends on position, it is called *local time*. Lorentz showed that the auxiliary fields $\mathbf{E}'$ and $\mathbf{B}'$ and the auxiliary charge density ρ' written as functions of the auxiliary space-time coordinates $(t', \mathbf{x}')$ satisfy Maxwell's equations:

(41)
$$\text{div}' \mathbf{E}' = \rho'/\varepsilon_0, \quad \text{curl}' \mathbf{B}' = \frac{1}{c^2}\frac{\partial \mathbf{E}'}{\partial t'},$$

$$\text{div}' \mathbf{B}' = 0, \quad \text{curl}' \mathbf{E}' = -\frac{\partial \mathbf{B}'}{\partial t'}.$$

When the factor l is set equal to one, what Lorentz showed, at least for static charge densities,[34] is that Maxwell's equations are invariant under what Poincaré (1906, 495) proposed to call textitLorentz transformations. For $l = 1$, the transformation formulae in eq. 40 for the fields $\mathbf{E}$ and $\mathbf{B}$ and for a static charge density ρ look exactly the same as in special relativity. The transformation formulae for the space-time coordinates do not. Bear in mind, however, that Lorentz did the transformation in two steps, given by eqs. 38 and 40, respectively. Schematically, we have:

(42) $\quad (t_0, \mathbf{x}_0, \mathbf{E}_0, \mathbf{B}_0, \rho_0) \rightarrow (t, \mathbf{x}, \mathbf{E}, \mathbf{B}, \rho) \rightarrow (t', \mathbf{x}', \mathbf{E}', \mathbf{B}', \rho')_{l=1}.$

Combining these two steps, we recover the familiar Lorentz transformation formulae. For the fields and the charge density, this is just a matter of replacing $(\mathbf{E}, \mathbf{B}, \rho)$ in eq. 40 by $(\mathbf{E}_0, \mathbf{B}_0, \rho_0)$ and setting $l = 1$. For the space-time coordinates, it takes only a minimal amount of algebra:

(43)
$$x' = \gamma x = \gamma(x_0 - vt_0), \quad y' = y = y_0, \quad z' = z = z_0,$$

$$t' = \frac{t}{\gamma} - \gamma\frac{v}{c^2}x = \frac{t_0}{\gamma} - \gamma\frac{v}{c^2}(x_0 - vt_0) = \gamma\left(t_0 - \frac{v}{c^2}x_0\right),$$

where in the second line we used that $1/\gamma + \gamma(v^2/c^2) = \gamma(1/\gamma^2 + \beta^2) = \gamma$.

The inverse of the transformation $(t_0, \mathbf{x}_0, \mathbf{E}_0, \mathbf{B}_0) \rightarrow (t', \mathbf{x}', \mathbf{E}', \mathbf{B}')$ for $l = 1$ is found by interchanging $(t_0, \mathbf{x}_0, \mathbf{E}_0, \mathbf{B}_0)$ and $(t', \mathbf{x}', \mathbf{E}', \mathbf{B}')$ and changing $\mathbf{v}$ to $-\mathbf{v}$. Doing the inversion for $l \neq 1$ also requires changing l to l^{-1}. The inverse of the transformation $(\mathbf{E}_0, \mathbf{B}_0) \rightarrow (\mathbf{E}', \mathbf{B}')$ for $l \neq 1$, for instance, is given by

(44)
$$\mathbf{E}_0 = \mathbf{E} = l^2\text{diag}(1, \gamma, \gamma)(\mathbf{E}' - \mathbf{v} \times \mathbf{B}'),$$

$$\mathbf{B}_0 = \mathbf{B} = l^2\text{diag}(1, \gamma, \gamma)\left(\mathbf{B}' + \frac{1}{c^2}\mathbf{v} \times \mathbf{E}'\right).$$

The transformation is symmetric only for $l = 1$. Unlike Lorentz before 1905, Poincaré and Einstein both looked upon the primed quantities as the quantities measured by the observer in the lab frame. In special relativity, the ether frame is just another inertial frame on a par with the lab frame. The situation for observers in these two frames will be fully symmetric only if $l = 1$. This was essentially the argument for both Poincaré and Einstein to set $l = 1$. As we shall see in the next section, Lorentz also ended up setting $l = 1$ but on the basis of a roundabout dynamical argument. For our purposes it is important that we leave the factor l undetermined for the time being.

The invariance of Maxwell's equations under the combination of transformations 38 and 40 allowed Lorentz to formulate what he called the *theorem of corresponding states*. This theorem says that for any field configuration in a frame at rest in the ether there is a corresponding field configuration in a frame moving through the ether such that the auxiliary fields $\mathbf{E}'$ and $\mathbf{B}'$ in the moving frame are the same functions of the auxiliary space and time coordinates $(t', \mathbf{x}')$ as the real fields $\mathbf{E}_0$ and $\mathbf{B}_0$ in the frame at rest of the real space and time coordinates $(t_0, \mathbf{x}_0)$. Lorentz was particularly interested in free field configurations (for which $\rho = 0$) describing patterns of light and darkness. Most experiments in optics eventually boil down to the observation of such patterns.

To describe a pattern of light and darkness it suffices to specify where the fields averaged over times that are long compared to the period of the light used vanish and where these averages are large. $\mathbf{E}'$ and $\mathbf{B}'$ are linear combinations of $\mathbf{E}$ and $\mathbf{B}$ (see eq. 40). They are large (small) when- and wherever $\mathbf{E}$ and $\mathbf{B}$ are. Since patterns of light and darkness by their very nature are effectively static, no complications arise from the x-dependence of local time. If it is light (dark) simultaneously at two points with coordinates $\mathbf{x}_0 = \mathbf{a}$ and $\mathbf{x}_0 = \mathbf{b}$ in some field configuration in a frame at rest in the ether, it will be light (dark) simultaneously at the corresponding points $\mathbf{x}' = \mathbf{a}$ and $\mathbf{x}' = \mathbf{b}$ in the corresponding state in a frame moving through the ether. In terms of the real coordinates these are the points $\mathbf{x} = (1/l)\text{diag}(1/\gamma, 1, 1)\mathbf{a}$ and $\mathbf{x} = (1/l)\text{diag}(1/\gamma, 1, 1)\mathbf{b}$. The pattern of light and darkness in a moving frame is thus obtained from its corresponding pattern in a frame at rest in the ether by contracting the latter by a factor γl in the direction of motion and a factor l in the directions perpendicular to the direction of motion. Examining the formula for the local time in eq. 40, one likewise sees that the periods of light waves in a moving frame are obtained by multiplying the periods of the light waves in the corresponding state at rest in the ether by a factor γ/l.

To account for the fact that these length-contraction and time-dilation effects in electromagnetic field configurations were never detected, Lorentz (1899) assumed that matter interacting with the fields (e.g., the optical components producing patterns of light and darkness) experiences these same effects. Lorentz thereby added a far-reaching physical assumption to his purely mathematical theorem of corresponding states. Elsewhere one of us has dubbed this assumption the *generalized contraction hypothesis* (Janssen,

1995, sec. 3.3; 2002b; Janssen and Stachel, 2004). It was through this hypothesis that Lorentz decreed a number of exemptions of the Newtonian laws that had jurisdiction over matter in his theory. The length-contraction and time-dilation rules to which matter and field alike had to be subject to account for the absence of any signs of ether drift are examples of such exemptions. The velocity dependence of mass is another (Janssen, 1995, sec. 3.3.6). This is the one that is important for our purposes.

Suppose an oscillating electron in a light source at rest in S_0 satisfies $\mathbf{F}_0 = m_0\mathbf{a}_0$. In the corresponding state in S the corresponding electron will then satisfy the same equation in terms of the auxiliary quantities, i.e.,

(45) $$\mathbf{F}' = m_0\mathbf{a}',$$

where $\mathbf{F}'$ is the same function of $(t', \mathbf{x}')$ as $\mathbf{F}_0$ is of $(t_0, \mathbf{x}_0)$, and where $\mathbf{a}' = d^2\mathbf{x}'/dt'^2$ and $\mathbf{a}_0 = d^2\mathbf{x}_0/dt_0^2$ are always the same at corresponding points in S and S_0. Lorentz assumed that motion through the ether affects all forces on the electron the same way it affects Coulomb forces[35]

(46) $$\mathbf{F}' = \frac{1}{l^2}\text{diag}(1, \gamma, \gamma)\mathbf{F}.$$

For the relation between the acceleration $\mathbf{a}'$ in terms of the auxiliary space and time coordinates and the real acceleration $\mathbf{a}$, Lorentz used the relation

(47) $$\mathbf{a}' = \frac{1}{l}\text{diag}(\gamma^3, \gamma^2, \gamma^2)\mathbf{a}.$$

In general, this relation is more complicated, but when the velocity $d\mathbf{x}_0/dt_0$ with which the electron is oscillating in S_0 is small, $d\mathbf{x}'/dt'$ (equal to $d\mathbf{x}_0/dt_0$ at the corresponding point in S_0) can be neglected and eq. 47 holds. A derivation of the general relation between $\mathbf{a}'$ and $\mathbf{a}$ was given by Planck (1906a) in the context of his derivation of the relativistic generalization of Newton's second law, a derivation mathematically essentially equivalent to Lorentz's 1899 derivation of the velocity dependence of mass, except that Planck only had to consider the special case $l = 1$.[36]

Lorentz probably arrived at eq. 47 through the following crude argument. If an electron oscillates around a fixed point in S with a low velocity and a small amplitude, the x-dependent term in the expression for local time can be ignored. In that case, we only need to take into account that $\mathbf{x}'$ differs from $\mathbf{x}$ by $l\,\text{diag}(\gamma, 1, 1)$ and that t' differs from t by l/γ (see eq. 40). This gives a quick and dirty derivation of eq. 47:

(48) $$\mathbf{a}' = \frac{d^2\mathbf{x}'}{dt'^2} = \left(\frac{\gamma}{l}\right)^2 l\,\text{diag}(\gamma, 1, 1)\frac{d^2\mathbf{x}}{dt^2} = \frac{1}{l}\text{diag}\left(\gamma^3, \gamma^2, \gamma^2\right)\mathbf{a}.$$

Inserting eqs. 46 and 47 into eq. 45, we find

(49)
$$\frac{1}{l^2}\text{diag}(l,\gamma,\gamma)\mathbf{F} = \frac{1}{l}\text{diag}\left(\gamma^3,\gamma^2,\gamma^2\right)m_0\mathbf{a}.$$

This can be rewritten as

(50)
$$\mathbf{F} = l\,\text{diag}\left(\gamma^3,\gamma,\gamma\right)m_0\mathbf{a}.$$

From this equation it follows that the oscillation of an electron in the moving source can only satisfy Newton's second law if the mass m of an electron with velocity $\mathbf{v}$ with respect to the ether (remember that the velocity of the oscillation itself was assumed to be negligible) differs from the mass m_0 of an electron at rest in the ether in precisely the following way:

(51)
$$m_{//} = l\gamma^3 m_0, \quad m_\perp = l\gamma m_0.$$

If $l = 1$, these are just the relativistic eqs. 36 and 37. It was Planck who showed in the paper mentioned above that these relations also obtain in special relativity.[37] Planck's interpretation of these relations was very different from Lorentz's. For Planck, as for Einstein, the velocity dependence of mass was part of a new relativistic mechanics replacing classical Newtonian mechanics. Lorentz wanted to retain Newtonian mechanics, even after he accepted in 1904 that there are no Galilean-invariant Newtonian masses or forces in nature. Consequently, he had to provide an explanation for the peculiar velocity-dependence of electron mass he needed to account for the absence of any detectable ether drift. In 1904, adapting Abraham's electron model, Lorentz provided such an explanation in the form of a specific model of the electron that exhibited exactly the velocity dependence of eq. 51 for $l = 1$.

4. ELECTROMAGNETIC ENERGY, MOMENTUM, AND MASS OF A MOVING ELECTRON

In this section we use Lorentz's theorem of corresponding states—or, in modern terms, the Lorentz invariance of Maxwell's equations—to calculate the energy, the momentum, and the Lagrangian for the field of a moving electron, conceived of as nothing but a surface charge distribution and its electromagnetic field. We then compute the longitudinal and the transverse mass of the electron.

We distinguish three different models. In all three the electron at rest in the ether is spherical. In Abraham's model it remains spherical when it is set in motion; in Lorentz's model it contracts by a factor γ in the direction of motion; and in the Bucherer-Langevin model it contracts by a factor $\gamma^{2/3}$ in the direction of motion but expands by a factor $\gamma^{1/3}$ in the directions

	moving electron	corresponding state stretch dimensions of moving system by $\mathrm{diag}(\gamma l, l, l)$
The rigid electron of Abraham (l arbitrary)	sphere (R, R, R)	ellipsoid $(\gamma l R, l R, l R)$
The contractile electron of Lorentz and Poincaré ($l = 1$)	ellipsoid $(R/\gamma, R, R)$	sphere (R, R, R)
The contractile electron of constant volume of Bucherer and Langevin ($l = \gamma^{-1/3}$)	ellipsoid $(R/\gamma^{2/3}, \gamma^{1/3} R, \gamma^{1/3} R)$	sphere (R, R, R)

FIGURE 1.   A moving electron according to the models of Abraham, Lorentz, and Bucherer-Langevin, and the corresponding states at rest in the ether.

perpendicular to the direction of motion so that its volume remains constant. Fig. 1 shows a moving electron according to these three models along with the corresponding states in a frame at rest in the ether. For Abraham's rigid electron the corresponding state is an ellipsoid; for the contractile electrons of Lorentz and Bucherer-Langevin it is a sphere.

In the corresponding state of a moving electron (in relativistic terms: in the electron's rest frame) there is no magnetic field. Hence $\mathbf{B}' = 0$ and eq. 44

gives:

$$\text{(52)} \qquad \mathbf{E} = l^2 \left(E_x', \gamma E_y', \gamma E_z' \right), \quad \mathbf{B} = \frac{\gamma l^2 v}{c^2} \left(0, -E_z', E_y' \right).$$

4.1. Energy

The energy of the electric and magnetic field is defined as

$$\text{(53)} \qquad U_{EM} = \int \left(\frac{1}{2} \varepsilon_0 E^2 + \frac{1}{2} \mu_0^{-1} B^2 \right) d^3 x.$$

For the field in eq. 52 with $B_x = 0$, it is given by

$$\text{(54)} \quad U_{EM} = \int \frac{1}{2} \varepsilon_0 E_x^2 d^3 x + \int \frac{1}{2} \varepsilon_0 \left(E_y^2 + E_z^2 \right) d^3 x + \int \frac{1}{2} \mu_0^{-1} \left(B_y^2 + B_z^2 \right) d^3 x.$$

Following Poincaré (1906, 523), we call these three terms A, B, and C. Using eq. 52 and $d^3 x = d^3 x' / \gamma l^3$, we find

$$A = \frac{l}{\gamma} \int \frac{1}{2} \varepsilon_0 E_x'^2 d^3 x' = \frac{l}{\gamma} A',$$

$$\text{(55)} \qquad B = l\gamma \int \frac{1}{2} \varepsilon_0 \left(E_y'^2 + E_z'^2 \right) d^3 x' = l\gamma B',$$

$$C = \frac{\mu_0^{-1} \gamma l v^2}{c^4} \int \frac{1}{2} \left(E_y'^2 + E_z'^2 \right) d^3 x' = l\gamma \beta^2 B',$$

where in the last step we used $c^2 = 1/\varepsilon_0 \mu_0$. If the corresponding state is spherical,

$$\text{(56)} \qquad B' = 2A' = \frac{2}{3} U_{EM}'.$$

It follows that for the models of Lorentz and Bucherer-Langevin:

$$\text{(57)} \qquad U_{EM} = l\gamma \left(\frac{1}{\gamma^2} + 2 + 2\beta^2 \right) A' = l\gamma \left(1 + \frac{1}{3} \beta^2 \right) U_{EM}',$$

where we used that $\gamma^{-2} = 1 - \beta^2$ and that $3A' = U_{EM}'$. Eq. 57 can also be written as

$$\text{(58)} \qquad U_{EM} = l\gamma \left(\frac{4}{3} - \frac{1}{3} (1 - \beta^2) \right) U_{EM}' = l \left(\frac{4\gamma}{3} - \frac{1}{3\gamma} \right) U_{EM}'.$$

4.2. Lagrangian

The Lagrangian can be computed the same way. We start from

$$(59) \qquad L_{EM} = \int \mathcal{L}_{EM} d^3x,$$

where $\mathcal{L}_{EM}$ is the Lagrange density defined as (note the sign)

$$(60) \qquad \mathcal{L}_{EM} \equiv \frac{1}{2}\mu_0^{-1}B^2 - \frac{1}{2}\varepsilon_0 E^2.$$

This quantity transforms as a scalar under Lorentz transformations as can be seen from its definition in manifestly Lorentz-invariant form:[38]

$$(61) \qquad \mathcal{L}_{EM} \equiv \frac{1}{4}\mu_0^{-1} F_{\mu\nu}F^{\mu\nu}.$$

It follows that $\mathcal{L}_{EM} = l^4 \mathcal{L}'_{EM}$ with $\mathcal{L}'_{EM} = -(1/2)\varepsilon_0 E'^2$. Eq. 59 thus gives:

$$(62) \qquad L_{EM} = \int l^4 \mathcal{L}'_{EM} \frac{d^3x'}{\gamma l^3} = -\frac{l}{\gamma}U'_{EM}.$$

4.3. Momentum

The electromagnetic momentum can also be computed in this way. For the field of the electron, the electromagnetic momentum density (see eq. 18) is:

$$(63) \qquad \mathbf{p}_{EM} = \varepsilon_0 \begin{pmatrix} E_y B_z - E_z B_y \\ -E_x B_z \\ E_x B_y \end{pmatrix} = \varepsilon_0 \gamma l^4 \frac{v}{c^2} \begin{pmatrix} \gamma(E_y'^2 + E_z'^2) \\ E_x' E_y' \\ E_x' E_z' \end{pmatrix}.$$

Because of symmetry (in all three models)

$$(64) \qquad \int p_{y_{EM}} d^3x = \int p_{z_{EM}} d^3x = 0.$$

For the x-component, we find

$$(65) \qquad P_{x_{EM}} = \frac{1}{\gamma l^3} \int p_{x_{EM}} d^3x' = \gamma l \frac{v}{c^2} \int \varepsilon_0 \left(E_y'^2 + E_z'^2 \right) d^3x' = \gamma l \frac{v}{c^2} 2B'$$

(see eq. 55). For the contractile electron (Lorentz and Bucherer-Langevin), $B' = (2/3)U'_{EM}$ (see eq. 56). In that case

$$(66) \qquad \mathbf{P}_{EM} = \frac{4}{3}\gamma l \left(\frac{U'_{EM}}{c^2} \right) \mathbf{v}.$$

This pre-relativistic equation will immediately strike anyone familiar with the basic formulae of special relativity as odd. Remember that from a relativistic point of view the energy U'_{EM} of the moving electron's corresponding state at rest in the ether is nothing but the energy $U_{0_{EM}}$ of the electron in its rest frame. Comparison of eq. 66 with $l = 1$ to $\mathbf{P} = \gamma m_0 \mathbf{v}$ (eq. 9) suggests that the rest mass of the electron is $m_0 = \frac{4}{3} U_{0\text{EM}}/c^2$. This seems to be in blatant contradiction to the equation everybody knows, $E = mc^2$. This is the notorious "4/3 puzzle" of the energy-mass relation of the classical electron. The origin of the problem is that the system we are considering, the self-field of the electron, is not closed and that its four-momentum consequently does not transform as a four-vector, at least not under the standard definition 12 of four-momentum. The solution to the puzzle is either to add another piece to the system so that the composite system is closed or to adopt the alternative Fermi-Rohrlich definition 14 (with a fixed unit vector n^μ) of the four-momentum of spatially extended systems.[39] As we shall see, the "4/3 puzzle" had already reared its ugly head before the advent of special relativity, albeit in a different guise.

4.4. Longitudinal and transverse mass

Substituting eqs. 58 and 66 for the energy and momentum of the field of a moving contractile electron into the expressions 26 and 31 for the electron's transverse and longitudinal mass, we find:

$$(67) \qquad m_{//} = \frac{dP_{\text{EM}}}{dv} = \frac{d(\gamma l v)}{dv} \frac{4}{3} \frac{U'_{\text{EM}}}{c^2},$$

$$(68) \qquad m_\perp = \frac{P_{\text{EM}}}{v} = \gamma l \frac{4}{3} \frac{U'_{\text{EM}}}{c^2},$$

$$(69) \qquad m_{//} = \frac{1}{v} \frac{dU_{\text{EM}}}{dv} = \frac{1}{v} \frac{d}{dv} \left(\frac{4\gamma l}{3} - \frac{l}{3\gamma} \right) U'_{\text{EM}}.$$

Several conclusions can be drawn from these equations. First, it turns out that eq. 67 only gives the velocity dependence of the longitudinal mass required by Lorentz's generalized contraction hypothesis for $l = 1$. Unfortunately, for $l = 1$, eq. 69 does not give the same longitudinal mass as eq. 67. One only obtains the same result for $l = \gamma^{-1/3}$. This is the value for the Bucherer-Langevin constant-volume contractile electron.

It is easy to prove these claims. Using eq. 35, we can write eq. 67 as

$$(70) \qquad m_{//} = \frac{dP_{\text{EM}}}{dv} = \left(\gamma^3 l + \gamma v \frac{dl}{dv} \right) \frac{4}{3} \frac{U'_{\text{EM}}}{c^2}.$$

From eqs. 68 and 70 it follows that the only way to ensure that $m_{//} = l\gamma^3 m_0$ and $m_\perp = l\gamma m_0$, as required by the generalized contraction hypothesis (see eq. 51), is to set the Newtonian mass equal to zero, to set $l = 1$, and to define the mass of the electron at rest in the ether as

$$(71) \qquad m_0 = \frac{4}{3} \frac{U'_{\text{EM}}}{c^2}$$

(which, from a relativistic point of view, amounts to the odd equation $E = \frac{3}{4}mc^2$). Eqs. 70 and 68 then reduce to

(72) $$m_{//} = \gamma^3 m_0, \quad m_\perp = \gamma m_0,$$

in accordance with eq. 51.

Lorentz (1904) had thus found a concrete model for the electron with a mass exhibiting exactly the velocity dependence that he had found in 1899. This could hardly be a coincidence. Lorentz concluded[40] that the electron was indeed nothing but a small spherical surface charge distribution, subject to a microscopic version of the Lorentz-FitzGerald contraction when set in motion, and that its mass was purely electromagnetic, i.e., the result of interaction with its self-field. This is Lorentz's version of the classical dream referred to by Pais in the passage we quoted in the introduction. The mass-velocity relations for Lorentz's electron model are just the relativistic relations 36–37. So it is indeed no coincidence that Lorentz found these same relations twice, first, in 1899, as a necessary condition for rendering ether drift unobservable (see eqs. 45–51) and then, in 1904, as the mass-velocity relations for a concrete Lorentz-invariant model of the electron. But the explanation is not, as Lorentz thought, that his model provides an accurate representation of the real electron; it is simply that the mass of *any* Lorentz-invariant model of *any* particle—whatever its nature and whatever its shape—will exhibit the exact same velocity dependence. This was first shown (for static systems) by Laue (1911a) and, to use Pais' imagery again, it killed Lorentz's dream.

Quite independently of Laue's later analysis, Lorentz's electron model appeared to be dead on arrival. The model as it stands is inconsistent. One way to show this is to compare expression 72 for the longitudinal mass $m_{//}$ derived from the electron's electromagnetic momentum to the expression for $m_{//}$ derived from its electromagnetic energy. These two calculations, it turns out, do not give the same result (Abraham, 1905, 188, 204).[41] Setting $l = 1$ in eq. 69 and using eq. 35, we find

(73) $$m_{//} = \frac{1}{v}\frac{4}{3}\frac{d\gamma}{dv}U'_{EM} - \frac{1}{3v}\frac{d}{dv}\left(\frac{1}{\gamma}\right)U'_{EM} = \gamma^3\frac{4}{3}\frac{U'_{EM}}{c^2} - \frac{1}{3v}\frac{d}{dv}\left(\frac{1}{\gamma}\right)U'_{EM}.$$

The first term in the last expression is equal to $m_{//}$ in eq. 70 for $l = 1$. Without even working out the second term, we thus see that momentum and energy lead to different expressions for the longitudinal mass of Lorentz's electron.

For the Bucherer-Langevin electron there is no ambiguity in the formula for its longitudinal mass. Inserting $l = \gamma^{-1/3}$ and eq. 71 into eq. 70, we find that the electromagnetic momentum of the Bucherer-Langevin electron gives:

$$(74) \qquad m_{//} = \frac{dP_{EM}}{dv} = \left(\gamma^{8/3} + \gamma v \frac{d\gamma^{-1/3}}{dv} \right) m_0 = \gamma^{8/3} \left(1 - \frac{1}{3}\beta^2 \right) m_0,$$

where in the last step we used eq. 35 in conjunction with

$$(75) \qquad \frac{d\gamma^{-1/3}}{dv} = -\frac{1}{3}\gamma^{-4/3}\frac{d\gamma}{dv} = -\frac{1}{3}\gamma^{5/3}\frac{v}{c^2}.$$

Inserting $l = \gamma^{-1/3}$ and eq. 71 into eq. 69, we find that its electromagnetic energy gives:

$$(76) \qquad m_{//} = \frac{1}{v}\frac{dU_{EM}}{dv} = \frac{c^2}{v}\frac{d}{d\gamma}\left(\gamma^{2/3} - \frac{1}{4}\gamma^{-4/3} \right) \frac{d\gamma}{dv}m_0.$$

Some simple gamma gymnastics establishes that eq. 76 reproduces eq. 74:[42]

$$(77) \qquad m_{//} = \frac{1}{v}\frac{dU_{EM}}{dv} = \gamma^{8/3} \left(1 - \frac{1}{3}\beta^2 \right) m_0.$$

So energy and momentum of the Bucherer-Langevin electron do indeed give the same longitudinal mass. The same is true for the Abraham electron, although the calculation is more involved and unimportant for our purposes.

One feature that the Abraham model and the Bucherer-Langevin model have in common and that distinguishes both models from Lorentz's is that the volume of the electron is constant. Hence, whatever forces are responsible for stabilizing the electron never do any work and can safely be ignored, as was done in the derivation of the basic equations 26 and 31 for longitudinal mass (see eqs. 16 and 29 and notes 23 and 27). This does not mean that no such forces are needed. In all three models, one is faced with the problem of the electron's stability. Abraham, however, argued that whereas Lorentz's contractile electron called for the explicit addition of non-electromagnetic stabilizing forces, he, Abraham, could simply take the rigidity of his own spherical electron as a given and proceed from there without ever running into trouble.

In the introduction of the 1903 exposition of his electron dynamics, Abraham (1903, 108–109) devoted two long paragraphs to the justification of this crucial assumption. He distinguished three sets of equations for the dynamics of the electron. We already encountered two of these, the "field equations" determining the self-field of the electron and the "fundamental dynamical equations" determining the motion of the electron in an external field. Logically prior to these, however, is what Abraham called the "basic kinematical equation," which "limits the freedom of motion of the electron."

This is the assumption that the electron always retains its spherical shape. Abraham tried to preempt the criticism he anticipated on this score:

> This basic kinematical hypothesis may strike many as arbitrary; invoking the analogy with ordinary electrically charged solid bodies, many would subscribe to the view that the truly enormous field strengths at the surface of the electron—field strengths a trillion times larger than those amenable to measurement—are capable of deforming the electron; that electrical and elastic forces on a spherical electron would be in equilibrium as long as the electron is at rest; but that the motion of the electron would change the forces of the electromagnetic field, and thereby the shape of the equilibrium state of the electron. This is not the view that has led to agreement with experiment. It also seemed to me that the assumption of a deformable electron is not allowed on fundamental grounds. The assumption leads to the conclusion that work is done either by or against the electromagnetic forces when a change of shape takes place, which means that in addition to the electromagnetic energy an internal potential energy of the electron needs to be introduced. If this were really necessary, it would immediately make an electromagnetic foundation of the theory of cathode and Becquerel rays, purely electric phenomena, impossible: one would have to give up on an electromagnetic foundation of mechanics right from the start. It is our goal, however, to provide a purely electromagnetic foundation for the dynamics of the electron. For that reason we are no more entitled to ascribe elasticity to the electron than we are to ascribe material mass to it. On the contrary, our hope is to learn to understand the elasticity of matter on the basis of the electromagnetic conception (Abraham, 1903, 108–109).

The suggestion that experimental data, presumably those of Kaufmann, supported his kinematics was wishful thinking on Abraham's part (cf. Miller, 1981, secs. 1.9 and 1.11). In support of his more general considerations— as an argument it is a textbook example of the genetic fallacy—Abraham proceeded to appeal to no less an authority than Heinrich Hertz:

> Hertz has convincingly shown that one is allowed to talk about rigid connections before one has talked about forces. Our dynamics of the electron does not talk about forces trying to deform the electron at all. It only talks about "external forces,"

which try to give [the electron] a velocity or an angular velocity, and about "internal forces", which stem from the [self-]
field of the electron and which balance these external forces.
Even these "forces" and "torques" are only auxiliary quantities defined in terms of the fundamental kinematic and electromagnetic concepts. The same holds for terms like "work,"
"energy," and "momentum." The guiding principle in choosing these terms, however, was to bring out clearly the analogy
between electromagnetic mechanics and the ordinary mechanics of material bodies (Abraham, 1903, 109).

Abraham submitted this paper in October 1902, almost three years before
the publication of Einstein's first paper on relativity. He can thus hardly
be faulted for basing his new electromagnetic mechanics on the old Newtonian kinematics. Minkowski would sneer a few years later that "approaching Maxwell's equation with the concept of a rigid electron seems to me the
same thing as going to a concert with your ears stopped up with cotton wool"
(quoted in Miller, 1981, sec. 12.4.5, 330). He made this snide comment during the 80th *Versammlung Deutscher Naturforscher und Ärzte* in Cologne
in September 1908, the same conference where he gave his now famous
talk "Space and Time" (Minkowski, 1909). His veritable diatribe against
the rigid electron, which he called a "monster" and "no working hypothesis
but a working hindrance," came during the discussion following a talk by
Bucherer (1908c), who presented data that seemed to contradict Abraham's
predictions for the velocity dependence of electron mass and support what
was by then no longer just Lorentz's prediction but Einstein's as well. It was
only decades later that these data were also shown to be inconclusive (Zahn
and Spees, 1938; quoted in Miller, 1981, 331).

Minkowski's comment suggests that we run Abraham's argument about
the kinematics of the electron in Minkowski rather than in Newtonian spacetime. We would then take it as a given that the electron has the shape of a
sphere in its rest frame, which implies that it will have the shape of a sphere
contracted in the direction of motion in any frame in which it is moving.
This, of course, is exactly Lorentz's model. This gives rise to a little puzzle.
The point of Abraham's argument in the passage we just quoted was that
by adopting rigid kinematical constraints we can safely ignore stabilizing
forces. His objection to Lorentz's model was that Lorentz *did* have to worry
about non-electromagnetic stabilizing forces or he would end up with two
different formulae for the longitudinal mass of his electron. How can these
two claims by Abraham be reconciled with one another? One's initial reaction might be that Abraham's kinematical argument does not carry over to

special relativity because the theory leaves no room for rigid bodies. That in and of itself is certainly true, but it is not the source of the problem. We could run the argument using some appropriate concept of an *approximately* rigid body (and as long as the electron is moving uniformly there is no problem whatsoever on this score). Abraham's argument that kinematic constraints can be used to obviate the need for discussion of the stability of the electron will then go through *as long as we use proper relativistic notions*. From a relativistic point of view, the analysis of Lorentz's model in this section is based on the standard non-covariant definition 12 of four-momentum. If we follow Fermi, Rohrlich and others and use definition 14 (with a fixed unit vector n^μ) instead, the ambiguity in the longitudinal mass of Lorentz's electron simply disappears. After all, under this alternative definition the combination of the energy and momentum of the electron's self-field transforms as a four-vector, even though it is an open system. This, in turn, guarantees—as we saw in eqs. 32–37 at the end of sec. 2—that energy and momentum give the same longitudinal mass. This shows that the ambiguity in the longitudinal mass of Lorentz's electron is not a consequence of the instability of the electron, but an artifact of the definitions of energy and momentum he used. We do not claim great originality for this insight. It is simply a matter of translating Rohrlich's analysis of the "4/3 puzzle" in special relativity (see the discussion following eq. 66) to a pre-relativistic setting.

4.5. Hamiltonian, Lagrangian, and generalized momentum

Poincaré (1906, 524) brought out the inconsistency of Lorentz's model in a slightly different way. He raised the question whether the expressions he found for energy, momentum, and Lagrangian for the field of the moving electron conform to the standard relations between Hamiltonian, Lagrangian, and generalized momentum. For an electron moving in the positive x-direction, these relations are

$$(78) \qquad U = \mathbf{P} \cdot \mathbf{v} - L = Pv - L, \quad P = P_x = \frac{dL}{dv}.$$

It turns out that the first relation is satisfied by both the Lorentz and the Bucherer-Langevin model, but that the second is satisfied only by the latter. Using eqs. 62 and 66 for P_{EM} and L_{EM}, respectively, we find

$$(79) \qquad P_{EM}v - L_{EM} = \left(\frac{4}{3} \gamma l \beta^2 + \frac{l}{\gamma} \right) U'_{EM} = \gamma l \left(\frac{4}{3} \beta^2 + 1 - \beta^2 \right) U'_{EM},$$

which does indeed reduce to the expression for U_{EM} found in eq. 57 for any value of l.

We now compute the conjugate momentum,

$$\text{(80)} \qquad \frac{dL_{\text{EM}}}{dv} = -U'_{\text{EM}} \frac{d}{dv}\left(\frac{l}{\gamma}\right) = U'_{\text{EM}}\left\{l\gamma\frac{v}{c^2} - \frac{1}{\gamma}\frac{dl}{dv}\right\},$$

where we used eq. 35 for $d\gamma/dv$. For the Lorentz model, with $l = 1$, this reduces to

$$\text{(81)} \qquad \frac{dL_{\text{EM}}}{dv} = \gamma\left(\frac{U'_{\text{EM}}}{c^2}\right)v,$$

which differs by the meanwhile familiar factor of 4/3 from the expression for P_{EM} read off from eq. 66 for $l = 1$. For the Bucherer-Langevin model, $l = \gamma^{-1/3}$ and with the help of eq. 75, we find:

$$\text{(82)} \qquad \frac{dL_{\text{EM}}}{dv} = \frac{U'_{\text{EM}}}{c^2}\left\{\gamma^{2/3}v + \frac{1}{3}\gamma^{2/3}v\right\} = \frac{4}{3}\gamma^{2/3}\left(\frac{U'_{\text{EM}}}{c^2}\right)v.$$

This agrees exactly with eq. 66 for $l = \gamma^{-1/3}$.

The relations 78 are automatically satisfied if $(U/c, \mathbf{P})$ transforms as a four-vector under Lorentz transformations. In that case, we have (see eq. 9):

$$\text{(83)} \qquad U = \gamma U_0, \quad P = \gamma\frac{U_0}{c^2}v.$$

Inserting this into $L = Pv - U$, we find

$$\text{(84)} \qquad L = \gamma U_0\beta^2 - \gamma U_0 = -\gamma U_0(1 - \beta^2) = -\frac{U_0}{\gamma},$$

which, in turn, implies that

$$\text{(85)} \qquad \frac{dL}{dv} = -U_0\frac{d}{dv}\left(\frac{1}{\gamma}\right) = \frac{U_0}{\gamma^2}\frac{d\gamma}{dv} = \gamma\frac{U_0}{c^2}v,$$

in accordance with eq. 83. This shows once again (cf. the discussion at the end of sec. 4.4) that the inconsistency in Lorentz's model can be taken care of by switching—in relativistic terms—from the standard definition 12 of four-momentum to the Fermi-Rohrlich definition 14 (with fixed n^μ). In that case the energy and momentum of the electron's self-field will satisfy eqs. 83–85 even though it is an open system.

5. POINCARÉ PRESSURE

In this section we give a streamlined version of the argument with which Poincaré (1906, 525–529) introduced what came to be known as "Poincaré pressure" to stabilize Lorentz's purely electromagnetic electron.[43]

The Lagrangian for the electromagnetic field of a moving electron can in all three models (Abraham, Lorentz, Bucherer-Langevin) be written as

$$(86) \qquad L_{EM} = \frac{\varphi(\vartheta/\gamma)}{\gamma^2 r}$$

(Poincaré, 1906, 525), where the argument ϑ/γ of the as yet unknown function φ is the 'ellipticity' (our term) of the "ideal electron" (Poincaré's term for the corresponding state of the moving electron). The ellipticity is the ratio of the radius of the "ideal electron" in the directions perpendicular to the direction of motion ($l\vartheta r$) and its radius ($\gamma l r$) in the direction of motion. This is illustrated in Fig. 2, which is the same as Fig. 1, except that it shows the notation Poincaré used to describe the three electron models.

For the Abraham electron the ellipticity is $1/\gamma$; for both the Lorentz and the Bucherer-Langevin electron it is 1. By examining the Lorentz case, we can determine $\varphi(1)$. Inserting $U_{0_{EM}} = e^2/8\pi\varepsilon_0\gamma r$, where γr is the radius of the electron at rest in the ether, into eq. 62 for the Lagrangian, we find:

$$(87) \qquad L_{EM_{Lorentz}} = -\frac{U_{0_{EM}}}{\gamma} = -\frac{e^2}{8\pi\varepsilon_0\gamma^2 r},$$

Comparison with the general expression for L_{EM} in eq. 86 gives:

$$(88) \qquad \varphi(1) = -\frac{e^2}{8\pi\varepsilon_0}.$$

Abraham (1902a, 37) found that the Lagrangian for his electron model has the form

$$(89) \qquad L_{EM_{Abraham}} = \frac{a}{r}\frac{1-\beta^2}{\beta}\ln\frac{1+\beta}{1-\beta}$$

(Poincaré, 1906, 526). Since $L_{EM} = \varphi(1)/r$ for $\beta = 0$ (in which case all three electron models coincide), it must be the case that $a = \varphi(1)$. From eqs. 86 and 89 it follows that

$$(90) \qquad \varphi(1/\gamma) = \gamma^2 r L_{EM_{Abraham}} = \frac{a}{\beta}\ln\frac{1+\beta}{1-\beta}.$$

The Lagrangian for the Lorentz model told us that $\varphi(1) = a = -e^2/8\pi\varepsilon_0$. The Lagrangian for the Abraham model allows us to determine $\varphi'(1)$. We start from eq. 90 and develop both the right-hand side and the argument $1/\gamma$ of φ on the left-hand side to second order in β. This gives (ibid.):

$$(91) \qquad \varphi\left(1-\frac{1}{2}\beta^2\right) = a\left(1+\frac{1}{3}\beta^2\right).$$

	real electron (in motion) dimensions: $(r, \vartheta r, \vartheta r)$	ideal electron (at rest) dimensions: $(\gamma l r, l \vartheta r, l \vartheta r)$
The rigid electron of Abraham $\vartheta = 1$ l arbitrary r constant	sphere (r, r, r)	ellipsoid $(\gamma l r, l r, l r)$
The contractile electron of Lorentz and Poincaré $\vartheta = \gamma$ $l = 1$ $\gamma r = $ constant	ellipsoid $(r, \gamma r, \gamma r)$	sphere $(\gamma r, \gamma r, \gamma r)$
The contractile electron of constant volume of Bucherer and Langevin $\vartheta = \gamma$ $l = \gamma^{-1/3}$ $\gamma l r = \gamma^{2/3} r = $ constant	ellipsoid $(r, \gamma r, \gamma r)$	sphere $(\gamma^{2/3} r, \gamma^{2/3} r, \gamma^{2/3} r)$

FIGURE 2.   A moving electron according to the models of Abraham, Lorentz, and Bucherer-Langevin, and the corresponding states at rest in the ether.

Now differentiate both sides:

$$(92) \qquad\qquad -\beta \varphi' \left(1 - \frac{1}{2}\beta^2\right) = \frac{2}{3}a\beta.$$

It follows that $\varphi'(1) = -(2/3)a$.

As Poincaré notes, all three electron models satisfy a constraint of the form

(93) $r = b\vartheta^{m},$

where b is a constant and where the exponent m depends on which model we consider. In the Abraham model $\vartheta = 1$ and r is a constant. Hence, $r = b$. In the Lorentz model, $\vartheta = \gamma$ and γr is a constant. It follows that $\vartheta r = b$, or $r = b\vartheta^{-1}$. In the Bucherer-Langevin model, $\vartheta = \gamma$ and $\gamma^{2/3} r$ is a constant. It follows that $\vartheta^{2/3} r = b$, or $r = b\vartheta^{-2/3}$. In other words, the values of m in the three models are

(94)
$$\begin{array}{c} \text{Abraham}: m = 0, \\ \text{Lorentz}: m = -1, \\ \text{Bucherer} - \text{Langevin}: m = -2/3. \end{array}$$

Substituting $r = b\vartheta^{m}$ into the general expression 86 for the Lagrangian, we find:

(95) $$L_{\text{EM}} = \frac{\varphi(\vartheta/\gamma)}{b\gamma^{2}\vartheta^{m}}.$$

Poincaré proceeds to investigate whether this Lagrangian describes a stable physical system. To this end, he checks whether $\partial L_{\text{EM}}/\partial\vartheta$ vanishes. It turns out that for $m = -2/3$ it does, but that for $m = -1$ it does not. Denote the argument of the function φ with $u \equiv \vartheta/\gamma$.

(96) $$\frac{\partial L_{\text{EM}}}{\partial\vartheta} = \frac{\varphi'(u)}{b\gamma^{3}\vartheta^{m}} - \frac{m\varphi(u)}{b\gamma^{2}\vartheta^{m+1}}.$$

This derivative vanishes if

(97) $$\varphi'(u) = \frac{\gamma m\varphi(u)}{\vartheta} = m\frac{\varphi(u)}{u}.$$

For the Lorentz and Bucherer-Langevin models $u = 1$, and this condition reduces to

(98) $$\varphi'(1) = m\varphi(1).$$

Inserting $\varphi(1) = a$ and $\varphi'(1) = -(2/3)a$, we see that the purely electromagnetic Lagrangian only describes a stable system for $m = -2/3$, which is the value for the Bucherer-Langevin electron. The Lorentz electron calls for an additional term in the Lagrangian.[44] The total Lagrangian is then given by the sum

(99) $$L_{\text{tot}} = L_{\text{EM}} + L_{\text{non-EM}}.$$

Like L_{EM}, L_{non-EM} is a function of ϑ and r. Treating these variables as independent, we can write the stability conditions for the total Lagrangian as

$$(100) \qquad \frac{\partial}{\partial \vartheta}(L_{EM} + L_{non-EM}) = 0, \quad \frac{\partial}{\partial r}(L_{EM} + L_{non-EM}) = 0.$$

Evaluating the partial derivatives of L_{EM} given by eq. 86,

$$(101) \qquad \frac{\partial L_{EM}}{\partial \vartheta} = \frac{\varphi'(u)}{\gamma^3 r}, \quad \frac{\partial L_{EM}}{\partial r} = -\frac{\varphi(u)}{\gamma^2 r^2},$$

and inserting the results into the stability conditions, we find

$$(102) \qquad \frac{\partial L_{non-EM}}{\partial \vartheta} = -\frac{\varphi'(u)}{\gamma^3 r}, \quad \frac{\partial L_{non-EM}}{\partial r} = \frac{\varphi(u)}{\gamma^2 r^2}.$$

Poincaré (1906, 528–529) continues his analysis without picking a specific model. We shall only do the calculation for the Lorentz model. So we no longer need subscripts such as in eqs. 87 and 89 to distinguish between the models of Abraham and Lorentz. For the Lorentz model $m = -1$, $\gamma = \vartheta$, $r = b/\vartheta$, and $u = 1$. Substituting these values into eqs. 102 and using that $\varphi(1) = a$ and $\varphi'(1) = -(2/3)a$, we find:

$$(103) \qquad \frac{\partial L_{non-EM}}{\partial \vartheta} = \frac{2a}{3b\vartheta^2}, \quad \frac{\partial L_{non-EM}}{\partial r} = \frac{a}{b^2}.$$

These equations are satisfied by a Lagrangian of the form

$$(104) \qquad L_{non-EM} = Ar^3\vartheta^2,$$

where A is a constant. Since $r^3\vartheta^2$ is proportional to the volume V of the moving electron, L_{non-EM} can be written as

$$(105) \qquad L_{non-EM} = P_{Poincaré}V,$$

where $P_{Poincaré}$ is a constant. We chose the letter P because this constant turns out be a (negative) pressure. To determine the constant A, we take the derivative of eq. 104 with respect to ϑ and r, and eliminate r from the results, using $r = b/\vartheta$:

$$(106) \qquad \frac{\partial L_{non-EM}}{\partial \vartheta} = 2Ar^3\vartheta = \frac{2Ab^3}{\vartheta^2}, \quad \frac{\partial L_{non-EM}}{\partial r} = 3Ar^2\vartheta^2 = 3Ab^2.$$

Comparison with eqs. 103 gives:

$$(107) \qquad A = \frac{a}{3b^4}.$$

Finally, we write L_{non-EM} in a form that allows easy comparison with $L_{EM} = a/\gamma^2 r$ (see eq. 86 with $\varphi(\vartheta/\gamma) = \varphi(1) = a$). Using eq. 107 along with $\vartheta = \gamma$

and $b = \gamma r$, we can rewrite eq. 104 as

$$(108) \qquad L_{\text{non-EM}} = \frac{a}{3b^4} r^3 \vartheta^2 = \frac{1}{3} \frac{a}{\gamma^2 r} = \frac{1}{3} L_{\text{EM}} = -\frac{1}{3} \frac{U_{0\text{EM}}}{\gamma},$$

where in the last step we used eq. 62 for $l = 1$. Using that the volume V_0 of Lorentz's electron at rest is equal to γV, we can rewrite this as:

$$(109) \qquad L_{\text{non-EM}} = -\frac{1}{3} \frac{U_{0\text{EM}}}{V_0} V.$$

Comparison with expression 105 gives:

$$(110) \qquad P_{\text{Poincaré}} = -\frac{1}{3} \frac{U_{0\text{EM}}}{V_0}$$

(Laue, 1911b, 164, eq. 171). Note that this so-called Poincaré pressure is negative. The pressure is present only inside the electron and vanishes outside (Poincaré, 1906, 537).[45] It can be written more explicitly with the help of the ϑ-step-function (defined as: $\vartheta(x) = 0$ for $x < 0$ and $\vartheta(x) = 1$ for $x \geq 0$). For an electron moving through the ether with velocity v in the x-direction, the Poincaré pressure in a co-moving frame (related to a frame at rest in the ether by a Galilean transformation) is:

$$(111) \qquad P_{\text{Poincaré}}(\mathbf{x}) = -\frac{1}{3} \frac{U_{0\text{EM}}}{V_0} \vartheta \left(R - \sqrt{\gamma^2 x^2 + y^2 + z^2} \right),$$

where R is the radius of the electron at rest. So there is a sudden drop in pressure at the edge of the electron, which is the only place where forces are exerted.[46] These forces serve two purposes. First, they prevent the electron's surface charge distribution from flying apart under the influence of the Coulomb repulsion between its parts. Second, as the region where $P_{\text{Poincaré}}(\mathbf{x})$ is non-vanishing always coincides with the ellipsoid-shaped region occupied by the moving electron, these forces make the electron contract by a factor γ in the direction of motion.

Relations 78 between Hamiltonian, Lagrangian and generalized momentum, only one of which was satisfied by Lorentz's original purely electromagnetic electron model, are both satisfied once $L_{\text{non-EM}}$ is added to the Lagrangian. Using the total Lagrangian,

$$(112) \qquad L_{\text{tot}} = L_{\text{EM}} + L_{\text{non-EM}} = \frac{4}{3} L_{\text{EM}} = -\frac{4}{3} \frac{U_{0\text{EM}}}{\gamma},$$

to compute the total momentum, we find:

$$(113) \qquad P_{\text{tot}} = \frac{dL_{\text{tot}}}{dv} = \frac{4}{3} \frac{dL_{\text{EM}}}{dv} = \frac{4}{3} \gamma \frac{U_{0\text{EM}}}{c^2} v,$$

where in the last step we used eq. 81. This is just the electromagnetic momentum P_{EM} found earlier (see eq. 66 for $l = 1$). With the help of these expressions for L_{tot} and P_{tot}, we can compute the total energy:

$$(114) \qquad U_{tot} = P_{tot}v - L_{tot} = \frac{4}{3}\gamma U_{0_{EM}}\beta^2 + \frac{4}{3}\frac{U_{0_{EM}}}{\gamma} = \frac{4}{3}\gamma U_{0_{EM}}.$$

The total energy is the sum of the electromagnetic energy (see eq. 58),

$$(115) \qquad U_{EM} = \frac{4}{3}\gamma U_{0_{EM}} - \frac{1}{3}\frac{U_{0_{EM}}}{\gamma},$$

and the non-electromagnetic energy,

$$(116) \qquad U_{non-EM} = \frac{1}{3}\frac{U_{0_{EM}}}{\gamma},$$

which is minus the product of the Poincaré pressure (see eq. 110) and the volume $V = V_0/\gamma$ of the moving electron. The total energy of the system at rest is

$$(117) \qquad U_{0_{tot}} = \frac{4}{3}U_{0_{EM}},$$

and its rest mass is $m_{0_{tot}} = U_{0_{tot}}/c^2$ accordingly. Eq. 113 can thus be rewritten as

$$(118) \qquad P_{tot} = \gamma\left(\frac{U_{0_{tot}}}{c^2}\right)v = \gamma m_{0_{tot}}v.$$

The troublesome factor 4/3 has disappeared.

The total energy and momentum transform as a four-vector under Lorentz transformations. In the system's rest frame its four-momentum is $P_{0_{tot}}^\mu = (U_{0_{tot}}/c, 0, 0, 0)$. In a frame moving with velocity v in the x-direction, it is

$$(119) \qquad P_{tot}^\mu = \Lambda^\mu_{\ \nu}P_{0_{tot}}^\nu = \left(\gamma\frac{U_{0_{tot}}}{c}, \gamma\beta\frac{U_{0_{tot}}}{c}, 0, 0\right),$$

in accordance with eqs. 114, 117, and 118. As we saw at the end of sec. 2, if $(U/c, \mathbf{P})$ transforms as a four-vector, it is guaranteed that energy and momentum lead to the same longitudinal mass. With Poincaré's amendment Lorentz's electron model may no longer be purely electromagnetic—at least it is fully consistent.

As we pointed out earlier, the problem that Abraham found in Lorentz's purely electromagnetic electron model (viz. that momentum and energy lead to different expressions for the longitudinal mass) returns in special relativity as the infamous "4/3 puzzle" of the mass-energy relation of the classical electron. Mathematically, these two problems are identical and the introduction of Poincaré pressure thus takes care of both. In the next section, we shall reintroduce Poincaré pressure à la Max Laue (1911a, 1911b) in his relativistic treatment of Lorentz's electron model.

Before we do so, we need to deal with a serious error committed by Poincaré (1906, 538) in his calculation of the transverse and longitudinal mass of the stabilized Lorentz electron. As a result of this error, Poincaré overestimated what he had accomplished in his paper.[47] The calculations in (Poincaré, 1906) that we have covered so far are all from section 6 of the paper. This section is phrased entirely in terms of energies, momenta, and Lagrangians. The consideration of mass is explicitly postponed (Ibid., 522). In section 4 we showed how Poincaré restated the problem of the ambiguity of the longitudinal mass of Lorentz's electron in terms of the model failing to satisfy one of the standard relations between Hamiltonian, Lagrangian, and generalized momentum (Ibid., 524; cf. sec. 4.5). In this section we traced the steps that Poincaré took in the remainder of section 6 to restore the validity of these relations for Lorentz's model (Ibid., 525–529). This is a completely unobjectionable way to proceed, from a pre-relativistic as well as from a relativistic point of view.[48]

In section 7 of his paper, Poincaré (1906, 531) finally introduces Abraham's definitions 26 of the electromagnetic longitudinal and transverse mass of the electron. And at the end of section 8, at the very end of his discussion of electron models and just before he turns to the problem of gravitation, he computes the mass of the electron in Lorentz's model, limiting himself to what he calls—in scare quotes—the ""experimental mass," i.e., the mass for small velocities" (Ibid., 538). He writes down the Lagrangian 86 for the special case of the Lorentz electron. Using that $\varphi(\vartheta/\gamma) = \varphi(1) = a$ (where $a = -e^2/8\pi\varepsilon_0$) and $\gamma r = b$ (with b the radius of the electron at rest in the ether), we arrive at the expression given by Poincaré at this point,

$$(120) \qquad L_{EM} = \frac{a}{b}\sqrt{1 - v^2/c^2},$$

except that Poincaré uses H instead of L_{EM} and sets $c = 1$. For small velocities, eq. 120 reduces to

$$(121) \qquad L_{EM} \approx \frac{a}{b}\left(1 - \frac{1}{2}\frac{v^2}{c^2}\right).$$

Poincaré concludes that for small velocities both the longitudinal and the transverse mass of the electron is given by a/b. Since a is negative, he must have meant $-a/b$. This is just a minor slip. Poincaré's result corresponds to $U_{0_{EM}}/c^2$,[49] which differs from the result that we found by the infamous factor of 4/3 (see eqs. 117–118). How did Poincaré arrive at his result? It is hard to see how he could have found this in any other way than the following.

Computing the electromagnetic momentum as the generalized momentum corresponding to the Lagrangian 121, one finds

(122)
$$P_{\text{EM}} = \frac{dL_{\text{EM}}}{dv} \approx -\frac{a}{b}\frac{v}{c^2}.$$

Inserting this result into definitions 26 for longitudinal and transverse mass, one arrives at:

(123)
$$m_{//} = \frac{dP_{\text{EM}}}{dv} \approx -\frac{a}{bc^2}, \quad m_{\perp} = \frac{P_{\text{EM}}}{v} \approx -\frac{a}{bc^2}.$$

This is just the result reported by Poincaré (recall that he set $c = 1$). However, we had no business using eq. 122! As Poincaré himself had pointed out in section 6 of his paper, in the case of the Lorentz model, the electromagnetic momentum P_{EM} is *not* equal to the generalized momentum dL_{EM}/dv. The relation $P = dL/dv$ only holds for the *total* momentum and the *total* Lagrangian. The total Lagrangian is 4/3 times the electromagnetic part. For low velocities it reduces to (cf. eq. 121):

(124)
$$L_{\text{tot}} \approx \frac{4}{3}\frac{a}{b}\left(1 - \frac{1}{2}\frac{v^2}{c^2}\right).$$

Replacing L_{EM} by L_{tot} in eqs. 122–123, we find that the low-velocity limit of the electron mass is 4/3 times $-a/b$ or 4/3 times $U_{0_{\text{EM}}}/c^2$, in accordance with what we found above. Unlike the minus sign of $-a/b$ that Poincaré lost in his calculation, the conflation of L_{EM} and L_{tot} has dire consequences. If we use L_{tot} it is immediately obvious that the mass of Lorentz's electron is not of purely electromagnetic origin, whereas if we use L_{EM} we are led to believe that it is. In fact, this is exactly what Poincaré claimed, both at the end of section 8 and in the introduction of his paper. In the introduction, he writes:

> If the inertia of matter is exclusively of electromagnetic origin, as is generally admitted since Kaufmann's experiment, and all forces are of electromagnetic origin (apart from this constant pressure that I just mentioned), the postulate of relativity may be established with perfect rigor. (Poincaré, 1906, 496)

Commenting on this passage, Miller (1973, 248) writes: "However the presence of these stresses [the Poincaré pressure] negates a purely electromagnetic theory of the electron's inertia." We agree. One has to choose between the "postulate of relativity" and mass being "exclusively of electromagnetic origin." Even Poincaré cannot have his cake and eat it too.[50]

6. THE RELATIVISTIC TREATMENT OF THE ELECTRON MODEL OF LORENTZ AS AMENDED BY POINCARÉ

From the point of view of Laue's relativistic continuum mechanics, the problem with Lorentz's fully electromagnetic electron is that it is not a closed system. The four-divergence of the energy-momentum tensor of its electromagnetic field does not vanish. Computing this four-divergence tells us what needs to be added to this energy-momentum tensor to obtain a closed system, i.e., a system with a total energy-momentum tensor such that $\partial_\nu T_{\text{tot}}^{\mu\nu} = 0$. Unsurprisingly, the part that needs to be added is just the energy-momentum tensor for the Poincaré pressure.

The energy-momentum tensor for the electromagnetic field is given by (Jackson, 1975, sec. 12.10)

$$(125) \qquad T_{\text{EM}}^{\mu\nu} = \mu_0^{-1} \left(F^\mu{}_\alpha F^{\alpha\nu} + \frac{1}{4}\eta^{\mu\nu} F_{\alpha\beta} F^{\alpha\beta} \right),$$

where $F^{\mu\nu}$ is the electromagnetic field tensor with components (ibid., 550):

$$(126) \qquad F^{\mu\nu} = \begin{pmatrix} 0 & -E_x/c & -E_y/c & -E_z/c \\ E_x/c & 0 & -B_z & B_y \\ E_y/c & B_z & 0 & -B_x \\ E_z/c & -B_y & B_x & 0 \end{pmatrix}.$$

Inserting the components of the field tensor into eq. 125 for the energy-momentum tensor, we recover the familiar expressions for the electromagnetic energy density (cf. eq. 53), (c times) the electromagnetic momentum density (cf. eq. 18), and (minus) the Maxwell stress tensor (cf. eq. 20).

$$T_{\text{EM}}^{00} = \frac{1}{2}\varepsilon_0 E^2 + \frac{1}{2}\mu_0^{-1} B^2 = u_{\text{EM}},$$

$$(127) \qquad \left(T_{\text{EM}}^{01}, T_{\text{EM}}^{02}, T_{\text{EM}}^{03} \right) = \left(T_{\text{EM}}^{10}, T_{\text{EM}}^{20}, T_{\text{EM}}^{30} \right) = c\varepsilon_0 \mathbf{E} \times \mathbf{B} = c\mathbf{p}_{\text{EM}},$$

$$T_{\text{EM}}^{ij} = -\varepsilon_0 \left(E^i E^j - \frac{1}{2}\delta^{ij} E^2 \right) - \mu_0^{-1} \left(B^i B^j - \frac{1}{2}\delta^{ij} B^2 \right) = -T_{\text{Maxwell}}^{ij}.$$

We calculate the four-divergence of the energy-momentum tensor for the electromagnetic field of Lorentz's electron in its rest frame. Lorentz invariance guarantees that if the four-divergence of the total energy-momentum tensor vanishes in the rest frame ($\partial_{0_\nu} T_{0_{\text{tot}}}^{\mu\nu} = 0$), it will vanish in all frames ($\partial_\nu T_{\text{tot}}^{\mu\nu} = 0$). In the rest frame, we have

$$(128) \qquad T_{0_{\text{EM}}}^{\mu\nu} = \begin{pmatrix} u_{0_{\text{EM}}} & 0 \\ 0 & -T_{0_{\text{Maxwell}}}^{ij} \end{pmatrix}.$$

Consider the four-divergence $\partial_{0_\nu} T^{\mu\nu}_{0_{EM}}$ of this tensor. Since the system is static, only the spatial derivatives, $\partial_{0_j} T^{\mu j}_{0_{EM}}$, give a contribution. Since $T^{0j}_{0_{EM}} = 0$, there will only be contributions for $\mu = i$. Using eq. 127, we can write these contributions as:[51]

(129) $$\partial_{0_j} T^{ij}_{0_{EM}} = -\partial_{0_j} T^{ij}_{0_{Maxwell}} = -\rho_0 E^i_0,$$

The charge density ρ_0 is the surface charge density $\sigma = e/4\pi R^2$ (where e is the charge of the electron and R the radius of the electron in its rest frame):

(130) $$\rho_0 = \sigma \delta(R - r_0),$$

where $r_0 \equiv \sqrt{x_0^2 + y_0^2 + z_0^2}$. Inside the electron there is no electric field (it is a miniature version of Faraday's cage); outside the field is the same as that of a point charge e located at the center of the electron. At $r_0 = R$, right at the surface of the electron, the field has a discontinuity. Its magnitude, E_0, jumps from 0 to $e/4\pi\varepsilon_0 R^2$. At this point we need to use the average of these two values (see, e.g., Griffith, 1999, 102–103). At $r_0 = R$ the field is thus given by

(131) $$E^i_{0_{r_0=R}} = \frac{e}{8\pi\varepsilon_0 R^2} \frac{x^i_0}{R} = \frac{\sigma}{2\varepsilon_0} \frac{x^i_0}{R}.$$

Substituting eqs. 131 and 130 into eq. 129, we find that the divergence of the energy-momentum tensor of the electron's electromagnetic field in its rest frame is:

(132) $$\partial_{0_\nu} T^{\mu\nu}_{0_{EM}} = \begin{cases} \mu = 0: & 0 \\ \mu = i: & -\dfrac{\sigma^2}{2\varepsilon_0} \dfrac{x^i_0}{R} \delta(R - r_0). \end{cases}$$

It vanishes everywhere except at the surface of the electron. To get a total energy-momentum tensor with a four-divergence that vanishes everywhere,

(133) $$\partial_{0_\nu} T^{\mu\nu}_{0_{tot}} = \partial_{0_\nu} \left(T^{\mu\nu}_{0_{EM}} + T^{\mu\nu}_{0_{non-EM}} \right) = 0,$$

we need to add the Poincaré pressure of eq. 111, which in the electron's rest frame is described by the energy-momentum tensor[52]

(134) $$T^{\mu\nu}_{0_{non-EM}} = -\eta^{\mu\nu} P_{Poincaré} \vartheta(R - r_0).$$

Using that $\eta^{ij}\partial_j\vartheta(R - r) = \delta(R - r)(x^i/R)$, we find that the four-divergence of this energy-momentum tensor is given by:

(135) $$\partial_{0_\nu} T^{\mu\nu}_{0_{non-EM}} = \begin{cases} \mu = 0: & 0 \\ \mu = i: & -P_{Poincaré} \dfrac{x^i_0}{R} \delta(R - r_0). \end{cases}$$

Inserting eq. 110 for the Poincaré pressure, using $U_{0_{EM}} = e^2/8\pi\varepsilon_0 R$, $V_0 = \frac{4}{3}\pi R^3$, and $\sigma = e/4\pi R^2$, we find:

$$(136) \quad P_{Poincaré} = -\frac{U_{0_{EM}}}{3V_0} = -\frac{e^2}{(8\pi\varepsilon_0 R)(4\pi R^3)} = -\frac{1}{2\varepsilon_0}\left(\frac{e}{4\pi R^2}\right)^2 = -\frac{\sigma^2}{2\varepsilon_0}.$$

This is the expression for the Poincaré pressure given, e.g., in (Lorentz, 1915, 214), (Schwinger, 1983, 376–377, eqs. (24) and (34)), and (Rohrlich, 1997, 1056, eq. (A.4)). Substituting this expression in eq. 135 and comparing the result with eq. 132, we see that the Poincaré pressure indeed ensures that the four-divergence of the electron's total energy-momentum tensor vanishes. The reader is invited to compare this straightforward and physically clearly motivated introduction of Poincaré pressure to (the streamlined version of) Poincaré's own derivation presented in sec. 5.

We now calculate the contributions of $T_{EM}^{\mu\nu}$ and $T_{non-EM}^{\mu\nu}$ to the electron's four-momentum. We begin with the contribution coming from the electron's electromagnetic field:

$$(137) \qquad P_{EM}^\mu = \frac{1}{c}\int T_{EM}^{\mu 0}d^3x.$$

Using that $T^{\mu\nu} = \Lambda^\mu_{\ \rho}\Lambda^\nu_{\ \sigma}T_0^{\rho\sigma}$ and that $d^3x = d^3x_0/\gamma$, we can rewrite this as

$$(138) \qquad P_{EM}^\mu = \frac{1}{c\gamma}\Lambda^\mu_{\ \rho}\Lambda^0_{\ \sigma}\int T_{0_{EM}}^{\rho\sigma}d^3x_0.$$

Eq. 128 tells us that there will only be contributions for $\rho\sigma = 00$ and $\rho\sigma = ij$. We denote these contributions as $P_{EM}^\mu(00)$ and $P_{EM}^\mu(ij)$.

For $P_{EM}^\mu(00)$ we have:

$$(139) \qquad P_{EM}^\mu(00) = \frac{1}{c\gamma}\Lambda^\mu_{\ 0}\Lambda^0_{\ 0}\int T_{0_{EM}}^{00}d^3x_0.$$

Since $\Lambda^\mu_{\ 0} = (\gamma, \gamma\beta, 0, 0)$ (see eq. 7) and the integral over $T_{0_{EM}}^{00}$ gives $U_{0_{EM}}$, this turns into:

$$(140) \qquad P_{EM}^\mu(00) = \left(\gamma\frac{U_{0_{EM}}}{c}, \gamma\frac{U_{0_{EM}}}{c^2}\mathbf{v}\right).$$

This is just the Lorentz transform of $P_{0_{EM}}^\mu = (U_{0_{EM}}/c, 0, 0, 0)$. It is the additional contribution $P_{EM}^\mu(ij)$, coming from $T_{0_{EM}}^{ij}$, that is responsible for the fact that the four-momentum of the electron's electromagnetic field does not transform as a four-vector.

For $P_{EM}^\mu(ij)$ we have:

$$(141) \qquad P_{EM}^\mu(ij) = \frac{1}{c\gamma}\Lambda^\mu_{\ i}\Lambda^0_{\ j}\int T_{0_{EM}}^{ij}d^3x_0.$$

The integrand is minus the Maxwell stress tensor in the electron's rest frame (see eq. 127):

$$(142) \qquad T_{0EM}^{ij} = -\varepsilon_0 \begin{pmatrix} E_{0_x}^2 - \frac{1}{2}E_0^2 & E_{0_x}E_{0_y} & E_{0_x}E_{0_z} \\ E_{0_y}E_{0_x} & E_{0_y}^2 - \frac{1}{2}E_0^2 & E_{0_y}E_{0_z} \\ E_{0_z}E_{0_x} & E_{0_z}E_{0_y} & E_{0_z}^2 - \frac{1}{2}E_0^2 \end{pmatrix}.$$

The integrals over the off-diagonal terms are all zero. The integrals over the three diagonal terms are equal to one another and given by

$$(143) \qquad \int \varepsilon_0 \left(\frac{1}{2}E_0^2 - \frac{1}{3}E_0^2 \right) d^3x_0 = \frac{1}{3}\int \frac{1}{2}\varepsilon_0 E_0^2 d^3x_0 = \frac{1}{3}U_{0EM}.$$

Since $(1/\gamma)\Lambda^\mu{}_1\Lambda^0{}_1 = (\gamma\beta^2, \gamma\beta, 0, 0)$ and $\Lambda^\mu{}_i\Lambda^0{}_i = 0$ for $i = 2, 3$ (see eq. 7), only the 11-component of eq. 141 is non-zero:

$$(144) \qquad P_{EM}^\mu(11) = \left(\frac{1}{3}\gamma\beta^2\frac{U_{0EM}}{c}, \frac{1}{3}\gamma\frac{U_{0EM}}{c^2}\mathbf{v} \right).$$

Adding eqs. 140 and 144, we find:

$$(145) \qquad P_{EM}^\mu = P_{EM}^\mu(00) + P_{EM}^\mu(11) = \left(\gamma\left(1 + \frac{1}{3}\beta^2\right)\frac{U_{0EM}}{c}, \frac{4}{3}\gamma\frac{U_{0EM}}{c^2}\mathbf{v} \right).$$

This, unsurprisingly, is exactly the result we found earlier for the energy and momentum of the electromagnetic field of Lorentz's electron (see eqs. 57 and 66 with $l = 1$ and $U_{EM}' = U_{0EM}$).

The calculation of the contributions to the four-momentum coming from $T_{non-EM}^{\mu\nu}$ is completely analogous to the calculation in eqs. 137–145. We start with:

$$(146) \qquad P_{non-EM}^\mu = \frac{1}{c\gamma}\Lambda^\mu{}_\rho\Lambda^0{}_\sigma \int T_{0non-EM}^{\rho\sigma} d^3x_0.$$

Since $T_{0non-EM}^{\mu\nu}$ is diagonal (see eq. 134), there will only be contributions when $\rho = \sigma$. Since $\Lambda^0{}_\mu = (\gamma, \gamma\beta, 0, 0)$, the only contributions will be for $\rho = \sigma = 0$ and $\rho = \sigma = 1$. We denote these by $P_{non-EM}^\mu(00)$ and $P_{non-EM}^\mu(11)$, respectively, and calculate them separately. Since $T_{0non-EM}^{00} = -P_{Poincaré}\vartheta(R - r_0)$ (see eq. 134) and $\int \vartheta(R - r_0)d^3x_0 = V_0$,

$$(147) \qquad \int T_{0non-EM}^{00} d^3x_0 = -P_{Poincaré}V_0 = \frac{1}{3}U_{0EM},$$

where we used eq. 110. Hence,

$$(148) \qquad P_{non-EM}^\mu(00) = \left(\gamma\frac{1}{3}\frac{U_{0EM}}{c}, \gamma\frac{1}{3}\frac{U_{0EM}}{c^2}\mathbf{v} \right),$$

which is just the Lorentz transform of $P_{0_{EM}}^{\mu} = \left(\frac{1}{3}U_{0_{EM}}/c, 0, 0, 0\right)$. Similarly, we find:

(149) $P_{non-EM}^{\mu}(11) = \left(-\frac{1}{3}\gamma\beta^2\frac{U_{0_{EM}}}{c}, -\frac{1}{3}\gamma\frac{U_{0_{EM}}}{c^2}\mathbf{v}\right).$

Comparing eq. 149 to eq. 144, we see that $P_{non-EM}^{\mu}(11)$ is exactly the opposite of $P_{EM}^{\mu}(11)$:

(150) $P_{EM}^{\mu}(11) + P_{non-EM}^{\mu}(11) = 0.$

This is a direct consequence of what is known as *Laue's theorem* (Miller, 1981, 352). This theorem (Laue, 1911a, 539) says that for a "complete [i.e., closed] static system" (*vollständiges statisches System*):

(151) $\int T_{0_{tot}}^{ij} d^3x_0 = 0.$

For the electron we have $T_{0_{tot}}^{ij} = T_{0_{EM}}^{ij} + T_{0_{non-EM}}^{ij}$. From eqs. 142–143 we read off that

(152) $\int T_{0_{EM}}^{ij} d^3x_0 = \begin{cases} i \neq j: & 0 \\ i = j: & \frac{1}{3}U_{0_{EM}}. \end{cases}$

In analogy with eq. 147, we find:

(153) $\int T_{0_{non-EM}}^{ij} d^3x_0 = \begin{cases} i \neq j: & 0 \\ i = j: & -\frac{1}{3}U_{0_{EM}}. \end{cases}$

Laue's theorem thus holds for this system, as it should, and eq. 150 is a direct consequence of this. Using eqs. 138 and 146, we find

(154) $P_{EM}^{\mu}(ij) + P_{non-EM}^{\mu}(ij) = \frac{1}{c\gamma}\Lambda^{\mu}{}_i\Lambda^0{}_j \int \left(T_{0_{EM}}^{ij} + T_{0_{non-EM}}^{ij}\right) d^3x_0,$

which by Laue's theorem vanishes, as is confirmed explicitly by eqs. 152–153. Since $P_{EM}^{\mu}(ij) = P_{non-EM}^{\mu}(ij) = 0$ except when $i = j = 1$, the sum of the 11-components considered in eq. 150 is equal to the sum of the ij-components.

Laue's theorem ensures that the four-momentum of a closed static system transforms as a four-vector. The total four-momentum of the electron is the sum of four terms (see eqs. 140, 144, 148, and 149):

(155) $P_{tot}^{\mu} = P_{EM}^{\mu}(00) + P_{non-EM}^{\mu}(00) + P_{EM}^{\mu}(ij) + P_{non-EM}^{\mu}(ij).$

The last two terms cancel each other because of Laue's theorem, and all that is left is:

(156) $P_{tot}^{\mu} = P_{EM}^{\mu}(00) + P_{non-EM}^{\mu}(00).$

Using eqs. 140 and 148 for these two contributions we recover eq. 119 for the total energy and momentum of the electron:

$$(157) \qquad P^\mu_{\text{tot}} = \left(\gamma \frac{4}{3} \frac{U_{0_{\text{EM}}}}{c}, \gamma \frac{4}{3} \frac{U_{0_{\text{EM}}}}{c^2} \mathbf{v} \right) = \left(\gamma \frac{U_{0_{\text{tot}}}}{c}, \gamma \frac{U_{0_{\text{tot}}}}{c^2} \mathbf{v} \right).$$

As we pointed out above (see eqs. 133–134 and note 52), we still have a closed system if we set the 00-component of $T^{\mu\nu}_{0_{\text{non-EM}}}$ to zero. This does not affect the result for $P^\mu_{\text{non-EM}}(ij)$, which only depends on the ij-components of $T^{\mu\nu}_{0_{\text{non-EM}}}$. $P^\mu_{\text{non-EM}}(00)$, however, will be zero if $T^{00}_{0_{\text{non-EM}}} = 0$ (see eq. 146). The total four-momentum will still be a four-vector but compared to eq. 157 the system's rest energy will be smaller by $\frac{1}{3} U_{0_{\text{EM}}}$:

$$(158) \qquad P^\mu_{\text{tot}} = P^\mu_{\text{EM}}(00) = \left(\gamma \frac{U_{0_{\text{EM}}}}{c}, \gamma \frac{U_{0_{\text{EM}}}}{c^2} \mathbf{v} \right).$$

To reiterate: if the stabilizing mechanism for the electron does *not* contribute to the energy in the rest frame but only to the stresses, $T^{00}_{0_{\text{non-EM}}} = 0$ and only the first term in eq. 156 contributes to the four-momentum. In this case, the electron's rest mass is $U_{0_{\text{EM}}}/c^2$ (see eq. 158). If the stabilizing mechanism *does* contribute to the energy in the electron's rest frame, $T^{00}_{0_{\text{non-EM}}} \neq 0$ and both terms in eq. 156 contribute to the four-momentum. If $T^{00}_{0_{\text{non-EM}}} = \frac{1}{3} (U_{0_{\text{EM}}}/V_0) \vartheta(R - r_0)$, as in Poincaré's specific model (see eqs. 134 and 110), the electron's rest mass is $\frac{4}{3} U_{0_{\text{EM}}}/c^2$ (see eq. 157).[53]

The arbitrariness of the Lorentz-Poincaré electron is much greater than the freedom we have in choosing the 00-component of the energy-momentum tensor for the mechanism stabilizing a spherical surface charge distribution. For starters, we can choose a (surface or volume) charge distribution of any shape we like—a box, a doughnut, a banana, etc. As long as this charge distribution is subject to the Lorentz-FitzGerald contraction, we can turn it into a system with the exact same energy-momentum-mass-velocity relations as the Lorentz-Poincaré electron by adding the appropriate non-electromagnetic stabilizing mechanism.[54] Of course, as the analysis in this section, based on (Laue, 1911a), shows, any closed static system will have the same energy-momentum-mass-velocity relations as the Lorentz-Poincaré electron, no matter whether it consists of charges, electromagnetic fields, and Poincaré pressure or of something else altogether. The only thing that matters is that whatever the electron is made of satisfies Lorentz-invariant laws. The restriction to *static* closed systems, moreover, is completely unnecessary. Any closed system will do.[55] In short, there is nothing we can learn about the nature and structure of the electron from studying its energy-momentum-mass-velocity relations.

Lorentz himself emphasized this in lectures he gave at Caltech in 1922. In a section entitled "Structure of the Electron" in the book based on these lectures and published in 1927, he wrote:

> The formula for momentum was found by a theory in which it was supposed that in the case of the electron the momentum is determined wholly by that of the electromagnetic field [...] This meant that the whole mass of an electron was supposed to be of electromagnetic nature. Then, when the formula for momentum was verified by experiment, it was thought at first that it was thereby proved that electrons have no "material mass." Now we can no longer say this. Indeed, the formula for momentum is a general consequence of the principle of relativity, and a verification of that formula is a verification of the principle and tells us nothing about the nature of mass or of the structure of the electron. (Lorentz, 1927, 125).

By 1922 this point was widely appreciated. In his famous review article on relativity, Pauli (1921, 82–83), for instance, wrote:

> It constituted a definite progress that Lorentz's law of the variability of mass could be derived from the theory of relativity without making any specific assumptions on the electron shape or charge distribution. Also nothing need be assumed about the nature of the mass: [the relativistic formula for the velocity-dependence of mass] is valid for every kind of ponderable mass [...] The old idea that one could distinguish between the "constant" true mass and the "apparent" electromagnetic mass, by means of deflection experiments on cathode rays, can therefore not be maintained.

7. FROM THE ELECTROMAGNETIC VIEW OF NATURE TO RELATIVISTIC CONTINUUM MECHANICS

Experiment was supposed to be the final arbiter in the debate over the electron models of Abraham, Lorentz-Poincaré, and Bucherer-Langevin. Later analysis, however, showed that the results of the experiments of Kaufmann and others were not accurate enough to decide between the different models. They only "indicated a large qualitative increase of mass with velocity" (Zahn and Spees, 1938).[56] All parties involved took these experiments much too seriously, especially when the data favored their own theories. Abraham hyped Kaufmann's results. Lorentz was too eager to believe Bucherer's results, while his earlier concern over Kaufmann's appears to have been somewhat disingenuous. Einstein's cavalier attitude toward Kaufmann's experiments stands in marked contrast to his belief in later results purporting to prove him right.

In Abraham's defense, it should be said that he could also be self-deprecating about his reliance on Kaufmann's data. At the 78th *Versammlung Deutscher Naturforscher und Ärzte* in Stuttgart in 1906, he got quite a few laughs when he joked: "When you look at the numbers you conclude from them that the deviations from the Lorentz theory are at least twice as big as mine, so you may say that the [rigid] sphere theory represents the reflection of β-rays twice as well as the relativity theory [by which Abraham meant Lorentz's electron model in this context]" (quoted in Miller, 1981, sec. 7.4.3, 221).

In 1906 Lorentz gave a series of lectures at Columbia University in New York, which were published in 1909. On the face of it, he seems to have taken Kaufmann's results quite seriously at the time. He wrote: "His [i.e., Kaufmann's] new numbers agree within the limits of experimental errors with the formulae given by Abraham, but [...] are decidedly unfavourable to the idea of a contraction such as I attempted to work out" (Lorentz, 1915, 212–213; quoted in Miller, 1981, sec. 12.4.1). Shortly before his departure for New York, he had told Poincaré the same thing: "Unfortunately my hypothesis of the flattening of electrons is in contradiction with Kaufmann's results, and I must abandon it. I am therefore at the end of my rope (*au bout de mon latin*)."[57] These passages strongly suggest that Lorentz took Kaufmann's results much more seriously than Einstein. Miller indeed draws that conclusion. Lorentz expert A. J. Kox, however, has pointed out to one of us (MJ) that Lorentz's reaction was probably more ambivalent (see also Hon, 1995, sec. 6). This is suggested by what Lorentz continues to say after acknowledging the problem with Kaufmann's data in his New York lectures: "Yet, though it seems very likely that we shall have to relinquish this idea altogether, *it is, I think, worth while looking into it somewhat more closely*" (Lorentz, 1915, 213; our italics). Lorentz then proceeds to discuss his idea *at length*.

In response to Kaufmann's alleged refutation of special relativity Einstein wrote in an oft-quoted passage:[58]

> Abraham's and Bucherer's theories of the motion of the electron yield curves that are significantly closer to the observed curve than the curve obtained from the theory of relativity. However, the probability that their theories are correct is rather

small, in my opinion, because their basic assumptions concerning ... the moving electron are not suggested by theoretical systems that encompass larger complexes of phenomena
(Einstein, 1907b, 439).

This is a fair assessment of Bucherer's theory. Whether it is also a fair assessment of Abraham's electromagnetic program is debatable. This will not concern us here. What we want to point out is that Einstein, like Abraham and
Lorentz, took the experimental data much more seriously when they went
his way. In early 1917, Friedrich Adler, detained in Vienna awaiting trial for
his assassination of the Austrian prime minister Count Stürgkh in November 1916, began sending Einstein letters and manuscripts attacking special
relativity.[59] He was still at it in the fall of 1918, when the exchange that is interesting for our purposes took place. Einstein wrote: "for a while Bucherer
advocated a theory that comes down to a different choice for l [see eq. 40 and
Fig. 1]. But a different choice for l is out of the question now that the laws
of motion of the electron have been verified with great precision."[60] From
his prison cell in Stein an der Donau Adler replied: "Now, I would be very
interested to hear, *which* experiments you see as definitively decisive about
the laws of motion of the electron. For as far as my knowledge of the literature goes, I have not found any claim of a final decision."[61] Adler went on
to quote remarks from Laue, Lorentz, and the experimentalist Erich Hupka,
spanning the years 1910–1915, all saying that this was still an open issue.[62]
In his response Einstein cited three recent studies (published between 1914
and 1917), which, he wrote, "have so to speak *conclusively shown* [*sicher
bewiesen*] that the relativistic laws of motion of the electron apply (as opposed to, for instance, those of Abraham)" (Einstein's emphasis).[63] Even
considering the context in which it was made, this is a remarkably strong
statement.

   Much more interesting than the agreement between theory and experiment or the lack thereof were the theoretical arguments that Abraham and
Lorentz put forward in support of their models. Lorentz was right in thinking
that it was no coincidence that his contractile electron exhibited exactly the
velocity dependence he needed to account for the absence of ether drift (see
the discussion following eq. 72). He could not have known at the time that
this particular velocity dependence is a generic feature of relativistic closed
systems. As the quotation at the end of sec. 6 shows, he did recognize this
later on. Abraham was right that fast electrons call for a new mechanics. His
new electromagnetic mechanics is much closer to relativistic mechanics than
to Newtonian mechanics. Like Lorentz, he just did not realize that this new
mechanics reflected a new kinematics rather than the electromagnetic nature

of all matter. Abraham at least came to accept that Minkowski space-time was the natural setting for his electromagnetic program.

Proceeding along similar lines as Abraham in developing his electromagnetic mechanics, we can easily get from Newtonian particle mechanics to relativistic continuum mechanics and back again. The first step is to read $\mathbf{F} = m\mathbf{a}$ as expressing momentum conservation (cf. the discussion following eq. 15 in sec. 2.2). In continuum mechanics, the differential form of the conservation laws is the fundamental law and the integral form is a derived law. In other words, the fundamental conservation laws are expressed in local rather than global terms. This reflects the transition from a particle ontology to a field ontology. Special relativity integrates the laws of momentum and energy conservation. These laws, of course, are Lorentz-invariant rather than Galilean-invariant. We thus arrive at the fundamental law of relativistic continuum mechanics, the Lorentz-invariant differential law of energy-momentum conservation, $\partial_\nu T^{\mu\nu} = 0$. To recap: there are four key elements in the transition from Newtonian particle mechanics based on $\mathbf{F} = m\mathbf{a}$ to relativistic continuum mechanics based on $\partial_\nu T^{\mu\nu} = 0$. They are (in no particular order): the transition from Galilean invariance to Lorentz invariance, the focus on conservation laws rather than force laws, the integration of the laws of energy and momentum conservation, and the transition from a particle ontology to a field ontology.

We now show how, once we have relativistic continuum mechanics, we recover Newtonian particle mechanics. Consider a closed system described by continuous (classical) fields such that the total energy-momentum tensor $T^{\mu\nu}_{\text{tot}}$ of the system can be split into a part describing a localizable particle (e.g., an electron à la Lorentz-Poincaré[64]) and a part describing its environment (e.g., an external electromagnetic field):

$$(159) \qquad T^{\mu\nu}_{\text{tot}} = T^{\mu\nu}_{\text{par}} + T^{\mu\nu}_{\text{env}}.$$

Using our fundamental law, $\partial_\nu T^{\mu\nu}_{\text{tot}} = 0$, integrated over space, we find

$$(160) \qquad 0 = \int \partial_\nu T^{\mu\nu}_{\text{tot}} d^3x = \int \partial_\nu T^{\mu\nu}_{\text{par}} d^3x + \int \partial_\nu T^{\mu\nu}_{\text{env}} d^3x.$$

As long as $T^{\mu\nu}_{\text{par}}$ drops off faster than $1/r^2$ as we go to infinity, Gauss's theorem tells us that

$$(161) \qquad \int \partial_i T^{\mu i}_{\text{par}} d^3x = 0.$$

For $\partial_\nu T^{\mu\nu}_{env}$ we can substitute minus the density f^μ_{ext} of the four-force acting on the particle. The spatial components of eq. 160 can thus be written as

(162) $$\frac{d}{dx_0} \int T^{i0}_{par} d^3x = \int f^i_{ext} d^3x.$$

The right-hand side gives the components of $\mathbf{F}_{external}$. Since

(163) $$P^\mu_{par} \equiv \frac{1}{c} \int T^{\mu 0}_{par} d^3x$$

and $x^0 = ct$, the left-hand side is the time derivative of the particle's momentum. Eq. 162 is thus equivalent to

(164) $$\frac{d\mathbf{P}_{par}}{dt} = \mathbf{F}_{ext}.$$

This equation has the same form (and the same transformation properties) as Abraham's electromagnetic equation of motion 21. In Abraham's equation, $\mathbf{P}_{par}$ is the electromagnetic momentum of the electron, and $\mathbf{F}_{ext}$ is the Lorentz force exerted on the electron by the external fields. Under the appropriate circumstances and with the appropriate identification of the Newtonian mass m, Abraham's electromagnetic equation of motion reduces to Newton's second law, $\mathbf{F} = m\mathbf{a}$ (see eq. 28). The same is true for our more general eq. 164. This equation, however, is not tied to electrodynamics. It is completely agnostic about the nature of both the particle and the external force. The only thing that matters is that it describes systems in Minkowski space-time, which obey relativistic kinematics. $\mathbf{P}_{par}$ and $\mathbf{F}_{ext}$, like Abraham's electromagnetic momentum and the Lorentz force, only transform as vectors under Galilean transformations in the limit of low velocities, where Lorentz transformations are indistinguishable from Galilean transformations. They inherit their transformation properties from $\partial_\nu T^{\mu\nu}_{par}$ and f^μ_{ext}, respectively, which transform as four-vectors under Lorentz transformations.

It only makes sense to split the total energy-momentum tensor $T^{\mu\nu}_{tot}$ into a particle part and an environment part, if the interactions holding the particle together are much stronger than the interactions of the particle with its environment. Typically, therefore, the energy-momentum of the particle taken by itself will very nearly be conserved, i.e.,

(165) $$\partial_\nu T^{\mu\nu}_{par} \approx 0.$$

This means that the particle's four-momentum will to all intents and purposes transform as a four-vector under Lorentz transformations and satisfy the relations for a strictly closed system (see eqs. 4–13):

(166) $$P^\mu_{par} \equiv \frac{1}{c} \int T^{\mu 0}_{par} d^3x \approx (\gamma m_0 c, \gamma m_0 \mathbf{v}).$$

Inserting $\mathbf{P}_{\mathrm{par}} = \gamma m_0 \mathbf{v}$ into eq. 164, we can reduce the problem in relativistic continuum mechanics that we started from in eq. 159 to a problem in the relativistic mechanics of point particles. In the limit of small velocities, such problems once again reduce to problems in the Newtonian mechanics of point particles.

To the best of our knowledge, this way of recovering particle mechanics from what might be called 'field mechanics' was first worked out explicitly in the context of general rather than special relativity (Einstein, 1918; Klein, 1918).[65] Relativistic continuum mechanics played a crucial role in the development of general relativity. For one thing, the energy-momentum tensor is the source of the gravitational field in general relativity.[66] Even before the development of general relativity, Einstein recognized the importance of relativistic continuum mechanics. In an unpublished manuscript of 1912, he wrote:

> The general validity of the conservation laws and of the law of the inertia of energy [...] suggest that [the symmetric energy-momentum tensor $T^{\mu\nu}$ and the equation $f^\mu = -\partial_\nu T^{\mu\nu}$] are to be ascribed a general significance, even though they were obtained in a very special case [i.e., electrodynamics]. We owe this generalization, *which is the most important new advance in the theory of relativity*, to the investigations of Minkowski, Abraham, Planck, and Laue (Einstein, 1987–2002, Vol. 4, Doc. 1, [p. 63]; our emphasis).

Einstein went on to give a clear characterization of relativistic continuum mechanics:

> To every kind of material process we want to study, we have to assign a symmetric tensor $(T_{\mu\nu})$ [...] Then [$f^\mu = -\partial_\nu T^{\mu\nu}$] must always be satisfied. The problem to be solved always consists in finding out how $(T_{\mu\nu})$ is to be formed from the variables characterizing the processes under consideration. If several processes can be isolated in the energy-momentum balance that take place in the same region, we have to assign to each individual process its own stress-energy tensor $(T_{\mu\nu}^{(1)})$, etc., and set $(T_{\mu\nu})$ equal to the sum of these individual tensors (ibid.).

As the development of general relativity was demonstrating the importance of continuum mechanics, developments in quantum theory—the Bohr model and Sommerfeld's relativistic corrections to it—rehabilitated particle mechanics, be it of Newtonian or relativistic stripe. As a result, relativistic continuum mechanics proved less important for subsequent developments in areas of physics other than general relativity than Einstein thought in 1912 and than our analysis in this paper suggests. The key factor in this was that it gradually became clear in the 1920s that elementary particles are point-like and not spatially extended like the electron models discussed in this paper. That special relativity precludes the existence of rigid bodies is just one of the problems such models are facing.

In hindsight, Lorentz, the guarded Dutchman, comes out looking much better than Abraham, his impetuous German counterpart. At one point, for instance, Lorentz (1915, 215) cautioned:

> In speculating on the structure of these minute particles we must not forget that there may be many possibilities not dreamt of at present; it may very well be that other internal forces serve to ensure the stability of the system, and perhaps, after all, we are wholly on the wrong track when we apply to the parts of an electron our ordinary notion of force (Lorentz, 1915, 215).

This passage is quoted approvingly by Pais (1972, 83). Even a crude operationalist argument of the young Wolfgang Pauli, which would have made his godfather Ernst Mach proud, can look prescient in hindsight. Criticizing the work of later proponents of the electromagnetic worldview in his review article on relativity, Pauli concluded:

> Finally, a conceptual doubt should be mentioned. The continuum theories make direct use of the ordinary concept of electric field strength, even for the fields in the interior of the electron. This field strength, however, is defined as the force acting on a test particle, and since there are no test particles smaller than an electron or a hydrogen nucleus the field strength at a given point in the interior of such a particle would seem to be unobservable by definition, and thus be fictitious and without physical meaning (Pauli, 1921, 206).

This moved Valentin Bargmann (1960, 189)—who had accompanied Einstein on his quest for a classical unified field theory, a quest very much in the spirit of Abraham's electromagnetic program—to write in the Pauli memorial volume:

> A physicist will feel both pride and humility when he reads Pauli's remarks today. In the light of our present knowledge the attempts which Pauli criticizes may seem hopelessly naïve, although it was certainly sound practice to investigate what the

profound new ideas of general relativity would contribute to the understanding of the thorny problem of matter (Bargmann, 1960, 189).

Putting such hagiography to one side, we conclude our paper by quoting and commenting on two oft-quoted passages that nicely illustrate some of the key points of our paper. The first is a brief exchange between Planck and Sommerfeld following a lecture by the former at the *Naturforscherversammlung* in Stuttgart on September 19, 1906.[67] Planck talked about "[t]he Kaufmann measurements of the deflectability of β-rays and their relevance for the dynamics of electrons." Abraham, Bucherer,[68] Kaufmann, and Sommerfeld all took part in the discussion afterwards. It was Planck who got to the heart of the matter:

> Abraham is right when he says that the essential advantage of the sphere theory would be that it be a purely electrical theory. If this were feasible, it would be very beautiful indeed, but for the time being it is just a postulate. At the basis of the Lorentz-Einstein theory lies another postulate, namely that no absolute translation can be detected. These two postulates, it seems to me, cannot be combined, and what it comes down to is which postulate one prefers. My sympathies actually lie with the Lorentzian postulate (Planck, 1906b, 761).

Whereupon Sommerfeld, pushing forty, quipped: "I suspect that the gentlemen under forty will prefer the electrodynamical postulate, while those over forty will prefer the mechanical-relativistic postulate" (Ibid.). The reaction of the assembled physicists to Sommerfeld's quick retort has also been preserved in the transcript of this session: "hilarity" (*Heiterkeit*). This exchange between Planck and Sommerfeld is perhaps the clearest statement in the contemporary literature of the dilemma that lies behind the choice between the electron models of Abraham and Lorentz. Physicists had to decide what they thought was more important, full relativity of uniform motion or the reduction of mechanics to electrodynamics. We find it very telling that in 1906 a leader in the field such as Sommerfeld considered the former the conservative and the latter the progressive option. Unlike Abraham, Lorentz, and Planck, however, Sommerfeld did not fully appreciate what was at stake.

First of all, his preference for the "electrodynamical postulate" was mainly because Lorentz's contractile electron was incompatible with superluminal velocities.[69] This can be inferred from a comment on Lorentz's electron model in (Sommerfeld, 1904c). In this paper—translated into Dutch by Peter Debye, Sommerfeld's student at the time (Eckert and Märker, 2000, 148), and communicated to the Amsterdam Academy of Sciences by Lorentz himself—Sommerfeld summarized and simplified his trilogy on electron theory in the proceedings of the Göttingen Academy (Sommerfeld, 1904a, 1904b, 1905a). He wrote:

> As is well-known, Lorentz, for very important reasons, has recently formulated the hypothesis that the shape of the electron is variable, i.e., that for every velocity the electron takes on the shape of a so-called "Heaviside ellipsoid." For velocities greater than that of light this hypothesis cannot be used; one can hardly speak of a "Heaviside hyperboloid" as the shape of the electron (Sommerfeld, 1904c, 433).

Sommerfeld's objections to Lorentz's program were thus not nearly as principled as Abraham's (cf. the passages from Abraham, 1903, quoted in sec. 4.4).

Moreover, from letters he wrote to Wien and Lorentz in November and December of 1906 (letters 102 and 103 in Eckert and Märker, 2000) it appears that Sommerfeld only became familiar with Einstein's work *after* the meeting in Stuttgart. On December 12, 1906, he wrote to Lorentz:

> Meanwhile I have also studied Einstein. It is remarkable to see how he arrives at the exact same results as you do (also with respect to his relative time) despite his very different epistemological point of departure. However, his deformed time, like your deformed electron, does not really sit well with me (Eckert and Märker, 2000, 258).

This passage suggests that Sommerfeld had not read (Einstein, 1905) before the 1906 *Naturforscherversammlung*. So Sommerfeld may not even have realized at the time that there was at least one gentleman well under forty, albeit one not in attendance in Stuttgart, who preferred the "mechanical-relativistic postulate," nor that the mechanics involved need not be Newtonian. By the time of the next *Naturforscherversammlung*, the following year in Dresden, Sommerfeld (1907), still only 39, had jumped ship and had joined the relativity camp (Battimelli, 1981, 150, note 30).[70] In an autobiographical sketch written in 1919, Sommerfeld ruefully looks back on this whole episode. Referring to the trilogy (Sommerfeld, 1904a, 1904b, 1905a), he wrote: "The last of these appeared in the critical year 1905, the birth year of relativity. These difficult and protracted studies, to which I originally attached great value, were therefore condemned to fruitlessness" (Sommerfeld, 1968, Vol. 4, 677).[71]

The second passage that we want to look at comes from Lorentz's important book *The Theory of Electrons*, based on his 1906 lectures in New

York and first published in 1909. Referring to Einstein and special relativity, Lorentz wrote

> His results concerning electromagnetic and optical phenomena (leading to the same contradiction with Kaufmann's results that was pointed out in §179[72]) agree in the main with those which we have obtained in the preceding pages, the chief difference being that Einstein simply postulates what we have deduced, with some difficulty and not altogether satisfactorily, from the fundamental equations of the electromagnetic field. (Lorentz, 1915, 229–230).

The parenthetical reference to "Kaufmann's results" suggests that the famous clause that concludes this sentence—"Einstein simply postulates what we have deduced [...] from the fundamental equations of the electromagnetic field"—refers, at least in part, to Lorentz's own struggles with the velocity dependence of electron mass.[73] The relativistic derivation of these relations is mathematically equivalent to Lorentz's 1899 derivation of them from the requirement, formally identical to the relativity principle, that ether drift can never be detected (see sec. 3, eqs. 45–51). From Lorentz's point of view, the relativistic derivation therefore amounted to nothing more than postulating these relations on the basis of the relativity principle. Lorentz himself had gone to the trouble of producing a concrete model of the electron such that its mass exhibited exactly the desired velocity-dependence (see sec. 4, eqs. 67–73). As we saw at the end of sec. 6, by 1922, if not much earlier, Lorentz had recognized that this had led him on a wild goose chase: "the formula for momentum [of which those for the velocity dependence of mass are a direct consequence] is a general consequence of the principle of relativity [...] and tells us nothing about the nature of mass or of the structure of the electron." This was Lorentz's way of saying what Pais said in the quotation with which we began this paper.

Acknowledgments: We want to thank Jon Dorling, Tony Duncan, Michael Eckert, Gordon Fleming, Klaus Frovin Jørgensen, Shaul Katzir, A. J. Kox, Christoph Lehner, John Norton, Jürgen Renn, Serge Rudaz, Rob Rynasiewicz, Tilman Sauer, Urs Schöpflin, Robert Schulmann, Matthias Schwerdt, John Stachel, Rick Swanson, and Scott Walter for helpful discussions, suggestions, and references. One of us (MJ) gratefully acknowledges support from the *Max-Planck-Institut für Wissenschaftsgeschichte* in Berlin.

University of Minnesota
USA

NOTES

[1]For discussion of and references to the experimental literature, we refer to (Miller, 1981), (Cushing, 1981), and (Hon, 1995). (Pauli, 1921, 83) briefly discusses some of the later experiments. See also (Gerlach, 1933), a review article on electrons first published in the late 1920s.

[2]Moreover, classical electron models have continued to attract attention from (distinguished) physicists (see note 12 below). In addition, the notion of "Poincaré pressure" introduced to stabilize Lorentz's electron (see below) resurfaced in a theory of Einstein (1919), which is enjoying renewed interest (Earman, 2003), as well as in other places (see, e.g., Grøn, 1985, 1988).

[3]See (Sommerfeld, 1904a, 1904b, 1904c, 1905a, 1905b) and (Herglotz, 1903). For a discussion of the development of Sommerfeld's attitude toward the electromagnetic program and special relativity, see (McCormmach, 1970, 490) and (Walter 1999a, 69–73, Forthcoming, sec. 3). On Minkowski and the electromagnetic program, see (Galison, 1979), (Pyenson, 1985, Ch. 4), (Corry, 1997), and (Walter, 1999a, 1999b, Forthcoming).

[4]This particular history of the electron is conspicuously absent, however, from the collection of histories of the electron brought together in (Buchwald and Warwick, 2001). One of us (MJ) bears some responsibility for that and hopes to make amends with this paper.

[5]See also (Darrigol, 2000). We refer to (Janssen, 1995, 2002b) for references to and discussion of earlier literature on this topic.

[6]For more recent commentary, see (Corry, 1999).

[7]As Born explains in introductory comments to the reprint of (Born, 1909b) in a volume with a selection of his papers (Born, 1965, Vol. 1, XIV–XV). For a brief discussion of the debate triggered by Born's work and references to the main contributions to this debate, see the editorial note, "Einstein on length contraction in the theory of relativity," in (Einstein, 1987–2002, Vol. 3, 478–480).

[8]For a brief discussion of this acrimonious exchange, see (Miller, 1981, sec. 1.13.1, especially notes 57 and 58).

[9]For brief discussions, see (Balazs, 1972, 29–30) and (Warwick, 2003, Ch. 8, especially 413–414). We also refer to Warwick's work for British reactions to the predominantly German developments discussed in our paper. See, e.g., (Warwick, 2003, 384) for comments by James Jeans on electromagnetic mass.

[10]For a brief discussion, see (Balazs, 1972, 30)

[11]See also (Cuvaj, 1968). We have benefited from (annotated) translations of Poincaré's paper by Schwartz (1971, 1972) and Kilmister (1970), as well as from the translation of passages from (Poincaré, 1905), the short version of (Poincaré, 1906), by Keswani and Kilmister (1983). A new translation of parts of (Poincaré, 1906) by Scott Walter will appear in (Renn, Forthcoming).

[12]See (Rohrlich, 1960, 1965, 1970, 1997). See also, e.g., (Fermi, 1921, 1922) [cf. note 20 below], (Wilson, 1936), (Dirac, 1938), (Kwal, 1949), (Caldirola, 1956), (Zink, 1966, 1968, 1971), (Pearle, 1982), (Schwinger, 1983) [in a special issue on

the occasion of Dirac's 80th birthday], (Comay, 1991), (Yaghjian, 1992), (Moylan, 1995), and (Hnizdo, 1997).

[13]The letter U rather than E is used for energy to avoid confusion with the electric field. We shall be using SI units throughout. For conversion to other units, see, e.g., (Jackson, 1975, 817–819).

[14]From $ds^2 = \eta_{\mu\nu}dx^\mu dx^\nu = (c^2 - v^2)dt^2$ it follows that $ds = c\sqrt{1 - v^2/c^2}dt = cdt/\gamma$.

[15]The energy-momentum tensor is typically symmetric. In that case, $T^{i0} = T^{0i}$, which means that the momentum density (T^{i0}/c) equals the energy flow density (cT^{0i}) divided by c^2. As was first noted by Planck, this is one way of expressing the inertia of energy, $E = mc^2$.

[16]Which is why $T^{\mu\nu}$ is also known as the stress-energy tensor or the stress-energy-momentum tensor.

[17]See (Rohrlich, 1965, 89–90, 279–281) or (Janssen, 1995, sec. 2.1.3) for the details of the proof, which is basically an application of the obvious generalization of Gauss's theorem (which says that for any vector field $\mathbf{A}$, $\oint \mathbf{A} \cdot d\mathbf{S} = \int \operatorname{div} \mathbf{A} \, d^3x$) from three to four dimensions.

[18]This way of writing P^μ was suggested to us by Serge Rudaz. See (Janssen, 2002b, 440–441, note; 2003, 47) for a more geometrical way of stating the argument below.

[19]As Gordon Fleming (private communication) has emphasized, the rest frame cannot always be uniquely defined. For the systems that will concern us here, this is not a problem. Following Fleming, one can avoid the arbitrary choice of n^μ altogether by accepting that the four-momentum of spatially extended systems is a hyperplane-dependent quantity.

[20]Some of Fermi's earliest papers are on this issue (Miller, 1973, 317). We have not been able to determine what sparked Fermi's interest in this problem. His biographer only devotes one short paragraph to it: "In January 1921, Fermi published his first paper, "On the Dynamics of a Rigid System of Electrical Charges in Translational Motion" [Fermi, 1921]. This subject is of continuing interest; Fermi pursued it for a number of years and even now it occasionally appears in the literature" (Segrè, 1970, 21).

[21](Rohrlich, 1965, 17) notes that Fermi's idea was forgotten and independently rediscovered at least three times, by W. Wilson (1936), by Bernard Kwal (1949), and then by Rohrlich himself (Rohrlich, 1960). This goes to show that John Stachel's meta-theorem—anything worth discovering once in general relativity has been discovered at least twice—also holds for special relativity. In the preface to the second edition of his textbook on special relativity, Aharoni (1965) cites (Rohrlich, 1960) as the motivation for some major revisions of the first edition, published in 1959. In this same preface, Aharoni lists Dirac (1938) and Kwal as rediscoverers. For a concise exposition of Rohrlich's work, see (Aharoni, 1965, sec. 5.5, 160–165).

[22]See (Abraham, 1902a, 25–26; 1903, 110). In both places, he cites (Poincaré, 1900) for the basic idea of ascribing momentum to the electromagnetic field. For

discussion, see (Miller, 1981, sec. 1.10), (Darrigol, 1995; 2000, 361), and (Janssen, 2003, sec. 3)

[23] In fact, another force, a stabilizing force $\mathbf{F}_{stab}$, needs to be added to keep the charges from flying apart under the influence of their Coulomb repulsion.

[24] See, e.g., (Lorentz, 1904a, sec. 7), (Abraham, 1905, sec. 5), (Jackson, 1975, 238–239), (Janssen, 1995, 56–58), (Griffith, 1999, 351–352). In special relativity, we would write eq. 19 as the integral over the spatial components of the Lorentz four-force density f^{μ}, which is equal to minus $\partial_{\nu} T_{EM}^{\mu\nu}$, the four-divergence of the energy-momentum tensor for the electron's self-field,

$$F_{self}^{i} = - \int \partial_{\nu} T_{EM}^{i\nu} d^3x,$$

with $T_{EM}^{i0} \equiv c\varepsilon_0 (\mathbf{E} \times \mathbf{B})^i$ and $T_{EM}^{ij} \equiv -T_{Maxwell}^{ij}$ (cf. eqs. 127 and 129 and note 51).

[25] This assumption may sound innocuous, but under the standard definition 12 of the four-momentum of spatially extended systems, the (ordinary three-)momentum of open systems will in general not be in the direction of motion. Because both Lorentz's and Abraham's electrons are symmetric around an axis in the direction of motion, the momentum of their self-fields is always in the direction of motion, even though these fields by themselves do not constitute closed systems. If a system has momentum that is not in the direction of motion, it will be subject to a turning couple trying to align its momentum with its velocity. Trouton and Noble (1903) tried in vain to detect this effect on a charged capaticor hanging from the ceiling of their laboratory on a torsion wire (cf. Janssen, 2002b, 440–441, note, and Janssen, 1995, especially secs. 1.4.2 and 2.2.5). Ehrenfest (1907) raised the question whether the electron would be subject to a turning couple if it were *not* symmetric around the axis in the direction of motion. Einstein (1907a) countered that the behavior of the electron would be independent of its shape. This exchange between Einstein and Ehrenfest is discussed in (Miller, 1981, sec. 7.4.4.). Laue (1911a) proved Einstein right (see also Pauli, 1921, 186–187). As with the capacitor in the Trouton-Noble experiment, the electromagnetic momentum of the electron is not the only momentum of the system. The non-electromagnetic part of the system also contributes to its momentum. Laue showed that the total momentum of a closed static system is always in the direction of motion. From a modern point of view this is because the four-momentum of a closed system (static or not) transforms as a four-vector under Lorentz transformations. The momenta of open systems, such as the subsystems of a closed static system, need not be in the direction of motion, in which case the system is subject to equal and opposite turning couples. A closed system never experiences a net turning couple. The turning couples on open systems, it turns out, are artifacts of the standard definition 12 of the four-momentum of spatially extended systems. Under the alternative Fermi-Rohrlich definition (see the discussion following eq. 14), there are no turning couples whatsoever (see Butler, 1968; Janssen, 1995; and Teukolsky, 1996).

[26] Substituting the momentum, $p = mv$, of Newtonian mechanics for P_{EM} in eq. 26, we find $m_{//} = m_{\perp} = m$.

[27] But recall that there should be an additional term, $\mathbf{F}_{stab}$, on the right-hand side of eq. 16 (see note 23).

[28] Substituting the kinetic energy, $U_{kin} = \frac{1}{2}mv^2$, of Newtonian mechanics for U_{EM} in eq. 31, we find $m_{//} = m$, in accordance with the result found on the basis of eq. 26 and $p = mv$ (see note 26).

[29] The converse is not true. For the electron model of Bucherer and Langevin (see sec. 4) $(U_{EM}/c, \mathbf{P}_{EM})$ is not a four-vector, yet U_{EM} and $\mathbf{P}_{EM}$ give the same longitudinal mass $m_{//}$ (see eqs. 74–77). The same is true for the Newtonian energy $U_{kin} = \frac{1}{2}mv^2$ and the Newtonian momentum $p = mv$ (see notes 26 and 28).

[30] The first relation follows from $\dfrac{d\gamma^{-2}}{dv} = \dfrac{d}{dv}\left(1 - \dfrac{v^2}{c^2}\right)$ or $-2\gamma^{-3}\dfrac{d\gamma}{dv} = -2\dfrac{v}{c^2}$; the second is found with the help of the first:

$$\frac{d(\gamma v)}{dv} = \gamma + v\frac{d\gamma}{dv} = \gamma + \gamma^3\beta^2 = \gamma^3(1 - \beta^2 + \beta^2) = \gamma^3.$$

[31] For more extensive discussion, see (Janssen, 1995, Ch. 3; 2002b; Janssen and Stachel, 2004).

[32] For the magnetic field it is the motion of charges with respect to the ether that matters, not the motion with respect to the lab frame.

[33] For the induced $\mathbf{E}$ and $\mathbf{B}$ fields it is the changes in the $\mathbf{B}$ and $\mathbf{E}$ fields at fixed points in the ether that matter, not the changes at fixed points in the lab frame.

[34] Lorentz only started using the relativistic transformation formula for non-static charge densities and for current densities in 1915 (Janssen, 1995, secs. 3.5.3 and 3.5.6).

[35] See (Lorentz, 1895, sec. 19–23) for the derivation of this transformation law and (Janssen, 1995, sec. 3.2.5) or (Zahar, 1989, 59–61) for a reconstruction of this derivation in modern notation.

[36] For an elegant and elementary exposition of Planck's derivation, see (Zahar, 1989, sec. 7.1, 227–237). The equations for the relation between $\mathbf{a}'$ and $\mathbf{a}$ can be found on p. 232, eqs. (2)–(4).

[37] Einstein (1905, 919) obtained $m_{\perp} = \gamma^2 m_0$ instead of $m_{\perp} = \gamma m_0$, the result obtained by Planck and Lorentz (for $l = 1$). The discrepancy comes from Einstein using $\mathbf{F}' = \mathbf{F}$ instead of $\mathbf{F}' = \text{diag}(1, \gamma, \gamma)\mathbf{F}$, the now standard transformation law for forces used by Lorentz and Planck (Zahar, 1989, 233). Einstein made it clear that he was well aware of the arbitrariness of his definition of force. When (Einstein, 1905) was reprinted in (Blumenthal, 1913), a footnote was added in which Einstein's original definition of force is replaced by the one of Lorentz and Planck. Recently a slip of paper came to light with this footnote in Einstein's own hand. This shows that the footnote was added by Einstein himself and not by Sommerfeld as suggested, e.g., by Miller (1981, 369, 391).

[38] See eq. 126 below for the relation between the (contravariant) electromagnetic field strength tensor $F^{\mu\nu}$ (and its covariant form $F_{\mu\nu} = \eta_{\mu\rho}\eta_{\nu\sigma}F^{\rho\sigma}$) and the components of $\mathbf{E}$ and $\mathbf{B}$.

[39]If the Fermi-Rohrlich definition is used, the relation $d^3x = d^3x'/\gamma l^3$ used in going from eq. 54 to eq. 55 no longer holds. Kwal (1949) clearly recognized that this is the source of the problem. In the abstract of his paper he wrote: "The appearance of the factor 1/3 in the expression for the total energy of the moving electron results from the simultaneous use in the calculation of a tensorial quantity (the energy-momentum tensor) and a quantity that is not [a tensor] (the volume element). The difficulty disappears with a tensorial definition of the volume element."

[40]This is an example of what one of us called a "common origin inference" or *COI* in (Janssen, 2002a). The example illustrates how easy it is to overreach with this kind of argument (for other examples see, ibid., 474, 491, 508).

[41]Cf. (Miller, 1981, sec. 1.13.2). Miller cites a letter of January 26, 1905, in which Abraham informed Lorentz of this difficulty. See also (Lorentz, 1915, 213).

[42]Carrying out the differentiation with respect to γ in eq. 76, we find:

$$m_{//} = \frac{c^2}{v}\left(\frac{2}{3}\gamma^{-\frac{1}{3}} + \frac{1}{3}\gamma^{-\frac{7}{3}}\right)\gamma^3\frac{v}{c^2}m_0,$$

where we used eq. 35 for $d\gamma/dv$. This in turn can be rewritten as

$$m_{//} = \gamma^{\frac{8}{3}}\left(\frac{2}{3} + \frac{1}{3}\gamma^{-2}\right)m_0 = \gamma^{\frac{8}{3}}\left(\frac{2}{3} + \frac{1}{3}\left(1 - \beta^2\right)\right)m_0 = \gamma^{\frac{8}{3}}\left(1 - \frac{1}{3}\beta^2\right)m_0.$$

[43]We are grateful to Serge Rudaz for his help in reconstructing this argument.

[44]Referring to (Poincaré, 1885, 1902a, 1902b), Scott Walter (Forthcoming, sec. 1) makes the interesting suggestion that "[s]olving the stability problem of Lorentz's contractile electron was a trivial matter for Poincaré, as it meant transposing to electron theory a special solution to a general problem he had treated earlier at some length: to find the equilibrium form of a rotating fluid mass."

[45]As Miller (1973, 300) points out, in the short announcement of his 1906 paper, Poincaré (1905, 491) mistakenly wrote that the electron "is under the action of constant external pressure" (Keswani and Kilmister, 1983, 352).

[46]For a detailed analysis of the completely analogous case of the forces on a capacitor in the Trouton-Noble experiment, see secs. 2.3.3 and 2.4.2 of (Janssen, 1995).

[47]We are grateful to Scott Walter for reminding us of this problem. We essentially follow the analysis of the problem by Miller (1973, 298–299), although we draw a slightly different conclusion (see note 48). Schwartz (1972, 871) translates the relevant passage from (Poincaré, 1906) but passes over the problem in silence.

[48]One might object, however, that our reading of Poincaré is too charitable. Poincaré certainly does not explicitly say, once he has derived the expression for Poincaré pressure at the end of section 6, that this restores the standard relations between Hamiltonian, Lagrangian, and generalized momentum in Lorentz's model. Yet we take this to be the rationale behind his calculations. Miller (1973, 248) is harder on Poincaré: "contrary to what is sometimes attributed to this paper [Poincaré, 1906], Poincaré never computed the counter term [our eq. 116] necessary to cancel

the second term on the right-hand side of [our eq. 115], nor did he reduce the factor of 4/3 in [the electromagnetic momentum] to unity [compare $\mathbf{P}_{EM}$ in eq. 66 to $\mathbf{P}_{tot}$ in eq. 113]." This is all true. Our rejoinder on behalf of Poincaré is that he did not need to do any of this to remove the inconsistency in Lorentz's model.

[49]Compare eq. 120 to eq. 87, which for small velocities reduces to

$$L_{EM} \approx -U_{0_{EM}}(1 - v^2/2c^2).$$

[50]Shaul Katzir (private communication) has suggested a more charitable interpretation of Poincaré's comments. Poincaré, Katzir suggests, recognized that the electron is not a purely electromagnetic system but believed that its mass is nonetheless given by its electromagnetic momentum through eqs. 26. For the specific model proposed by Poincaré this is not true. The non-electromagnetic piece he added to stabilize Lorentz's electron does contribute to the electron's mass, giving a total mass of $(4/3)U_{0_{EM}}/c^2$. As we shall see in sec. 6, however, it is possible to add a stabilizing piece that does not contribute to the electron's mass (see our remarks following eq. 158).

[51] The derivation of eq. 129 is essentially just the reverse of the derivation of eq. 17 and can be pieced together from the passages we cited for the latter (see note 24). From a relativistic point of view, eq. 129 is immediately obvious since the (four-)gradient of the energy-momemtum tensor gives minus the density of the (four-)force acting on the system (see, e.g., Pauli, 1921, 126, eq. (345)). The right-hand side of eq. 129 is minus the Lorentz force density in the absence of a magnetic field (ibid., 85, eq. (225)).

[52] $T^{00}_{0_{non-EM}}$ can be any function of the spatial coordinates and the system will still be closed. Of course, this component needs to be chosen in such a way that $T^{\mu\nu}_{non-EM}$ continues to transform as a tensor. We ensure this by changing definition 134 to:

$$T^{\mu\nu}_{0_{non-EM}} \equiv -\eta^{\mu\nu}P_{Poincaré}\vartheta(R - r_0) + f(\mathbf{x}_0)\frac{u_0^\mu u_0^\nu}{c^2},$$

where $u^\mu = \gamma(c, \mathbf{v})$ is the electron's four-velocity. The function $f(\mathbf{x}_0)$ can be chosen arbitrarily as long as the energy density is positive definite everywhere. Hence, it must satisfy the condition $f(\mathbf{x}_0) \geq 0$ outside the electron and the condition $f(\mathbf{x}_0) \geq P_{Poincaré}$ inside. If we choose $f(\mathbf{x}_0) = P_{Poincaré}\vartheta(R - r_0)$, the definition above becomes

$$T^{\mu\nu}_{0_{non-EM}} \equiv -\left(\eta^{\mu\nu} - \frac{u_0^\mu u_0^\nu}{c^2}\right)P_{Poincaré}\vartheta(R - r_0),$$

in which case $T^{00}_{0_{non-EM}} = 0$. This definition was proposed by Schwinger (1983, 379, eqs. (42)–(43)).

[53]As Rohrlich (1997, 1056), following (Schwinger, 1983, 374, 379), put it: "The argument over whether m_{es}[equal to $U_{0_{EM}}/c^2$ in our notation] or $m_{ed} = 4m_{es}/3$ is the "right" answer is thus resolved: [. . .] it depends on the model; either value as well as any value in between is possible [as are values greater than m_{ed}; cf. note 52 above]. But in all cases, one obtains a four-vector for the stabilized charged sphere". Which

situation obtains cannot be decided experimentally. The rest mass of the electron can be determined, but that value can be represented by U_0/c^2, by $4U_0/3c^2$, or by some other value by adjusting the radius of the electron, for instance, which cannot be determined experimentally.

[54]This stabilizing system will not be as simple as the Poincaré pressure for the Lorentz-Poincaré electron. Without the spherical symmetry of this specific model, eq. 134 for the non-electromagnetic part of the energy-momentum tensor will be more complicated. See (Janssen, 1995, sec. 2.3.3, especially eq. (2.96)) for another simple example, the stabilizing mechanism for the surface charge distribution on a plate capacitor, worked out with the help of Tony Duncan.

[55]See the discussion following eq. 13 and (Janssen, 2003, 46–47).

[56]Quoted in (Miller, 1981, 331). In a review article about electrons originally published in the late 1920s, Walter Gerlach still claimed that the experiments of Bucherer and others decided in favor of the relativistic formula for the velocity dependence of the electron mass. Gerlach concluded: "Today there is therefore no reason to doubt the correctness of the results of the investigations of Bucherer, Wolz, Schaefer, and Neumann that *the experimentally observed velocity-dependence of the electron mass agrees, within the margins to be expected from the sources of error inherent in the method, only with the Lorentz-Einstein theory of the electron*" (Gerlach, 1933, 81). In a footnote, he adds: "Also note in this context the corresponding corroboration on the basis of [De Broglie] "wavelength"-measurements of electrons of different velocity by Ponte [1930]." Inspired by Zahn and Spees, (Rogers et al., 1940) repeated the experiment of the 1910s with sufficient accuracy to distinguish the relativistic prediction from Abraham's. Despite this result, (Faragó and Jánossy, 1957), in a subsequent review of the experimental confirmation of the relativistic formula for the velocity dependence of electron mass, essentially concurred with Zahn and Spees (Battimelli, 1981, 149; note 63 explains the reason for our qualification).

[57]Lorentz to Poincaré, March 8, 1906 (see Miller, 1981, sec. 12.4.1, for the quotation, and pp. 318–319 for a reproduction of the letter in facsimile).

[58]See, e.g., (Holton, 1988, 252–253), (Miller, 1981, sec. 12.4.3), (Hon, 1995, 208), and (Janssen, 2002a, 462, note 9).

[59]See Adler to Einstein, March 9, 1917 (Einstein, 1987–2002, Vol. 8, Doc. 307). In 1909 Adler had supported Einstein's candidacy for a post at the University of Zurich for which both of them had applied (see Einstein to Michele Besso, April 29, 1917 (Einstein, 1987–2002, Vol. 8, Doc. 331)). Einstein reciprocated in 1917 by drafting a petition on behalf of a number of Zurich physicists asking the Austrian authorities for leniency in Adler's case, even as Adler was busying himself with a critique of his benefactor's theories (see the letter to Besso quoted above). A draft of Einstein's petition is reproduced in facsimile in (Renn, 2005, 317). Adler's father, the well-known Austrian social democrat Victor Adler, considered using his son's railings against relativity for an insanity defense. His son, however, was determined to stand by his critique of relativity, even if it meant ending up in front of the firing squad. Adler was in fact sentenced to death but it was clear to all involved that

he would not be executed. The death sentence was commuted to eighteen years in prison on appeal and Adler was pardoned immediately after the war. This bizarre story is related in (Fölsing, 1997, 402–405). For an analysis of the psychology behind Adler's burning martyrdom, see (Ardelt, 1984).

[60] Einstein to Adler, September 29, 1918 (Einstein, 1987–2002, Vol. 8, Doc. 628; translation here and in the following are based on Ann M. Hentschel's).

[61] Adler to Einstein, October 12, 1918 (Einstein, 1987–2002, Vol. 8, Doc. 632; Adler's emphasis).

[62] Cf., however, the quotation from Lorentz in note 72 below.

[63] Einstein to Adler, October 20, 1918 (Einstein, 1987–2002, Vol. 8, Doc. 636). Two of the studies cited by Einstein involved the deflection of fast electrons as in the experiments of Kaufmann, Bucherer, and others; the third—by Karl Glitscher (1917), a student of Sommerfeld—used the fine structure of spectral lines to distinguish between the relativistic and the Abraham prediction for the velocity dependence of the electron mass. The experiment is not mentioned in the review article on electrons by Gerlach (1933), but Faragó and Jánossy (1957, sec. 2) review it very favorably. They write: "Analyzing the available experimental material, we have come to the conclusion that it is the fine-structure splitting in the spectra of atoms of the hydrogen type which give [sic] the only high-precision confirmation of the relativistic law of the variation of electron mass with velocity" (Faragó and Jánossy, 1957, 1417; quoted in Hon, 1995, 197).

[64] In general we need the fields associated with the particle to be sharply peaked around the worldline of the particle, a four-dimensional 'world-tube.'

[65] Einstein and Felix Klein corresponded about this issue in 1918 (Einstein, 1987–2002, Vol. 8, Docs. 554, 556, 561, 566, and 581). See also Hermann Weyl to Einstein, November 16, 1918 (Einstein, 1987–2002, Vol. 8, Doc. 657). A precursor to this approach can be found in (Einstein and Grossmann, 1913, sec. 4), where Einstein pointed out that the geodesic equation, which governs the motion of a test particle in a gravitational field, can be obtained by integrating $T^{\mu\nu}_{;\nu} = 0$—the vanishing of the covariant divergence of $T^{\mu\nu}$, the general-relativistic generalization of $\partial_\nu T^{\mu\nu} = 0$—over the 'worldtube' of the corresponding energy-momentum tensor for pressureless dust ("thread of flow" [Stromfaden] is the term Einstein used). This argument can also be found in the so-called Zurich Notebook (Einstein, 1987–2002, Vol. 4, Doc. 10, [p. 10] and [p. 58]). For analysis of these passages, see (Norton, 2000, Appendix C) and "A Commentary on the Notes on Gravity in the Zurich Notebook" in (Renn, Forthcoming, sec. 3 and 5.5.10; the relevant pages of the notebook are referred to as '5R' and '43L').

[66] See (Renn and Sauer, Forthcoming) for extensive discussion of the role of the energy-momentum tensor in the research that led to general relativity.

[67] This exchange is also discussed, for instance, in (Miller, 1981, sec. 7.3.4), (McCormmach, 1970, 489–490), and (Jungnickel and McCormmach, 1986, 249–250).

[68] Understandably, Bucherer took exception to the fact that Planck only discussed the electron models of Lorentz and Abraham (Planck, 1906b, 760).

[69]For brief discussions of the debate over superluminal velocities in the years surrounding the advent of special relativity, see (Miller, 1981, 110–111, note 57) and the editorial note, "Einstein on Superluminal Signal Velocities," in (Einstein, 1987–2002, Vol. 5, 56–60).

[70]See (Walter, 1999a, sec. 3.1) for a more charitable assessment of the development of Sommerfeld's views.

[71]We are grateful to Michael Eckert for alerting us to this passage and for providing us with the date of this part of Sommerfeld's autobiographical sketch.

[72] In the second edition, Lorentz added the following footnote at this point: "Later experiments [. . .] have confirmed [eq. 37] for the transverse electromagnetic mass, so that, in all probability, the only objection that could be raised against the hypothesis of the deformable electron and the principle of relativity has now been removed" (Lorentz, 1915, 339).

[73]For more extensive discussion of this passage, see (Janssen, 1995, sec. 4.3).

REFERENCES

Abraham, M. (1902a). Dynamik des Elektrons, *Königliche Gesellschaft der Wissenschaften zu Göttingen. Mathematisch-physikalische Klasse, Nachrichten* pp. 20–41.

Abraham, M. (1902b). Prinzipien der Dynamik des Elektrons, *Physikalische Zeitschrift* **4**: 57–63.

Abraham, M. (1903). Prinzipien der Dynamik des Elektrons, *Annalen der Physik* **10**: 105–179.

Abraham, M. (1904a). Die Grundhypothesen der Elektronentheorie, *Physikalische Zeitschrift* **5**: 576–579.

Abraham, M. (1904b). Zur Theorie der Strahlung und des Strahlungsdruckes, *Annalen der Physik* **14**: 236–287.

Abraham, M. (1904c). Kritik der Erwiderung des Hrn. W. Wien, *Annalen der Physik* **14**: 1039–1040. Response to (Wien, 1904c).

Abraham, M. (1905). *Theorie der Elektrizität,* Vol. 2, *Elektromagnetische Theorie der Strahlung*, Teubner, Leipzig.

Abraham, M. (1909). Zur Elektrodynamik bewegter Körper, *Circolo Matematico di Palermo, Rendiconti* **28**: 1–28.

Aharoni, J. (1965). *The Special Theory of Relativity,* 2nd ed., Dover, New York.

Ardelt, R. G. (1984). *Friedrich Adler. Probleme einer Persönlichkeitsentwicklung um die Jahrhundertwende*, Österreichischer Bundesverlag, Vienna.

Balazs, N. L. (1972). The Acceptability of Physical Theories: Poincaré versus Einstein, *in* L. O'Raifeartaigh (ed.), *General Relativity. Papers in Honour of J. L. Synge*, Clarendon Press, Oxford, pp. 21–34.

Bargmann, V. (1960). Relativity, *in* M. Fierz and V. F. Weisskopf (eds), *Theoretical Physics in the Twentieth Century. A Memorial Volume to Wolfgang Pauli*, Interscience Publishers, New York, pp. 187–198.

Battimelli, G. (1981). The Electromagnetic Mass of the Electron: A Case Study of a Non-Crucial Experiment, *Fundamenta Scientiae* **2**: 137–150.

Blumenthal, O. (ed.) (1913). *Das Relativitätsprinzip. Eine Sammlung von Abhandlungen*, Teubner, Leipzig.

Born, M. (1909a). Die träge Masse und das Relativitätsprinzip, *Annalen der Physik* **28**: 571–584.

Born, M. (1909b). Die Theorie des starren Elektrons in der Kinematik des Relativitätsprinzips, *Annalen der Physik* **30**: 1–56 ("Berichtigung," ibid. 840).

Born, M. (1910). Über die Definition des starren Körpers in der Kinematik des Relativitätsprinzips, *Physikalische Zeitschrift* **11**: 233–234.

Born, M. (1965). *Ausgewählte Abhandlungen*, Vol. 2, Vandenhoeck & Ruprecht, Göttingen.

Bucherer, A. H. (1904). *Mathematische Einführung in die Elektronentheorie*, Teubner, Leipzig.

Bucherer, A. H. (1905). Das deformierte Elektron und die Theorie des Elektromagnetismus, *Physikalische Zeitschrift* **6**: 833–834.

Bucherer, A. H. (1907). On a New Principle of Relativity in Electromagnetism, *Philosophical Magazine* **13**: 413–420.

Bucherer, A. H. (1908a). On the Principle of Relativity and on the Electromagnetic Mass of the Electron. A Reply to Mr. E. Cunningham, *Philosophical Magazine* **15**: 316–318. Response to (Cunningham, 1907).

Bucherer, A. H. (1908b). On the Principle of Relativity. A Reply to Mr. E. Cunningham, *Philosophical Magazine* **16**: 939–940. Response to (Cunningham, 1908).

Bucherer, A. H. (1908c). Messungen an Becquerelstrahlen. Die experimentelle Bestätigung der Lorentz-Einsteinschen Theorie, *Physikalische Zeitschrift* **9**: 755–762.

Buchwald, J. Z. and Warwick, A. (eds) (2001). *Histories of the Electron. The Birth of Microphysics*, The MIT Press, Cambridge.

Butler, J. W. (1968). On the Trouton-Noble Experiment, *American Journal of Physics* **36**: 936–941.

Caldirola, P. (1956). A New Model of Classical Electron, *Supplemento al Volume III, Serie X del Nuovo Cimento* pp. 297–343.

Comay, E. (1991). Lorentz Transformations of Electromagnetic Systems and the 4/3 Problem, *Zeitschrift für Naturforschung A* **46**: 377–383.

Corry, L. (1997). Hermann Minkowski and the Postulate of Relativity, *Archive for History of Exact Sciences* **51**: 273–314.

Corry, L. (1999). From Mie's Electromagnetic Theory of Matter to Hilbert's Unified Foundations of Physics, *Studies in History and Philosophy of Modern Physics* **30**: 159–183.

Cunningham, E. (1907). On the Electromagnetic Mass of a Moving Electron, *Philosophical Magazine* **14**: 538–547.

Cunningham, E. (1908). On the Principle of Relativity and the Electromagnetic Mass of the Electron. A Reply to Dr. A. H. Bucherer, *Philosophical Magazine* **16**: 423–428. Response to (Bucherer, 1908a).

Cushing, J. T. (1981). Electromagnetic Mass, Relativity, and the Kaufmann Experiments, *American Journal of Physics* **49**: 1133–1149.

Cuvaj, C. (1968). Henri Poincaré's Mathematical Contributions to Relativity and the Poincaré Stresses, *American Journal of Physics* **36**: 1102–1113.

Damerow, P., Freudenthal, G., McLaughlin, P. and Renn, J. (2004). *Exploring the Limits of Preclassical Mechanics. A Study of Conceptual Development in Early Modern Science: Free Fall and Compounded Motion in the Work of Descartes, Galileo, and Beeckman.* 2nd ed., Springer, New York.

Darrigol, O. (1995). Henri Poincaré's Criticism of Fin de Siècle Electrodynamics, *Studies in History and Philosophy of Modern Physics* **26**: 1–44.

Darrigol, O. (2000). *Electrodynamics from Ampère to Einstein*, Oxford University Press, Oxford.

Dirac, P. A. M. (1938). Classical Theory of Radiating Electrons, *Proceedings of the Royal Society* (London) *Series A* **167**: 148–169.

Earman, J. (2003). The Cosmological Constant, the Fate of the Universe, Unimodular Gravity, and All That, *Studies in History and Philosophy of Modern Physics* **34**: 559–577.

Eckert, M. and Märker, K. (eds) (2000). *Arnold Sommerfeld. Wissenschaftlicher Briefwechsel. Band 1: 1892-1918*, Deutsches Museum Verlag für Geschichte der Naturwissenschaften und der Tecknik, Berlin, Diepholz, München.

Ehrenfest, P. (1906). Zur Stabilitätsfrage bei den Bucherer-Langevin-Elektronen, *Physikalische Zeitschrift* **7**: 302–303.

Ehrenfest, P. (1907). Die Translation deformierbarer Elektronen und der Flächensatz, *Annalen der Physik* **23**: 204–205.

Einstein, A. (1905). Zur Elektrodynamik bewegter Körper, *Annalen der Physik* **17**: 891–921. Reprinted in (Einstein, 1987–2002), Vol. 2, Doc. 23.

Einstein, A. (1907a). Bemerkungen zu der Notiz von Hrn. Paul Ehrenfest: "Die Translation deformierbarer Elektronen und der Flächensatz", *Annalen der Physik* **23**: 206–208. Reprinted in (Einstein, 1987–2002), Vol. 2, Doc. 44.

Einstein, A. (1907b). Über die vom Relativitätsprinzip geforderte Trägheit der Energie, *Annalen der Physik* **23**: 371–384. Reprinted in (Einstein, 1987–2002), Vol. 2, Doc. 45.

Einstein, A. (1918). Der Energiesatz in der allgemeinen Relativitätstheorie, *Preußische Akademie der Wissenschaften* (Berlin). *Sitzungsberichte* pp. 448–459. Reprinted in (Einstein, 1987–2002), Vol. 7, Doc. 9.

Einstein, A. (1919). Spielen Gravitationsfelder im Aufbau der materiellen Elementarteilchen eine wesentliche Rolle?, *Preußische Akademie der Wissenschaften* (Berlin). *Sitzungsberichte* pp. 349–356. Reprinted in (Einstein, 1987–2002), Vol. 7, Doc. 17.

Einstein, A. (1987–2002). *The Collected Papers of Albert Einstein.* 8 Vols. Edited by J. Stachel et al., Princeton University Press, Princeton.

Einstein, A. and Grossmann, M. (1913). *Entwurf einer verallgemeinerten Relativitätstheorie und einer Theorie der Gravitation*, Teubner, Leipzig. Reprinted in (Einstein, 1987–2002), Vol. 4, Doc. 13.

Faragó, P. S. and Jánossy, L. (1957). Review of the Experimental Evidence for the Law of Variation of the Electron Mass with Velocity, *Il Nuovo Cimento Ser. 10* **5**: 1411–1436.

Fermi, E. (1921). Sulla dinamica di un sistema rigido di cariche elettriche in moto translatorio, *Nuovo Cimento* **22**: 199–207. Reprinted in (Fermi, 1962–1965), Vol. 1, pp. 1-7.

Fermi, E. (1922). Über einen Widerspruch zwischen der elektrodynamischen und der relativistischen Theorie der elektromagnetischen Masse, *Physikalische Zeitschrift* **23**: 340–344.

Fermi, E. (1962–1965). *The Collected Works of Enrico Fermi.* 2 Vols. Edited by E. Segrè, University of Chicago Press, Accademia Nazionale dei Lincei, Chicago, Rome.

Feynman, R. P., Leighton, R. B. and Sands, M. (1964). *The Feynman Lectures on Physics,* Vol. II, *Mainly Electromagnetism and Matter*, Addison-Wesley, Reading, MA.

Fölsing, A. (1997). *Albert Einstein. A Biography*, Viking, New York.

Galison, P. L. (1979). Minkowski's Space-Time: From Visual Thinking to the Absolute World, *Historical Studies in the Physical Sciences* **10**: 85–121.

Gerlach, W. (1933). Elektronen, *in* H. Geiger (ed.), *Handbuch der Physik.* 2nd ed. Vol. 22, Part 1, *Elektronen-Atome-Ionen*, Springer, Berlin, pp. 1–89.

Glitscher, K. (1917). Spektroskopischer Vergleich zwischen den Theorien des starren und des deformierbaren Elektrons, *Annalen der Physik* **52**: 608–630.

Goldberg, S. (1970). The Abraham Theory of the Electron: The Symbiosis of Experiment and Theory, *Archive for History of Exact Sciences* **7**: 7–25.

Griffith, D. J. (1999). *Introduction to Electrodynamics,* 3rd ed., Prentice Hall, Upper Saddle River, New Jersey.

Grøn, Ø. (1985). Repulsive Gravitation and Electron Models, *Physical Review D* **31**: 2129–2131.

Grøn, Ø. (1988). Poincaré Stress and the Reissner-Nordström Repulsion, *General Relativity and Gravitation* **20**: 123–129.

Herglotz, G. (1903). Zur Elektronentheorie, *Königliche Gesellschaft der Wissenschaften zu Göttingen. Mathematisch-physikalische Klasse. Nachrichten* pp. 357–382.

Herglotz, G. (1910). Über den vom Standpunkt des Relativitätsprinzips aus als 'starr' zu bezeichnenden Körper, *Annalen der Physik* **31**: 393–415.

Herglotz, G. (1911). Über die mechanik des deformierbaren körpers vom standpunkte der relativitätstheorie, *Annalen der Physik* **36**: 493–533.

Hnizdo, V. (1997). Hidden Momentum and the Electromagnetic Mass of a Charge and Current Carrying Body, *American Journal of Physics* **65**: 55–65.

Holton, G. (1988). Mach, Einstein, and the Search for Reality, *Thematic Origins of Scientific Thought: Kepler to Einstein,* rev. ed., Harvard University Press, Cambridge, pp. 237–277.

Hon, G. (1995). Is the Identification of Experimental Error Contextually Dependent? The Case of Kaufmann's Experiment and Its Varied Reception, *in* J. Z. Buchwald (ed.), *Scientific Practice. Theories and Stories of Doing Physics*, University of Chicago Press, Chicago, pp. 170–223.

Jackson, J. D. (1975). *Classical Electrodynamics,* 2nd ed., John Wiley & Sons, New York.

Jammer, M. (1997). *Concepts of Mass in Classical and Modern Physics*, Dover, New York.

Janssen, M. (1995). *A Comparison between Lorentz's Ether Theory and Einstein's Special Theory of Relativity in the Light of the Experiments of Trouton and Noble*, Ph.D. Thesis, University of Pittsburgh.

Janssen, M. (2002a). COI Stories: Explanation and Evidence in the History of Science, *Perspectives on Science* **10**: 457–522.

Janssen, M. (2002b). Reconsidering a Scientific Revolution: the Case of Lorentz versus Einstein, *Physics in Perspective* **4**: 421–446.

Janssen, M. (2003). The Trouton Experiment, $E = mc^2$, and a Slice of Minkowski Space-Time, *in* J. Renn, L. Divarci and P. Schröter (eds), *Revisiting the Foundations of Relativistic Physics: Festschrift in Honor of John Stachel*, Kluwer, Dordrecht, pp. 27–54.

Janssen, M. and Stachel, J. (2004). *The Optics and Electrodynamics of Moving Bodies.* Preprint 265, Max Planck Institute for History of Science, Berlin.

Jungnickel, C. and McCormmach, R. (1986). *Intellectual Mastery of Nature: Theoretical Physics from Ohm to Einstein.* Vol. 2. *The Now Mighty Theoretical Physics, 1870-1925*, University of Chicago Press, Chicago.

Keswani, G. H. and Kilmister, C. W. (1983). Intimations of Relativity: Relativity Before Einstein, *British Journal for the Philosophy of Science* **34**: 343–354.

Kilmister, C. W. (1970). *Special Theory of Relativity*, Pergamon, London.

Klein, F. (1918). Über die Integralform der Erhaltungssätze und die Theorie der räumlich-geschlossenen Welt, *Königliche Gesellschaft der Wissenschaften zu Göttingen. Nachrichten* pp. 394–423.

Kragh, H. (1999). *Quantum Generations. A History of Physics in the Twentieth Century*, Princeton University Press, Princeton.

Kwal, B. (1949). Les expressions de l'énergie et de l'impulsion du champ électromagnétique propre de l'électron en mouvement, *Le Journal de Physique et le Radium*, Série VIII, Tome X, pp. 103–104.

Langevin, P. (1905). La physique des électrons, *Revue Générale des Sciences Pures et Appliquées* **16**: 257–276.

Laue, M. (1911a). Zur Dynamik der Relativitätstheorie, *Annalen der Physik* **35**: 524–542. Reprinted in (Laue, M. von, 1961), Vol. 1, pp. 135–153.

Laue, M. (1911b). *Das Relativitätsprinzip*, Vieweg, Braunschweig.

Laue, M. von (1961). *Gesammelte Schriften und Vorträge.* 3 Vols, Friedrich Vieweg und Sohn, Braunschweig.

Levi-Civita, T. (1907). Sulla massa elettromagnetica, *Nuovo Cimento* **16**: 387–412. Reprinted in (Levi-Civita, 1956), Vol. 2, pp. 587–613.

Levi-Civita, T. (1909). Teoria asintotica delle radiazioni elettriche, *Rendiconti della Reale Accademia dei Lincei* **18**: 83–93. Reprinted in (Levi-Civita, 1956), Vol. 3, pp. 81–92.

Levi-Civita, T. (1956). *Opere Matematiche. Memorie e Note.* 4 Vols., Edited by F. Giordani et al. , Nicola Zanichelli Editore, Bologna.

Lorentz, H. A. (1895). *Versuch einer Theorie der electrischen und optischen Erscheinungen in bewegten Körpern*, Brill, Leiden. Reprinted in (Lorentz, 1934-39), Vol. 5, 1–138.

Lorentz, H. A. (1899). Vereenvoudigde theorie der electrische en optische verschijnselen in lichamen die zich bewegen, *Koninklijke Akademie van Wetenschappen te Amsterdam. Wis- en Natuurkundige Afdeeling. Verslagen van de Gewone Vergaderingen* (1898-99) **7**: 507–522. Slightly revised English translation: Simplified Theory of Electrical and Optical Phenomena in Moving Bodies. *Koninklijke Akademie van Wetenschappen te Amsterdam. Section of Sciences. Proceedings* **1** (1898–99): 427–442. Slightly revised French translation: Théorie simplifiée des phénoménes électriques et optiques dans des corps en mouvement. *Archives Néerlandaises des Sciences Exactes et Naturelles* **7** (1902): 64–80. French translation reprinted in (Lorentz, 1934-39), Vol. 5, 139–155.

Lorentz, H. A. (1900). Über die scheinbare Masse der Elektronen, *Physikalische Zeitschrift* **2**: 78–80 (79–80: discussion). Reprinted in (Lorentz, 1934-39), Vol. 3, 113–116.

Lorentz, H. A. (1904a). Weiterbildung der Maxwellschen Theorie. Elektronentheorie, *in* A. Sommerfeld (ed.), *Encyclopädie der mathematischen Wissenschaften, mit Einschluss iher Anwendungen*, Vol. 5, *Physik*, Part 2, Teubner, 1904–1922, Leipzig, pp. 45–288.

Lorentz, H. A. (1904b). Electromagnetische verschijnselen in een stelsel dat zich met willekeurige snelheid, kleiner dan die van het licht, beweegt, *Koninklijke Akademie van Wetenschappen te Amsterdam. Wis- en Natuurkundige Afdeeling. Verslagen van de Gewone Vergaderingen* (1903-04) **12**: 986–1009. Translation: Electromagnetic Phenomena in a System Moving with Any Velocity Smaller Than That of Light. *Koninklijke Akademie van Wetenschappen te Amsterdam. Section of Sciences. Proceedings* **6** (1903–04): 809–831. Translation reprinted in (Lorentz, 1934-39), Vol. 5, 172–197.

Lorentz, H. A. (1905). *Ergebnisse und Probleme der Elektronentheorie. Vortrag, gehalten am 20. Dezember 1904 im Elektrotechnischen Verein zu Berlin*, Springer, Berlin. Reprinted in (Lorentz, 1934-39), Vol. 8, 76-124.

Lorentz, H. A. (1915). *The Theory of Electrons and Its Applications to the Phenomena of Light and Radiant Heat.* 2d ed., Teubner, 1915, Leipzig.

Lorentz, H. A. (1927). *Problems of Modern Physics. A Course of Lectures Delivered in the California Institute of Technology*, Ginn, Boston.

Lorentz, H. A. (1934-39). *Collected Papers.* 9 Vols., Edited by P. Zeeman and A. D. Fokker, Nijhoff, The Hague:.

McCormmach, R. (1970). H. A. Lorentz and the Electromagnetic View of Nature, *Isis* **61**: 459–497.

Mie, G. (1912a). Grundlagen einer Theorie der Materie. (Erste Mitteilung.), *Annalen der Physik* **37**: 511–534.

Mie, G. (1912b). Grundlagen einer Theorie der Materie. (Zweite Mitteilung.), *Annalen der Physik* **39**: 1–40.

Mie, G. (1913). Grundlagen einer Theorie der Materie. (Dritte Mitteilung, Schluß.), *Annalen der Physik* **40**: 1–66.

Miller, A. I. (1973). A Study of Henri Poincaré's "Sur la dynamique de l'électron", *Archive for the History of Exact Sciences* **10**: 207–328. Reprinted in facsimile in (Miller, 1986).

Miller, A. I. (1981). *Albert Einstein's Special Theory of Relativity. Emergence (1905) and Early Interpretation (1905-1911)*, Addison-Wesley, Reading, Mass. Reprinted in 1998 (Springer, New York). Page references are to this reprint.

Miller, A. I. (1986). *Frontiers of Physics: 1900-1911*, Birkhäuser, Boston.

Minkowski, H. (1908). Die Grundgleichungen für die elektromagnetischen Vorgänge in bewegten Körpern, *Königliche Gesellschaft der Wissenschaften zu Göttingen. Mathematisch-physikalische Klasse. Nachrichten* pp. 53–111. Reprinted in (Minkowski, 1967), Vol. 2, 352-404.

Minkowski, H. (1909). Raum und Zeit, *Physikalische Zeitschrift* **20**: 104–11.

Minkowski, H. (1967). *Gesammelte Abhandlungen von Hermann Minkowski.* Edited by D. Hilbert, Chelsea, New York. Originally published in two volumes, Leipzig, 1911.

Moylan, P. (1995). An Elementary Account of the Factor 4/3 in the Electromagnetic Mass, *American Journal of Physics* **63**: 818–820.

Norton, J. D. (1992). Einstein, Nordström and the Early Demise of Scalar, Lorentz-Covariant Theories of Gravitation, *Archive for the History of Exact Sciences* **45**: 17–94.

Norton, J. D. (2000). 'Nature is the Realisation of the Simplest Conceivable Mathematical Ideas': Einstein and the Canon of Mathematical Simplicity, *Studies in History and Philosophy of Modern Physics* **31**: 135–170.

Pais, A. (1972). The Early History of the Theory of the Electron: 1897-1947, *in* A. Salam and E. P. Wigner (eds), *Aspects of Quantum Theory*, Cambridge University Press, Cambridge, pp. 79–93.

Pais, A. (1982). *'Subtle is the Lord . . . ' The Science and the Life of Albert Einstein*, Oxford University Press, Oxford.

Pauli, W. (1921). Relativitätstheorie, *in* A. Sommerfeld (ed.), *Encyklopädie der mathematischen Wissenschaften, mit Einschluß ihrer Anwendungen.* Vol. 5*, Physik*, Part 2, Teubner, 1904–1922, Leipzig, pp. 539–775. Translation with introduction and supplementary notes: *Theory of Relativity*. G. Field, trans. London: Pergamon, 1958. Reprinted: New York: Dover, 1981. German original (plus German translations of introduction and supplementary notes of English edition) reprinted as *Relativitätstheorie*. Edited by D. Giulini. Berlin: Springer, 2000. Page references are to English translation.

Pearle, P. (1982). Classical Electron Models, *in* D. Teplitz (ed.), *Electromagnetism: Paths to Research*, Plenum Press, New York, pp. 211–295.

Planck, M. (1906a). Das Prinzip der Relativität und die Grundgleichungen der Mechanik, *Deutsche Physikalische Gesellschaft. Verhandlungen* **8**: 136–141. Reprinted in (Planck, 1958), Vol. 2, pp. 115-120.

Planck, M. (1906b). Die Kaufmannschen Messungen der Ablenkbarkeit der β-Strahlen in ihrer Bedeutung für die Dynamik der Elektronen, *Deutsche Physikalische Gesellschaft. Verhandlungen* **8**: 418–432. Reprinted in *Physikalische Zeitschrift* **7** (1906): 753–759 (760–761: discussion); and in (Planck, 1958), Vol. 2, pp. 121-135.

Planck, M. (1908). Bemerkungen zum Prinzip der Aktion und Reaktion in der allgemeinen Dynamik, *Deutsche Physikalische Gesellschaft. Verhandlungen* **6**: 728–732. Reprinted in (Planck, 1958), Vol. 2, pp. 215–219.

Planck, M. (1958). *Physikalische Abhandlungen und Vorträge,* 3 Vols., Friedrich Vieweg und Sohn, Braunschweig.

Poincaré, H. (1885). Sur l'équilibre d'une masse fluide animée d'un mouvement de rotation, *Acta Mathematica* **7**: 259–380.

Poincaré, H. (1900). La théorie de Lorentz et le principe de réaction, *Archives Néerlandaises des Sciences Exactes et Naturelles* **2**: 252–278. Reprinted in (Poincaré, 1934–54), Vol. 9, 464-493.

Poincaré, H. (1902a). *Figures d'équilibre d'une masse fluide*, C. Naud, Paris.

Poincaré, H. (1902b). Sur la stabilité de l'équilibre des figures piriformes affectées par une masse fluide en rotation, *Philosophical Transactions of the Royal Society A* **198**: 333–373.

Poincaré, H. (1905). Sur la dynamique de l'électron, *Comptes Rendus de l'Académie des Sciences* **140**: 1504–1508. Short version of (Poincaré, 1906). Reprinted in (Poincaré, 1934–54), Vol. 9, 489–493. English translation in (Keswani and Kilmister, 1983).

Poincaré, H. (1906). Sur la dynamique de l'électron, *Rendiconti del Circolo Matematico di Palermo* **21**: 129–175. Reprinted in (Poincaré, 1934–54), Vol. 9, 494–550. Page references are to this reprint. English translation of secs. 1–4 and 9 in (Kilmister, 1970), and of secs. 6–8 in (Miller, 1973).

Poincaré, H. (1934–54). *Œuvres de Henri Poincaré.* 11 Vols., Gauthiers-Villars, Paris.

Ponte, M. (1930). Recherches sur la diffraction des électrons. Analyse électronique, *Annales de physique* **13**: 395–452.

Pyenson, L. (1985). *The Young Einstein. The Advent of Relativity*, Adam Hilger, Bristol and Boston.

Renn, J. (ed.) (2005). *Dokumente eines Lebensweg/Documents of a Life's Pathway*, Wiley-VCH, Weinheim.

Renn, J. (ed.) (Forthcoming). *The Genesis of General Relativity. Sources and Interpretation.* 4 Vols., Springer, Dordrecht.

Renn, J. and Sauer, T. (Forthcoming). A Pathway out of Classical Physics. To appear in (Renn, Forthcoming).

Rogers, M. M., Reynolds, A. W. and Rogers Jr, F. T. (1940). A Determination of the Masses and Velocities of Three Radium B Beta-Particles, *Physical Review* **57**: 379–383.

Rohrlich, F. (1960). Self-energy and the Stability of the Classical Electron, *American Journal of Physics* **28**: 639–643.

Rohrlich, F. (1965). *Classical Charged Particles: Foundations of Their Theory*, Addison-Wesley, Reading, MA.

Rohrlich, F. (1970). Electromagnetic Energy, Momentum, and Mass, *American Journal of Physics* **38**: 1310–1316.

Rohrlich, F. (1973). The Electron: Development of the First Elementary Particle Theory, *in* J. Mehra (ed.), *The Physicist's Conception of Nature*, Reidel, Dordrecht, pp. 331–369.

Rohrlich, F. (1997). The Dynamics of a Charged Sphere and the Electron, *American Journal of Physics* **65**: 1051–1056.

Schwartz, H. M. (1971). Poincaré's Rendiconti Paper on Relativity. Part I, *American Journal of Physics* **39**: 1287–1294.

Schwartz, H. M. (1972). Poincaré's Rendiconti Paper on Relativity. Part II, *American Journal of Physics* **40**: 862–872.

Schwarzschild, K. (1903a). Zur Elektrodynamik. I. Zwei Formen des Prinzips der kleinsten Action in der Elektronentheorie, *Königliche Gesellschaft der Wissenschaften zu Göttingen. Mathematisch-physikalische Klasse. Nachrichten* pp. 126–131.

Schwarzschild, K. (1903b). Zur Elektrodynamik. II. Die elementare elektrodynamische Kraft, *Königliche Gesellschaft der Wissenschaften zu Göttingen. Mathematisch-physikalische Klasse. Nachrichten* pp. 132–141.

Schwarzschild, K. (1903c). Zur Elektrodynamik. III. Ueber die Bewegung des Elektrons, *Königliche Gesellschaft der Wissenschaften zu Göttingen. Mathematisch-physikalische Klasse. Nachrichten* pp. 245–278.

Schwinger, J. (1983). Electromagnetic Mass Revisited, *Foundations of Physics* **13**: 373–383.

Segrè, E. (1970). *Enrico Fermi. Physicist*, University of Chicago Press, Chicago.

Sommerfeld, A. (1904a). Zur Elektronentheorie. I. Allgemeine Untersuchung des Feldes eines beliebig bewegten Elektrons, *Königliche Gesellschaft der Wissenschaften zu Göttingen. Mathematisch-physikalische Klasse. Nachrichten* pp. 99–130. Reprinted in (Sommerfeld, 1968), Vol. 2, pp. 39–70.

Sommerfeld, A. (1904b). Zur Elektronentheorie. II. Grundlage für eine allgemeine Dynamik des Elektrons, *Königliche Gesellschaft der Wissenschaften zu Göttingen. Mathematisch-physikalische Klasse. Nachrichten* pp. 363–439. Reprinted in (Sommerfeld, 1968), Vol. 2, pp. 71–147.

Sommerfeld, A. (1904c). Vereenvoudigde afleiding van het veld van, en de krachten werkende op een electron bij willekeurige beweging, *Koninklijke Akademie van Wetenschappen te Amsterdam. Wis- en Natuurkundige Afdeeling. Verslagen van de Gewone Vergaderingen* **13**: 431–452. Reprinted in (Sommerfeld, 1968), Vol. 2, pp. 17–38. English translation: Simplified Deduction of the Field and the Forces on an Electron, Moving in Any Given Way. *Koninklijke Akademie van Wetenschappen te Amsterdam. Section of Sciences.* Proceedings **7**: 346–367.

Sommerfeld, A. (1905a). Zur Elektronentheorie. III. Ueber Lichtgeschwindigkeits- und Ueberlichtgeschwindigkeits-Elektronen, *Königliche Gesellschaft der Wissenschaften zu Göttingen. Mathematisch-physikalische Klasse. Nachrichten* pp. 201–235. Reprinted in (Sommerfeld, 1968), Vol. 2, pp. 148–182.

Sommerfeld, A. (1905b). Über die Mechanik der Elektronen, *in* A. Krazer (ed.), *Verhandlungen des Dritten Internationalen Mathematiker-Kongresses in Heidelberg vom 8. bis 13. August 1904*, Teubner, Leipzig, pp. 417–432. Reprinted in (Sommerfeld, 1968), Vol. 2, pp. 1–16.

Sommerfeld, A. (1907). Ein Einwand gegen die Relativtheorie der Elektrodynamik und seine Beseitigung, *Physikalische Zeitschrift* **8**: 841–842. Reprinted in (Sommerfeld, 1968), Vol. 2, pp. 183–184.

Sommerfeld, A. (1910a). Zur Relativitätstheorie I. Vierdimensionale Vektoralgebra, *Annalen der Physik* **32**: 749–776. Reprinted in (Sommerfeld, 1968), Vol. 2, pp. 189–216.

Sommerfeld, A. (1910b). Zur Relativitätstheorie II. Vierdimensionale Vektoranalysis, *Annalen der Physik* **33**: 649–689. Reprinted in (Sommerfeld 1968), Vol. 2, pp. 217–257.

Sommerfeld, A. (1968). *Gesammelte Schriften.* 4 Vols. Edited by F. Sauter, Vieweg, Braunschweig.

Teukolsky, S. A. (1996). The Explanation of the Trouton-Noble Experiment Revisited, *American Journal of Physics* **64**: 1104–1106.

Trouton, F. T. and Noble, H. R. (1903). The Mechanical Forces Acting On a Charged Electric Condenser Moving Through Space, *Philosophical Transactions of the Royal Society* **202**: 165–181.

Walter, S. (1999a). Minkowski, Mathematicians, and the Mathematical Theory of Relativity, *in* H. Goenner, J. Renn, J. Ritter and T. Sauer (eds), *The Expanding Worlds of General Relativity*, Birkhäuser, Boston, pp. 45–86.

Walter, S. (1999b). The non-Euclidean Style of Minkowskian Relativity, *in* J. Gray (ed.), *The Symbolic Universe: Geometry and Physics, 1890–1930*, Oxford University Press, Oxford, pp. 91–127.

Walter, S. (Forthcoming). Breaking in the 4-Vectors: The Four-Dimensional Movement in Gravitation, 1905–1910. To appear in (Renn, Forthcoming).

Warwick, A. (2003). *Masters of Theory. Cambridge and the Rise of Mathematical Physics*, University of Chicago Press, Chicago.

Wien, W. (1900). Über die Möglichkeit einer elektromagnetischen Begründung der Mechanik, *Archives Néerlandais des Sciences Exactes et Naturelles* **2**: 96–107. Reprinted in *Annalen der Physik* **5** (1901): 501–513.

Wien, W. (1904a). Über die Differentialgleichungen der Elektrodynamik für bewegte Körper Part I, *Annalen der Physik* **13**: 641–662.

Wien, W. (1904b). Über die Differentialgleichungen der Elektrodynamik für bewegte Körper Part II, *Annalen der Physik* **13**: 663–668.

Wien, W. (1904c). Erwiderung auf die Kritik des Hrn. M. Abraham, *Annalen der Physik* **14**: 635–637. Response to (Abraham, 1904b).

Wien, W. (1904d). Poyntingscher Satz und Strahlung, *Annalen der Physik* **15**: 412–414. Response to (Abraham, 1904c).

Wilson, W. (1936). The Mass of a Convected Field and Einstein's mass-energy law, *The Proceedings of the Physical Society* (Cambridge) **48**: 736–740.

Yaghjian, A. D. (1992). *Relativistic Dynamics of a Charged Sphere. Updating the Lorentz-Abraham Model*, Springer, Berlin.

Zahar, E. (1989). *Einstein's Revolution: A Study in Heuristic*, Open Court, La Salle, IL.

Zahn, C. T. and Spees, A. A. (1938). A Critical Analysis of the Classical Experiments on the Variation of Electron Mass, *Physical Review* **53**: 511–521.

Zink, J. W. (1966). Electromagnetic Mass of the Classical Electron, *American Journal of Physics* **34**: 211–215.

Zink, J. W. (1968). On the Abraham-Lorentz Electron, *American Journal of Physics* **36**: 639–640.

Zink, J. W. (1971). Relativity and the Classical Electron, *American Journal of Physics* **39**: 1403–1404.

JEREMY GRAY

ENRIQUES: POPULARISING SCIENCE AND THE PROBLEMS OF GEOMETRY

ABSTRACT

Federigo Enriques' book *Problemi della Scienza* of 1906 (English translation *Problems of Science*, 1914) is a prolonged attempt to define and distinguish between facts and theories in order to analyse what constitutes reality. It is a positivist, somewhat anti-Kantian work, and it can be read as a long, if one-sided, conversation between Enriques and Poincaré. In this talk I shall draw out the views, sometimes in agreement, sometimes in conflict, of these two men, and then discuss how they illuminate issues in the philosophy of geometry that were of considerable contemporary importance, and which explain why Enriques' work was taken up in the United States.

1. INTRODUCTION

The Italian geometer Federigo Enriques (1871–1946) made his name in the 1890s as the creator, with his friend Guido Castelnuovo, of the first full theory of algebraic surfaces. This dovetailed very neatly with the more analytic work of Picard generalising Riemann's ideas to complex functions of two variables, and established him internationally as one of the leading mathematicians of his generation. In 1894 he became a professor of geometry at Bologna. He wrote up his lectures on projective geometry there and published them as a book (*Lezioni di geometria proiettiva*) in 1898. They are based on a set of six axioms, and when the next year David Hilbert published his own account of geometry from an axiomatic point of view Felix Klein suggested that Enriques's book be translated into German. The translation was published in 1903, with an introduction by Klein. On the strength of this work and his work in algebraic geometry he was asked by Klein to write the article on the principles of geometry for the *Encyclopädie der Mathematischen Wissenschaften*, which he did – the essay is dated 1907. Klein shared Enriques' views on the importance of intuition in geometry, and also his liking for the psycho-physical explanations of the acquisition of geometrical knowledge, such as Wundt proclaimed; Klein and Wundt had been colleagues at Leipzig in the early 1880s.[1]

His treatment of geometry in *Prinzipien der Geometrie* reveals several of the traits of his contemporary popularising works. He took a deep interest in the history of geometry, and began by describing Euclid's *Elements*, noting how its definitions, postulates and axioms differed from more modern ones. They were intended to be self-evident, they were not presumed to follow

135

V.F. Hendricks, K.F. Jørgensen, J. Lützen and S.A. Pedersen (eds.), Interactions: Mathematics, Physics and Philosophy, 1860-1930, pp. 135–154.

logically from the definitions, and they were in any case incomplete, as writers like Pasch had noted. More recent presentations of geometry, Enriques observed, regarded geometric objects as belonging either to the usual intuitive space, or to physical space with which we become acquainted through experience, or to some abstract space made intelligible by abstraction or generalisation. As part of this more sophisticated approach, there had been a rise in the standards of mathematical rigour, initiated by Pasch and taken up by Peano and Hilbert. This had led to questions about the arbitrariness of axioms and their independence, which were connected to contemporary work on logic. Finally, Enriques noted that these developments raised questions about the philosophy of geometry. Geometry could be approached either empirically, through its connections to experience, or from a logical and formal standpoint, as the Italians Peano, Padoa, and Pieri had done, paying particular attention to the number of the primitive (undefined) ideas. Enriques, as we shall see, had strong views about the proper relation of these two approaches, but he confined them to his popular writings.

By 1907 Enriques had become the Italian mathematician of stature who was involved in issues of mathematics education and philosophy. He thereby leant the authority he had earned as a research mathematician in a difficult, important but, to the general public, obscure domain to speculations of a quite different kind, targeted at a different audience where the criteria for success, and even influence, are much more fickle. His major work in this connection is the *Problemi della Scienza* of 1906 which, after some delays, was translated into English as *Problems of Science* and published in 1914.

With this book he joined the select group of scientists of any kind whose work reached out to the general public. Of course, Enriques did not simply switch roles, abandon mathematics and take up philosophy and the popularisation of science. Throughout the period we are considering he continued to do mathematics, specifically algebraic geometry. But he also moved towards logic and education and the writing of textbooks, all of which broadened his involvement in Italian intellectual life and also facilitated his move towards the public platform abroad. In the years before the First World War, Enriques was an editor of the journal *Scientia* and published many articles there. A number of these were collected and reworked for publication as a book in 1912, *Scienza e Razionalismo*, and the topics range from the value of science, through philosophical discussions of rationalism and empiricism (including a history of rationalism from the Eleatics to Leibniz) and of rationalism and historicism (in which he grappled with Hegel) to topics on the relation of science and religion, the existence of God, and the nature of reality, as well as, rather less grandly, the classification of the sciences and the implications of non-Euclidean geometry for the philosophy of Kant. His work invites comparison with that of his distinguished predecessor, Henri Poincaré, and, as will become apparent, Enriques had significant areas of agreement and one specific disagreement with Poincaré over the nature of geometry.

Erudite popularising work, such as Poincaré and Enriques supplied, was topical. The best of it remains in circulation, and in the form of Open Court and Dover publications has perhaps eclipsed later attempts to deal with similar themes.[2] Much of it was originally published by Paul Carus, who ran the Open Court Publishing House in Chicago and edited the *Monist*, including essays by Ernst Mach, and Poincaré, Hilbert's *Foundations of Geometry*, and eventually Enriques's *Problems of Science* and many others. It came out of a time when the modern profession of mathematics was defining itself in something like the form it still has, a period marked by the first International Congresses, which began in the period before the First World War. The Mathematicians met first in Zürich in 1897, then in Paris in 1900, in Heidelberg in 1904, in Rome in 1908, and Cambridge (England) in 1912, while the peripatetic philosophers went to Paris, Geneva, Heidelberg, and in 1912 to Bologna, where Enriques presided. These were occasions for the Italian mathematical community to see that it ranked perhaps second, behind the mighty Germans but ahead of the French. The Congresses offered images of the Italians in various ways: at Paris, for example, as logicians. Peano and his followers were prominent at both the International Congress of Philosophers and the International Congress of Mathematicians that followed it. Naturally, as an eloquent spokesman for this community, Enriques benefited from the depth and range of his countrymen's work, and we shall see that he was indeed regarded as drawing on that collective wisdom.

A major issue for mathematicians and their audience alike was the question of non-Euclidean geometry and the nature of space. Mathematicians had come up with new geometries, and their audience wanted to know what that meant, what these other geometries are, and how we can tell which is true, even though that question is entirely abstract – no-one was out there anxiously measuring parallaxes. There was a psychological aspect to this question, for in the century after Kant psychologists had turned the investigation of the preliminaries for any kind of thought into the study of how people think. What are concepts, what is it to know something? What sort of an intellectual activity is the study of geometry, regarded as the elementary appreciation of space? And, mindful of the error into which Kant had fallen, psychologists took care to make sure that their theories allowed human beings to discover non-Euclidean geometry. The pioneer in this regard was the energetic, if unsystematic, Wilhelm Wundt. Enriques was enraptured by Wundt's writings, which seemed to speak directly to his philosophical interests in the 1890s, and whom he described to his friend Castelnuovo in these terms 'the most marvellous philosophical, physiological, psychological, mathematical, etc. intelligence... Read the *Logik* of Wundt, at least that part about the methods of mathematics, and think that it is a physiologist who writes this, a physiologist who does not fear to scale the steep slopes of Kantian conceptions to illuminate from above the great course of all of science'.[3] It can surely be thought that Enriques was formulating a similar ambition even as he wrote these words. Finally, there was also a logical aspect to questions of this kind, in which mathematical ideas were taken to be rooted in logic or pure thought (with greater or lesser attention to how this came about).[4]

By 1900 was a widespread feeling, going well beyond the small circle of experts, that in logic as in geometry, much had happened since the Greeks, and much of that recently. Had mathematics been purely technical, had psychology defined itself as the study of measurable mental processes, had logic remained the driest kind of organised common sense, none of Enriques' public career would have been possible. Instead, as the public could easily discover, the experts had their own problems too. It is because interest in these matters was shared between the public and the experts, who both sensed that more was at stake than merely academic issues that Enriques had a platform on which to stand.

As is well-known, Moritz Pasch's *Vorlesungen über neuere Geometrie* (1882) is the first book in which a thorough reworking of geometry is proposed. Thereafter a line runs through Hilbert and into modern axiomatics. In an important paper of 1936 Ernest Nagel argued that the reformulation of geometry was an important source of modern logic. The kernel of his insight was that while duality in projective geometry puts points and lines on the plane on an exactly equal footing, intuition must prefer points. So mathematicians were forced away from intuition as the basis of geometry, and towards formalism and thence logic. I would add that non-Euclidean geometry further promoted this tendency. It is the geometry that raises the question of the nature of space, and with it the embarrassing problem of explaining why mathematicians had been so wrong about geometry for so long. Nagel's example of the formalist geometer was, of course, Hilbert but in this he was unfair, in ways that affect our understanding of Enriques. Indeed, as we shall see, the insight of Nagel owes a lot to the original work of Enriques.

As several scholars have noted recently, the study of geometry from an axiomatic point of view was taken up most eagerly not in Germany but in Italy. Italian mathematicians found unexpected properties of projective planes,[5] investigated a geometry in which, unlike Pasch's, the Archimedean axiom is false,[6] showed that Desargues' Theorem is automatic in projective spaces of dimensions 3 or more, but came to suspect that there may be two-dimensional projective spaces in which the theorem is false.[7] As these examples make clear, Italian mathematicians did not constrain their projective geometry to the facts of everyday experience. Pieri, unlike Pasch, completely abandoned any intention of formalising what is given in experience. Instead, as he wrote in (1895), he treated projective geometry "in a purely deductive and abstract manner, ... , independent of any physical interpretation of the premises". Primitive terms, such as line segments, "can be given any significance whatever, provided they are in harmony with the postulates which will be successively introduced" (Quoted in Bottazzini *Storia*, III.I, 276). In Pieri's presentation of plane projective geometry *(Principii della Geometria di Posizione* (1899b)) nineteen axioms were put forward, (typically: any two lines meet). For a time, Italian work travelled well; to England,[8] France,[9] and America.[10] It seems that in the early years of the 20th Century Pieri's ideas met with a greater degree of acceptance than is commonly recognised today.

So we see that there was a receptive public, and an active group of specialists. The result was a series of popularisations. In France the authority was Poincaré; in Germany, Ernst Mach; in Italy, Enriques. History of mathematics also benefited: the book by Engel and Stäckel was published in 1895, the history of non-Euclidean geometry by Roberto Bonola was published in Italian in 1906, and in a posthumous English translation, with a preface by Enriques, by the Open Court, in 1912.

2. ENRIQUES' PROBLEMS OF SCIENCE

Enriques' *Problems of Science* is difficult both to evaluate and to present. Some philosophical works become classics – which is not to say that they are widely or carefully read. They acquire a certain status, they are regularly kept in print, anthologised, and discussed in the subsequent literature steadily thereafter. Poincaré's popular essays are a good example, most recently re-packaged with an introduction by the late Stephen Jay Gould, no less, which says more about Gould's status as popular scientist extraordinaire than it does about the Poincaré's work. Enriques' *Problems of Science*, however, probably lives in that limbo of older works which are to be found on the open shelves of good university libraries, but not often consulted or argued with. It may be that 1914 was a particularly inauspicious year for such a book to come out, and that after the War intellectual issues had moved on to other things. It is indeed a book that had crystallised in the author's mind between 1901 and 1906 and discusses contemporary science without a single reference to Einstein. There is in fact something stale about the presentation of contemporary science altogether. It reads as if Enriques has picked it all up second-hand (as I suppose he did) whereas Poincaré's essays read as if he had got his hands dirty (as, theoretically, he had). Only the numerous passing references to research on psychology, physiology and the mechanisms of the brain stand out as fresh. However, whatever the reasons for its marginal status, it is an extremely interesting book.

The book addresses a familiar conundrum, then and now: the relation between symbols and their meanings, between syntax and semantics. The philosophical problem of science is to explain how talk about mathematical and scientific concepts makes sense. Enriques was a realist, in the sense that he believed it made sense to talk of objects and not merely our ideas about objects, but the way he defended and articulated such talk shows that he held with unusual vigour a viewpoint not unusual in his day, that human knowledge is acquired historically, both by the intellectual community down the ages and by each individual as he or she grows up. This process of knowledge acquisition, he observed, often inverts the 'logical' order, according to which (as d'Alembert had argued in the *Encyclopédie*, for example) geometry precedes mechanics because geometry is about some, but not all, of the concepts needed in mechanics. In Kant's *Critique of Pure Reason*, this appears as the *a priori* status of geometrical knowledge. Enriques, in contrast, argued the various sciences develop together, and geometry emerges 'as a part of physics, which has attained a high degree of perfection by virtue of the simplicity, generality, and relative independence of the relations included in it'. (*Problems of Science*, p. 181). It is crucial to Enriques' whole approach that this process of development is, insofar as it is meaningful, necessarily and forever incomplete.

One focus of this paper is on the comparison Enriques drew between himself and Poincaré (I am not aware that Poincaré ever thought to reply). Enriques regarded his own position on issues as being in some sense positivist, and when he disagreed with Poincaré he found the Frenchman to be transcendental. The touchstone, for Enriques, of a transcendental position was that it invoked ideas of a completed infinite process, such as abstracting away all irrelevant features (and not just more and more). A transcendental sense of space, he said, was offered when someone spoke of the inside of a ball of infinite radius, a concept which he regarded as meaningless and obtained by illegitimately passing to the limit of larger and larger balls. So for example, in discussing what is meant by the geometric terms 'point', 'straight line', and 'plane', Enriques could admit that there were no such objects in reality; we do not learn what the words mean by encountering examples of them (as we do elephants or ants). They are instead abstractions (p. 177): 'they serve as *symbols* to express certain relations of position amongst bodies, relations which are stated by means of the propositions of geometry'. Thus far, Enriques agreed with Poincaré. But that does not mean, said Enriques, that the propositions of geometry do not apply to any real fact but are merely, as Poincaré would have them, conventions by which we express physical facts. This is because these symbols find an 'approximate correspondence' in the physical world 'in certain objects for which they stand.'

The concept of a straight line, for example, is derived from the movement of solid bodies rotating about an axis; from the free motion of a particle unaffected by anything else; and from the path of light in a homogeneous medium. Each type of experience leads to a distinct system of geometry: the first to a metrical geometry, the third to a projective geometry. The agreement between the various experiences (each picks out the same object as a straight line) allows us to bring them together in a single geometric representation. Upon enquiry, this agreement is a symmetry in the phenomena that implies the homogeneity of space.

In two famous papers Poincaré had argued that attempts to decide if the geometry of space was Euclidean or non-Euclidean were bound to fail, because they required that straight lines be instantiated in some physical form.[11] In the first paper he gave the example of a metal plate cooling according to a particular formula as one moves away from the centre.[12] Geodesics in the plate will appear to us to be curved, because paths that tend towards the centre are measured with warmer, and therefore longer, measuring rods, and so fewer are needed. Poincaré chose the formula so the geometry on the plate is non-Euclidean geometry, and the geodesics appear to us as arcs of circles perpendicular to the edge of the plate. However, creatures living in the plate will regard the geodesics as straight. Such creatures, he argued, would not have the same geometry as ours. But this does not mean that we can say the space is Euclidean, for, as Poincaré argued in the very next essay, one is always free to assert that the physical object that instantiates the geodesic is not, after all, straight. Let us suppose that we put our trust in the paths of rays of light, and find that space seems to be described by non-Euclidean geometry. Then, said Poincaré, 'we should have a choice between two conclusions: we could give up Euclidean geometry, or modify the laws of optics, and suppose that light is not rigorously propagated in a straight line. It is needless to add that every one would look upon this solution as the more advantageous.'[13]

So Poincaré's conventionalist philosophy rests upon a supposed inability to distinguish physical from geometrical properties. But, said Enriques, we are denied in Poincaré's account any way of saying that the changes in length of rulers in the plate is due to changes in temperature, because our experience of heat is that it is a localised phenomenon to which different bodies respond differently. Because we cannot say the things that characterise what we say about heat, heat is not playing the role of a physical concept in the Poincaré model, but a geometrical one. The same is true of light rays, which demonstrably depart from straightness in inhomogeneous mediums. Therefore, Enriques concluded (p. 178), 'in this other world, geometry would be *really* and not merely apparently different from ours.' To think otherwise is to make the contrast between appearance and reality into something transcendental.

If in some strange way large regions of space were to appear to us to be more and more non-Euclidean, as measured by the behaviour of light rays, then Enriques would have argued as follows. Either we discover some way in which the light rays are being pulled out of their predicted path, by some process we can at least quantify and perhaps even control, or we do not, after exhaustive searching. On the first alternative, a new physical process has been discovered, and our idea of geometry is left unchanged. On the second alternative, we would have to say that geometry was to be altered.

A number of deep philosophical issues arise in a summary as brief as this, and in the course of the book *Problems of Science* Enriques touched on a number of them. One might object that all talk of Space was meaningless, because there is nothing (no object) to which the word refers. Enriques addressed this objection directly, and set out his position that: 'to Kant's thesis denying the existence of a real object corresponding to the word "space", we, together with Herbart, shall oppose the view that "spatial relations" are real. And to the nominalism recently maintained by Poincaré, which declares that these relations do not possess a real significance *absolutely independent of bodies*, we oppose a more precise estimate of the sense in which geometry is a *part of physics*.' (p. 174). This derived from his sense that there is 'an actual physical significance belonging to the *spatial relations or to the positions of bodies*, whose totality may well be denoted by the word "space".' These spatial relations therefore, he concluded, give us knowledge of reality.

Poincaré, Enriques reminded his readers with several quotations from *Science et hypothèse*, had stopped at the experience of the mutual relation of bodies – thus being a nominalist in Enriques' sense of the term, that being the antithesis of realism. Enriques believed that to talk about space was not to pass to talk of some (novel) object known transcendentally, but to talk of real relations between real bodies. Claims about space were claims about all the real (and presumably possible) relations between bodies, and we have seen that in discussing these claims Enriques would have argued that homogeneity was a property of space. Knowledge claims about space – many such claims, anyway – reduce to claims about measurements. Enriques gave two examples, of which the isosceles theorem was the first (p. 182). Following Klein's lectures[14] Enriques argued that the real meaning of the isosceles theorem is that if the side lengths differ by less than a given amount, ε, then the base angles differ by less than an amount depending on that ε in a specified way. In this way the theorems of geometry can be turned into statements about bodies. The second example was a discussion of how claims about the angle sums of non-Euclidean triangles could be tested astronomically, and shown to depend, within stated limits of accuracy, upon rather infeasibly large regions of space (p. 192).

What of the organisation of ideas in geometry, and of the evident fact that mathematicians had at their disposal at least two geometries of space (Euclidean and non-Euclidean geometry)? Enriques' opinion of axioms in geometry and the hypothetico-deductive side of the subject was possibly unexpected. He quoted Sartorius's remark (*Gauss zum Gedächtnis*, p. 81) that 'Gauss regarded geometry as a logical structure, only in case the theory of parallels were conceded as an axiom' only to disagree with it. In his opinion, none of the postulates of geometry had the character of a logical axiom, and all of the definitions of the fundamental entities of geometry were logically defective but instead made assumptions about reality. So in Enriques' opinion, geometry is not a matter of writing down some axioms (plausible or not, but in any case mutually consistent) and then reasoning purely logically. It is concealed talk about physically possible systems.[15] Different systems of postulates, forming various geometries, 'express *different physical hypotheses*' (p. 197).[16] But, because the process of constructing geometric concepts had, by 1906, gone far beyond any close link with sense data, Enriques felt compelled to explain how strikingly abstract concepts can have a certain objectivity, which he did by asserting that these abstract concepts give 'a *possible* representation of reality' (p. 191).[17]

Enriques held a number of interesting views on the nature of knowledge that derived from his emphasis on the process of the acquisition of knowledge. He held that what is known was subject to a continual process of revision, so that what may be 'known' at one time may be found to be false later on. Unlike Popper later, he did not regard this fallibility as the characteristic feature of scientific knowledge, and indeed the example he had in mind was the exactitude of geometry, but he did write: 'There is no reason however why it may not be *disproved* if untrue' (p. 184). His fallibilism derived from his already noted preference for meaningful discourse; Enriques was no formalist, unlike a number of his Italian contemporaries, or even Hilbert. Empirical verification may be definitive if it comes up with a negative result (a counter-example), but is merely probable when it comes up with a positive (confirmatory) result (p. 155). He distinguished between explicit and implicit hypotheses in a theory (the explicit hypotheses are those particular to the theory in question, the implicit hypotheses are those needed to connect the theory to the object of study, and may well be the explicit hypotheses of logically prior theories) and wrote:

> The progress of science is a process of *successive approximations,* in which new and *more precise, more probable and* more *extended inductions* result from partially verified *deductions,* and from those contradictions that correct the implicit hypotheses.
>
> In this process certain primary and general concepts, such as those of geometry and mechanics, give us some guiding principles that are but slightly variable if not *absolutely* fixed. Therefore we should turn our attention to these concepts in order to explain their actual value and their psychological origin. (p. 166).

Concepts, he observed (p. 117) come in two kinds: those appropriate to a certain physical reality, and those which are not tied down in that way. The second kind, Enriques regarded as psychological. Logic he regarded as operating in a meaning-independent way, on psychologised concepts. This is not the place to discuss Enriques' theory of how the brain works, or to venture a comparison with his ideas and those of modern cognitive science. Let us merely note that, in keeping with his fallibilism, Enriques practised what has been called 'meaning finitism': the idea that the meaning of a term of concept is established upon only finitely many instances and is therefore necessarily vague (see his discussion of how one learns the meaning of legal terms, pp. 113 and 119). Infinite classes are defined (p. 129) by considering the conceivable objects falling under a finite number of headings.

3. ENRIQUES AS A PHILOSOPHER

The mathematical community has evolved sophisticated ways of reading Enriques' work in algebraic geometry, much of which in any case is either correct or easy (these days) to put right. The same is not true of his writing as a philosopher or populariser. He held a subtle position, according to which knowledge is inseparable from the means of knowing, logic from psychology. This is a position on the nature of knowledge that was original and sophisticated, and even at times a little frightening in its implications for mathematics. This part of the dialogue has become a dead language, and with it Enriques's most original contribution lapses into the seemingly archaic. To restore him would require more than a keen philosophical sensibility and a well-stocked sharpness of mind (his own qualities); it requires a public eager for issues and willing to approach them in something like his way.

Restoration, however, is not the historian's task. It is Enriques' place in the period from 1900 to 1914 that concerns us here amid the constellation of ideas: mathematics, physics, philosophy, and their interactions. Enriques' allegiance was to Helmholtz, as he made clear in several places. Helmholtz is praised among all scientific men (p. 48) for offering 'the clearest insight into the office which epistemology ought to fulfil in the service of science', specifically for saying that all sorts of scientific questions lead to epistemological problems. Enriques therefore set himself the task of deriving a positive theory of knowledge, freed of philosophic controversies; a task he admitted that required the work of the entire scientific community.

His *Problems of Science* is full of discussions of positivism, in opposition to Kantianism, in metaphysics, in physics, in biology, and in the form of historical and sociological positivism.[18] His version of positivism set out to have both an objective and a subjective element. In the hard sciences (not Enriques's term) the subjective element reflects the way of representing and still more of acquiring the facts; in the social setting the subjective aspect becomes part of the facts to be explained. The task of scientific or positive epistemology becomes that of fixing standards that correspond to our conception of objective reality (p. 46) so as not to be lead into errors of the senses. Add to that the progress in mental interpretation, and concepts of reality move from crude to scientific facts – but how? What is objective, what subjective, and what is arbitrary?

Enriques focussed on the problems of logic and the growth of concepts, specifically those of geometry and mechanics. Logic, he wrote, might represent the ideal method of scientific construction, but positive epistemology points out its actual method (p. 47). For logic might include the methods of proof, but positive epistemology also discusses the method of discovery. It emerges in the course of a 72-page discussion of the problems of logic that whenever possible Enriques sought to subsume logic within psychology. Proper attention to the theme of psychologism in the period would require a book, but we must at least note that it was a central and dominating premise in the *Problems of Science*, and occupied many of its pages.[19]

By examining the situation in geometry ever since Plücker, in which various sorts of objects are treated as points in a space, Enriques came to the paradoxical conclusion that the most familiar definitions are no definitions at all, but the more obscure ones are. The usual definitions of a straight line, for example, emerges as a description or definition in the psychological sense, intended to recall certain images. More abstract definitions do however, define, for example, a system of postulates gives an implicit definition of the objects it refers to. The appropriate methods for reasoning about such things being necessarily abstract or logical, in this way Enriques came to something like the insight that Nagel was to spell out 30 years later. The difference is that while Enriques argued that the high level of abstraction forced geometers to argue more-or-less formally, he believed that nonetheless there was a core of meaning in the subject without which it was a sterile activity. Nagel, on the other hand, perhaps more sensitive to the developments inspired by Hilbert's work, but forgetful of Enriques' Italian contemporaries, placed more emphasis on the purely formal reasoning and the counter-intuitive (and therefore non-intuitive) nature of the objects.

In this connection, it is worth citing a lengthy passage also largely quoted by Avellone et al. in their recent paper.[20] It comes from the Appendix Enriques wrote in 1898 to his book on projective geometry.

> We have tried to show how projective geometry refers to intuitive concepts, psychologically well defined, and for that reason we have never missed an opportunity to show the agreement between deduction and intuition. On the other hand, however, we have warned that all deductions are based only on those propositions immediately inferred from intuition, which are stated as postulates.
>
> From this point of view geometry looks like a logical organism, in which the elementary concepts of 'point', 'line' and 'plane' (and those defined through these) are simply elements of some primitive logical relations (postulates) and of other logical relations that are then deduced (the theorems). The intuitive content of these concepts is totally irrelevant. From this observation originates a very important principle that informs all of modern geometry: *the principle of replaceability of geometrical elements.*
>
> Let us consider some concepts, defined in whatever way, that are conventionally identified with the names of 'point',

'line', and 'plane'. Let us assume that they verify the logical relations enounced by the postulates of projective geometry. All the theorems of such a geometry will still be meaningful and valid when we want to no longer consider them not as expressing intuitive relations between 'points', 'lines' and 'planes', but instead as relations between the given concepts, which are conventionally given those names.

In other words, *projective geometry can be considered as an abstract science, and it can therefore be given interpretations different from the intuitive one, by stating that its elements (points, lines, planes) are concepts determined in whatever way satisfy the logical relations expressed by the postulates.*

A first corollary of this general principle is the law of duality of space.

This shows quite clearly how Enriques held what might be called, with the example of Niels Bohr's dictum on quantum mechanics in our minds, a complementarity principle about geometry. In Enriques's view, projective geometry was simultaneously a matter of logic and an intuitive discipline. In the 1890s and all the way to the publication of the English edition of *Problems of Science* Enriques wished to argue that projective geometry, however recondite, was not ultimately or merely a formal system of rules, but was grounded in intuition and, as he came to argue, in the fundamentals of human psychology.

As Avellone, Brigaglia and Zappulla make clear, Enriques had opposition to his views within the Italian community, most notably Vailati. Vailati's allegiance was to Peano and the group of mathematicians and logicians who could most accurately claim to represent the logical point of view within mathematics. He even wrote his essays in Peano's ideographical style, thus further acknowledging the debt he was happy to admit he owed Pieri. In particular, as Avellone et al. show, Vailati saw geometry as a purely logical subject concerned with postulates and deductions. The truth of mathematical statements and the existence of objects conforming to them was not the concern of the mathematician. Underpinning this disagreement was a difference of opinion about the scope, and indeed the nature, of logic. Vailati took the strict formal view, whereas Enriques advocated a complicated ongoing relation between logic and psychology. It was, of course, the formal side that attracted Bertrand Russell, and which he was inspired by Peano to adopt when the met at the International Congress of Philosophers in Paris in 1900.

It was, however the psychologistic aspect that drew American philosophers to Enriques' book , and helps account for its American success.

4. THE AMERICAN RECEPTION

The *Problems of Science* was well received abroad. It was reviewed in *Mind* (**17**, 1908) where it was hailed as 'Probably the most comprehensive study that has appeared in recent years on the concepts on which modern science is built'. The reviewer regarded the grasp of modern science, traditional philosophy, and psychology as such as is 'rarely found united in one mind'. It was translated into German, and in two parts into French. The publication of the first part, on logic, was the occasion for H.M. Sheffer at Harvard to call for an English translation (*Philosophical Review*, **19**, 1910, 462–3). The translator of the second, geometrical part, introduced it as offering a middle way between Mach and Poincaré.

As Sheffer may well have known, the English translation was by then under way. It was made by Katherine Royce, the wife of Josiah Royce, who was the senior professor of philosophy at Harvard and had taught Sheffer. Royce had met Enriques at the International Congress of Philosophers in Heidelberg in 1908 and been impressed by him.[21] He wrote to J.M. Cattell (15 Oct. 1908, *Letters*, [1970]) that Enriques's book 'has the advantage over Poincaré's of going deeper into modern logical problems', and that 'as the book of a modern geometer and a notable representative of the great Italian school of logic, it would occupy a novel place in the literature'. Paul Carus of the Open Court agreed to publish the translation, but various administrative difficulties prevented Royce from finishing his wife's translation (as he had agreed to do) before the end of 1913. By then, however, Royce hoped in his Introductory Note to the book, it might be useful in combating the recent rise of anti-intellectualism which Royce feared would prefer easy, dramatic answers to patient, critical thought.

Royce argued that although Enriques's reputation was founded on his treatise on projective geometry and his essay on the foundations of geometry, as a philosopher he said much that pragmatists could accept. This was all the more surprising because the Italian book had been published in 1906, while the vogue for pragmatism had not begun until Heidelberg in 1908. Indeed, Enriques's form of pragmatism was largely original, many-sided, and judicious. Royce went on to comment on many aspects of Enriques' diverse yet synthesizing approach, before concluding by welcoming the book above all as a treatise on methodology.

Royce himself is an interesting figure. Born in California in 1855 he became a professor at Harvard in 1882. There he became an intimate friend of

William James, the distinguished pragmatist and psychologist, and of George Santayana. From philosophy and theology, he turned to logic, and in 1905 he published a lengthy paper on the close connection he saw between logic and geometry, and which cites work in the axiomatic style by Huntingdon (a mathematician at Harvard) and Veblen at Princeton. Royce's breadth is reminiscent of Wundt, who had impressed him as a young man: he studied mathematics, and his definition of philosophy embraced psychology as well as logic. he also had a sense of pragmatism. All these naturally pre-disposed him to the writings of Enriques. In so far as they were typical of their day – and they were a mainstream trend – they indicate that in America intellectual circles Enriques had a ready-made audience.[22]

Enriques, it is clear, was given credit for a point of view that is broader than his own. He was taken to represent a school to which he barely belonged, and his work was assimilated to traditions that it may not deliberately belong to. Enriques' interests in logic had nothing to do with the severe formalism of Peano, and everything to do with psychology. But both are issues in the popular perception of mathematics, and Enriques was taken to speak for both. Royce, for example, saw no problems in harmonising Enriques' ideas with any others he (Royce) happens to support, and others were equally imaginative, or simple careless, in their reading. When a reviewer writes in 1908 that 'Reality means, according to Enriques, the correspondence of sensations with expectation; a reality existing by itself, independent of our experience, is simply an absurdity' we find ourselves jolted into disagreement. Enriques indeed wrote '"The correspondence of the sensations with the expectation" always constitutes the true character of reality'. (*Problems*, p. 56). But he immediately went on: 'Reality as we think of it would not cease *to exist in itself*, even if all communication between our minds and the external world were broken off' (p. 56, italics in original). Then, to avoid the position he called transcendental, which in his view lead to a sceptical idealism about an unknowable phantasm before our eyes, Enriques explained that the term 'in itself' referred to our inability to modify our sensations as we wish. So, at the end of a lengthy discussion, Enriques defined the real as an *'invariant in the correspondence between volition and sensation'* (p. 65, italics in original). He did not dismiss externally existing reality as absurd, but his reviewer had no trouble in summarising him that way because such philosophical views were typical of the idealist philosophy then current.

5. CONCLUSION

Poincaré, Ernst Mach, and Enriques, probably in descending order, offer more than a series of philosophical positions worth arguing with to this day.

They present the historian with the spectacle of three major scientists and mathematicians deeply committed to reaching a large audience with probing and difficult essays. The practice of science led, for them, to questions about the nature of knowledge. If Enriques and Poincaré differed at some points, and Enriques was more of a realist or positivist, all three agreed at other points, notably the importance of psychology and the way knowledge is acquired. Their popular success tells us something more: that that mix of topics was a congenial one at the time. Rather than report a conclusion, let me urge a question: what does this confluence of leading scientists and ideas tell us about the place of science in the society of its day?

Open University
USA

NOTES

[1] See, for example, Klein's remarks about 'optical' and 'mechanical' properties of space in his Evanston Colloquium Lectures, (1894, 87).

[2] Paul Carus, who ran the Open Court Publishing House in Chicago and edited the *Monist*, published essays by Ernst Mach, and Poincaré, Hilbert's *Foundations of Geometry*, and eventually of Enriques's *Problems of Science* and many others.

[3] Enriques to Castelnuovo, 4 May 1896 ' ... il più meraviglioso ingegno filosofico, fisiologo, psicologo, matematico ecc. ... Leggi la ≪Logik≫ del Wundt quella parte almeno che riguarda i metodi della matematica, e pensa che è un fisiologo che scrive così: un fisiologo che non teme di salire l'erta della concezione kantiana per illuminare dall'alto il gran corso di tutte le scienze!.' in Bottazzini, Conte, and Gario, P., (1996, 261).

[4] The best-known example is in the philosophy of arithmetic, where Frege stood at one extreme in his resolute separation of concepts from their mode of acquisition, and at the other extreme one might place Poincaré, who roundly denigrated attempts to derive the integers from abstract sets.

[5] Fano (1892).

[6] Veronese (1891).

[7] Peano (1894, 73).

[8] See the citations in A. N. Whitehead's (1906).

[9] Couturat in his (1905).

[10] See J. W. Young (1911).

[11] Poincaré, L'expérience et la géométrie, in Poincaré (1902b, 95–110).

[12] Poincaré, L'espace et la géométrie, in Poincaré (1902b, 77–94).

[13] Poincaré, Experiment and geometry, in Poincaré (Dover, 72–88, quote appears on p. 73).

[14] Klein (1901).

[15]These need not be even remotely likely to be correct, and may certainly be false, for example the Clifford-Klein space forms.

[16]Enriques also noted that quite different axiom systems may describe the same geometry (*qua* relations of position of bodies) and gave the example of non-Archimedean geometry of Veronese and Hilbert, which, to an infinite degree of approximation, gave theorems identical with the usual Euclidean geometry (p. 198).

[17]He also admitted that points in one space may be objects in another, for example circles in the plane may be regarded as points in a space of all coplanar circles. But this recognition of a very significant development in the mathematics of the preceding 50 years did not affect his philosophy of geometry.

[18]There are similarly lengthy discussions of reality, and facts, but let us not wander too far.

[19]For such a book, focussed on the German university scene, see Kusch (1995).

[20]Avellone et al. (2002). The quotation from Enriques (1898, 376–7) will be found on p. 394. My translation differs slightly from the one in Avellone et al. (2002).

[21]Carus, Royce and Catherine Ladd Franklin were involved in setting up the International Congress of Philosophers; see Kennedy (1980).

[22]Cassius J. Keyser, the Adrian professor at Columbia University New York, was another in Enriques's audience. His *Human Worth of Rigorous Thinking* , 1st edition 1916, 2nd 1925, praises Enriques for his logical remarks, the 'important, much neglected and little understood subject of definition, its nature, variations and function, in the light of the recent literature, especially the suggestive handling of the matter by Enriques in his *Problems of Science*' (p. 192).

REFERENCES

Avellone, M., Brigaglia, A. and Zappulla, C. (2002). The Foundations of Projective Geometry in Italy from De Paolis to Pieri, *Archive for History of Exact Sciences* **56**: 363–425.

Bottazzini, U. (1988). Fondamenti dell'aritmetica e della geometria, *Storia della scienza moderna e contemporanea* p. III.I. UTET.

Bottazzini, U., Conte, A. and Gario, P. (1996). *Riposte Armonie, Lettere di Federigo Enriques a Guido Castelnuovo*, Bollati Boringhieri.

Couturat, L. (1905). *Principes des mathématiques*, Alcan, Paris.

Enriques, F. (1898). *Lezioni di geometria proiettiva*, Zanichelli, Bologna.

Enriques, F. (1906). *Problemi della Scienza*, Zanichelli, Bologna.

Enriques, F. (1907). Prinzipien der Geometrie, *Encyclopädie der Mathematischen Wissenschaften*, Teubner, Leipzig, p. III.1.1.

Enriques, F. (1912). *Scienza e Razionalismo*, Zanichelli, Bologna.

Enriques, F. (1914). *Problems of Science*, Open Court, Chicago. English translation by K. Joyce.

Enriques, F. (1922). *Per la storia della logica*, Zanichelli, Bologna. English translation: J. Rosenthal, 1929, *The Historic Development of Logic*, Holt, New York.

Fano, G. (1892). Sui postulati fondamentali della geometria proiettiva, *Giornale di Matematiche* **30**: 106–131.

Hilbert, D. (1899). *Grundlagen der Geometrie*, Teubner, Leipzig. Many subsequent editions.

Hilbert, D. (1971). *Foundations of Geometry*, Open Court, Illinois. 10th English edition: translation of the second German edition by L. Unger.

Kennedy, H. C. (1980). *Peano; The Life and works of Giuseppe Peano*, Reidel, Holland and Boston USA.

Klein, C. F. (1894). *Lectures on Mathematics, The Evanston Colloquium Lectures*, Macmillan and Co, New York.

Klein, C. F. (1901). *Anwendung der Differential- und Integral-Rechnung*.

Kusch, M. (1995). *Psychologism; A Case Study in the Sociology of Philosophical Knowledge*, Rouledge, London.

Marchisotto, E. A. (1993). Mario Pieri and his contributions to geometry and foundations of mathematics, *Historia Mathematica* **20**: 285–303.

Pasch, M. (1882). *Vorlesungen über neuere Geometrie*, Teubner, Leipzig.

Peano, G. (1889). I principii di geometria logicamente espositi. Rep. in *Opere scelte*, **2**, Rome, 1958, 56–91.

Peano, G. (1894). Sui fondamenti della Geometria, *Rivista di matematiche* **4**: 51–90.

Pieri, M. (1895). Sui principi che reggiono la geometria di posizione, *Atti Accademia Torino* **30**: 54–108.

Pieri, M. (1899a). Della geometria elementare come sistema ipotetico-deduttivo; monografia del punto e del mote, *Memorie della R. Academia delle scienze di Torino*.

Pieri, M. (1899b). I Principii della Geometria di Posizione, composti in sistema logico deduttivo, *Memorie della Reale Accademia delle Scienze di Torino (2)* **48**: 1–62.

Poincaré, H. (1887). Sur les hypothèses fondamentales de la géométrie, *Bulletin de la Société Mathématique de France* **15**: 203–216. in *Oeuvres*, **11**, 79–91.

Poincaré, H. (1898b). On the foundations of geometry, *The Monist* **9**: 1–43. Reprinted in Ewald, Vol. **2**, pp. 982–1011.

Poincaré, H. (1902a). Les fondements de la géométrie, *Bulletin de la Société Mathématique de France* **26**: 249–272. in *Oeuvres*, **11**, 92–113.

Poincaré, H. (1902b). *La science et l'hypothèse*, Flammarion, Paris. Reprint 1968, ed. J. Vuillemin.

Poincaré, H. (1912). L'Espace et le temps, *Derniers Pensées* pp. 97–109.

Poincaré, H. (1913). *Derniers Pensées*, Flammarion. Often reprinted.

Poincaré, H. (1997). *Three Supplementary Essays on the Discovery of Fuchsian Functions*, Akademie Verlag, Berlin and Blanchard, Paris. ed. with an introductory essay by J.J. Gray and S.A. Walter.

Poincaré, H. (Dover). *Science and Hypothesis*, Walter Scott Publishing Co. English translation of (Poincaré, 1902b) by W.J. Greenstreet, Walter Scott Publishing Co. Ltd. Dover reprint 1952.

Veronese, G. (1891). *Fondamenti di geometria a più dimensini e a più specie di unità rettilinee, espositi in forma elementare*, Padova.

Whitehead, A. N. (1906). *The Axioms of projective Geometry*, Cambridge University Press Tract nr. 4, Cambridge.

Young, J. W. (1911). *Fundamental Concepts of Algebra and Geometry (2nd ed. 1925)*, MacMillan, New York.

ULRICH MAJER

HILBERT'S AXIOMATIC APPROACH TO THE FOUNDATIONS OF SCIENCE—A FAILED RESEARCH PROGRAM?

1. INTRODUCTION

There is almost totally agreement that it is the aim of science 'to make pre-dictions about future events on the basis of prior experiences obtained by accidental observations and intentional experiment'.[1] There is still very im-pressive agreement that this aim can be achieved only by developing *theories* in a comprehensive sense, which means that they encompass large areas of natural phenomena and processes of a "law-like" nature. There is, however, already less agreement with respect to the question *'what theories are'*, what the logical status of theories is. Are they 'general propositions' or 'variable hypothesis'? Do they entail 'existential' assumptions or should theories be strictly free of any existential commitment?[2] But there is again consider-able agreement – at least since Kant – that 'real *scientific* theories of nature' must be stated in the language of mathematics. Yet there is remarkably little agreement in which way this shall be done, and in particular in which form a theory shall be *represented*: in an *axiomatic* or in a *constructive* way, as a *system of axioms* and their logical consequences or as a *collection of models* and rules for their construction?

Of course the two modes of presentation form no strict alternatives. There exist other modes of presentation, e.g. by writing out the fundamental equations of a physical theory or by giving a description of the 'intended models' of a theory. Most scientists decide the question of representation of a theory quite pragmatically using sometimes one sometimes another mode of presentation depending on their intellectual taste, their education and sci-entific training and other psychological and sociological preferences.

But, as a matter of fact, most scientists – except a small group of mathe-matical physicists – do not like the axiomatic mode of presentation. They prefer instead a less formal way of presentation, which focuses more on the *physical* content than on the *mathematical* structure of a theory. In the subse-quent paper I will investigate the reasons for this opinion. In particular I will ask where this opinion comes from, and what the cause for the widespread *antipathy* against the axiomatic presentation of physical theories is. The pur-suit of these questions may seem superfluous, but it is not—at least not to my mind. The answer may help to overcome the (unjustified) prejudice against the axiomatic approach in general and an axiomatic presentation of physical theories in particular, and this is, in my view, an essential precondition for a satisfactory solution of the foundational problems of modern physics in the form of a unified quantum-field-theory.

155

V.F. Hendricks, K.F. Jørgensen, J. Lützen and S.A. Pedersen (eds.), Interactions: Mathematics, Physics and Philosophy, 1860-1930, pp. 155–184.
© *2006 Springer.*

Because prejudices are often difficult to overcome, particularly if they are just a matter of taste, I'll proceed very carefully and divide my considerations into three parts. In the first part I will discusses a serious *misapprehension* of the axiomatic approach that is almost as old as 'modern axiomatics' itself. Modern axiomatic, as we know it today, had become established primarily[3] by Hilbert and his school at the turn of the 19th to 20th century. It will turn out that part of the misapprehension goes back to a number of unhappy formulations and a somewhat misleading explanation of the so-called 'axiomatic method' by Hilbert himself. This had the unfortunate consequence that the 'axiomatic point of view' was received with reservation and mistrust by many mathematicians. With the example of geometry I will try to correct the misapprehension.

In the second part I'll give a sketch of Hilbert's work in physics from 1900 to 1930. This will be crucial for a clearer and more distinctive understanding of the axiomatic point of view, in particular what Hilbert meant by the application of the 'axiomatic method' in physics. Most of Hilbert's extended work in physics has not been published for a number of reasons, which are hardly understandable from today's point of view. This had the double effect that a considerable part of his work in physics was not received by his fellow physicists and that that part of his work, which had been published, was in part misinterpreted regarding its methodological significance for our scientific understanding of nature.

In the third part I'll discuss in the light of the first two chapters some critical objections to the axiomatic approach in physics, which have been put forward by Hermann Weyl; this part is divided in two sections. In the first I present the position of Weyl and his objections against an axiomatic presentation of physical theories.[4] In the second section I try to defend Hilbert's axiomatic approach in physics against Weyl's methodological objections.

2. HILBERT'S AXIOMATIC POINT OF VIEW AND THE FOUNDATIONS OF GEOMETRY

In 1899 Hilbert published a work, entitled *Grundlagen der Geometrie* (GG) as part of a 'Festschrift' for the opening ceremony of the Gauss-Weber memorial. The publication was seen as a surprise, because nobody [outside Göttingen] had expected a book like this from Hilbert, who had become known as number theoretician by his 'Zahlbericht'. But, needless to say, for the insider this was no surprise at all. Hilbert had worked on the foundations of geometry already for several years. Indeed, the first work on the foundations of geometry dates back to 1893 when he was still in Königsberg. The publication in 1899 was the first harvest of his efforts over many years to come to grips with the foundations of geometry. And this was no easy task, because the *cognitive* situation in geometry was rather screwed and impenetrable, when Hilbert started his inquiries of geometry in ca. 1890.

Geometry had become separated into a confusing number of different branches and competing research programs like projective and synthetic, descriptive and applied, constructive and analytic geometry including differential geometry and perhaps most importantly Euclidean and non-Euclidean geometry. A large number of interesting yet unconnected results had been achieved and last but not least a number of important books had appeared as for example von Staudt's *Geometrie der Lage* (1847) and Pasch's *Vorlesungen über neuere Geometrie* (1882).

Regarding the genesis of the *Grundlagen der Geometrie* two points are important in retrospective. First when Hilbert started to lecture on geometry in 1891 he didn't know in which *direction* he should steer his investigations.[5] But once he had read Pasch's book it became quickly clear to him what his principal goal was: he wanted to revive *Euclid's axiomatic point of view*, which had been badly neglected during the eighteenth and nineteenth century, and bring geometry in this manner to a new logical perfection.[6] This was a long way, of course, and Hilbert didn't know immediately how to achieve this goal. This brings me to the second point. It's important to note that Hilbert in the beginning of his inquiries of the foundations of geometry did not dispose about that what he himself later called the 'axiomatic method'. This method emerged bit by bit *during* his investigations and was first explicitly stated and explained *after* GG had been published. (The label "axiomatic method" itself occurs for the first time in the lecture *Foundations of Euclidean Geometry* (1898/99) immediately preceding the publication of GG, but isn't used in GG itself, although the 'new approach' to geometry, what is meant by this label, is present throughout the whole book.)

In spite of these circumstances, I think, it does not come as a surprise that the real significance of Hilbert's efforts to revive Euclid's axiomatic approach was neither correctly apprehended nor fully understood. The main misapprehension[7] is roughly this: Most scientists are somehow convinced that Hilbert's main achievement in GG is a completely *new* understanding of what it means 'to set up an axiomatic system' of geometry [or some other discipline], namely *not to rely* on the (intuitive) *meaning* of expressions like 'point', 'straight line' and 'between', and to assert some evident truth regarding the meaning of these expressions, like Euclid apparently did, but instead to set up a self-consistent system of otherwise arbitrarily chosen axioms, expressed by meaningless symbols and to see *which* meanings can be given to the symbols. The problem with this understanding is that it is in one respect *correct* yet at the same time extremely *misleading* because it confuses or mingles two aspects in one, which have to be distinguished very sharply.

We have already said that Hilbert himself held Euclid's approach to geometry in the highest esteem and saw his own efforts regarding a 'new foundation' of geometry as a *continuation* and *improvement* of Euclid's axiomatic approach and not as a dismissal. This is clearly stated in the introduction to GG:

> Die Geometrie bedarf – ebenso wie die Arithmetik – zu ihrem folgerichtigen Aufbau nur weniger und einfacher Grundsätze. Diese Grundsätze heißen *Axiome* der Geometrie. Die Aufstellung der Axiome der Geometrie und die Erforschung ihres Zusammenhangs ist eine Aufgabe, die seit Euklid in zahlreichen vortrefflichen Abhandlungen der mathematischen Literatur sich erörtert findet. Die bezeichnete Aufgabe läuft auf die logische Analyse unserer räumlichen Anschauung hinaus.
>
> Die vorliegende Untersuchung ist ein neuer Versuch, für die Geometrie ein *vollständiges* und *möglichst einfaches* System von Axiomen aufzustellen und aus denselben die wichtigsten geometrischen Sätze in der Weise abzuleiten, daß dabei die Bedeutung der verschiedenen Axiomgruppen und die Tragweite der aus den einzelnen Axiomen zu ziehenden Folgerungen klar zu Tage tritt.[8]

According to my understanding this quotation shows beyond doubt that something *must be wrong* with the 'received view': that Hilbert *separated* himself from Euclid's axiomatic approach and aimed at a *completely new* conception of axiomatics. We have to find out first *what* is wrong with the received view and second *why* is it at the same time so attractive and apparently convincing that it pervades the interpretation of Hilbert's work up to the present day. In order to do so I introduce a *terminological distinction*, which does not occur in this explicit manner in Hilbert's writings, but is none the less utterly important, and underlies *implicitly* Hilbert's whole axiomatic procedure in GG (as elsewhere). The distinction is roughly this.

According to Hilbert's point of view (and contrary to the popular opinion) geometrical expressions like 'point', 'straight line' and 'between' are not completely meaningless symbols but have a certain *intuitive content* (or meaning) and this content has to be captured and represented by a *system of axioms* in a certain logical order and perfection, which I'll explain in a moment. Any such presentation of a specific content by an axiom-system I call an 'axiomatic representation' of the content in question, quite independently of the epistemological question what the source of the content is. This stipulation sounds obvious enough, but it has to be strictly distinguished from a *second* component of the axiomatic approach, the so-called 'axiomatic method'.

What does Hilbert mean by this expression? A real concise answer is not easy, because it needs a lot of technical notations, but it suffices for now to stress the following two points: (1) The 'axiomatic method' is *not identical* to the 'axiomatic point of view', it forms only *one component* of it; the other component is the already defined axiomatic presentation of a certain content. (2) The 'axiomatic method' is not a goal in itself but *a means to an end*, a 'logical tool' in order to accomplish a certain goal, and this goal is the axiomatic representation of a specific content such that the resulting system of axioms satisfies three conditions: (a) it should be *complete* with respect to the specific content; (b) it should be as *simple* as possible, which means that it entails no *superfluous* axioms, and (c) it has to be *"logically perspicuous"* or *distinct*, such that one can recognice which sentences follow from which axioms, and which sentences are logically *independent* from a certain (sub-)system of axioms.

Although it can hardly be denied in spite of the quotation above that *both* components are essential for an adequate understanding of Hilbert's axiomatic point of view there exists a strong tendency among mathematicians to ignore the first component in favour of the second and to equip the axiomatic method with miraculous abilities that it does not and cannot possess. I will now explain why this misapprehension of Hilbert's view is obviously so attractive that it survived all objections up to the present day, although it had been severely criticised from the very beginning first by Frege and later by Weyl—to mention only the most prominent critics. I see three reasons.

The first is a biographical one. Because Hilbert wanted to convince his contemporaries of the axiomatic point of view, i.e., its *superiority* over competing research programs regarding the foundations of geometry (such as the Erlanger Program of Felix Klein) he emphasised the *novel aspects* of his approach and passed over in silence what seemed anyway obvious. And this was, as a matter of fact, the traditional viewpoint of Euclid, whereas the superiority of the axiomatic method had to be demonstrated at every opportunity.

The second reason for ignoring the first component in favour of the second, is merely a psychological one. Most readers, I surmise, were eager to understand the extraordinary *success* of Hilbert's GG. This could be nothing else but the new axiomatic method. Although the term 'axiomatic method' does not occur in GG the method as such is present throughout the entire book. Consequently, most readers try to figure out – 'on their own account' so to speak because Hilbert had missed to present an explicit formulation or explanation of the new method – what the *essence* of the axiomatic method was. G. Frege is a good example. He tried very hard to come to grips with Hilbert's axiomatic approach. Although, or just because, Frege himself shared Euclid's point of view, he could not recognise the first component in Hilbert's GG (i.e., an axiomatic presentation of geometry) and interpreted instead Hilbert's axiomatic approach as resting solely and exclusively on the axiomatic method. Of the latter he gave a very interesting, but to my mind inadequate interpretation, which does not accord with Hilbert's own intentions. It is, however, beyond the scope of this essay to discuss Frege's logical interpretation of the axiomatic approach in detail.[9]

The third reason (for the neglect of the first component in favor of the second) is an *objective* one. It's connected with the methodological difficulty to combine the old viewpoint of Euclid with the new axiomatic method. At first glance this combination seems impossible because the new method seems to *contradict* the old viewpoint according to which the axioms of geometry cannot be chosen *arbitrarily* (because they have to represent a certain content) whereas an arbitrary choice of axioms seems to be exactly the *point* of the new axiomatic method.[10] Consequently, it seems as if one has to make an *exclusive* decision between the two components: *either* Euclid's old representative approach *or* the new axiomatic method, not both. But, needless to say, the contradiction is only apparent.

Correctly understood, there is no contradiction at all between the two components. On the contrary, there is a perfect harmony, an ideal *supplementation* of the first component by the second, which led (after a decade of research) to an essentially *improved* axiomatic representation of geometry— a representation, which Euclid could not achieve, because he did not know the axiomatic method, but had instead to *rely* on his geometrical 'intuition' and the 'construction' of figures.[11] Let me finish this chapter with a short schematic description how the 'interplay' between the two components works, how the axiomatic method indeed supplements the axiomatic representation of geometry.

The first point one has to become clear about is the circumstance that the axiomatic method is not an end in itself but only a *means to an end* to improve the axiomatic representation of a certain content, in our case that of geometry, according to the three criteria of completeness, simplicity and logical perspicuity or distinctness. Hence, the axiomatic method cannot be applied to an 'unarticulated' theory, to a "blank sheet of paper", but presupposes for its application a certain knowledge of facts articulated in a number of 'Grundsätze' or axioms, which – in a next step – can be logically analysed and eventually improved by means of the axiomatic method. This sounds like "Münchhausen's trick", but it is not because the axiomatic method is in the first line a procedure to check the logical dependence or independence of already articulated axioms and sentences: If an axiom turns out to be 'dependent' on the remaining set of axioms of an axiom-system it has either to be dropped completely or to be replaced by a weaker one.

The second point is equally important as the first, because it concerns the 'procedure' by which in particular the 'logical independence' of sentences is tested. Without this procedure the axiomatic method could never accomplish, what it should accomplish, namely a *control* and improvement of the degree to which an axiomatic presentation does or does not accord with the three criteria. The procedure by which Hilbert achieved this control is basically the same as what we today call 'model theory'.

In 1900 there was, of course, no model theory—at least not in the modern sense of an established logical discipline. Hilbert had to invent first something like this. One can say that his greatest achievement in a technical sense is the invention of a *precursor* of model theory: In GG Hilbert introduced *number fields*, in order to prove the logical *independence* of certain axioms, as e.g. the axiom of continuity, from other axioms, e.g. the axioms of congruence. The principal idea of such an independence-proof is well known; schematically stated it's this: In order to prove the independence of a sentence B from the axioms $A_1, \ldots, A_n$ one has to show that the united system of axioms '$A_1, \ldots, A_n$ and non-B' possess a model. This means that B cannot be a logical consequence of $A_1, \ldots, A_n$ because there is a structure, in which all sentences $A_1, \ldots, A_n$ and non-B are fulfilled; this in turn means that the system '$A_1, \ldots, A_n$ and non-B' is consistent, provided the model is consistent. Hence, the core of the axiomatic method is the proof of the *relative consistency* of $A_1, \ldots, A_n$ and non-B with respect to a model.

A final remark before I turn to the main question, whether the axiomatic point of view also makes sense in physics: The reader should be aware of three points:

First the question how to proof the dependence or independence of axioms and sentences from a given system of axioms was a logical mess before Hilbert had published GG. At least there existed no *systematic* treatment of the whole issue. Second, once the issue of independence-proofs by *number fields* had been clarified sufficiently Hilbert could use the technique to improve the axiomatic representation of geometry step by step according to the three aims of simplicity, distinctness and completeness until an 'optimal' axiomatic presentation of geometry (relative to the criteria) had been found.

Each step in this procedure has basically two parts: First, a systematic variation of single axioms in an axiomatic presentation, and second the proof that the resulting axiom system is consistent *relative* to an appropriate number field. Both parts together constitute that what Hilbert calls the axiomatic method. And third, as already mentioned, the axiomatic method says nothing regarding the source of our geometrical knowledge in the first place, i.e., where our geometrical knowledge does come from, it is merely a method of 'logical analysis', not in itself a *source* of knowledge. After this clarification let's turn to the main question:

- Is there any reason why the axiomatic point of view – correctly understood – cannot be applied successfully in the natural sciences, in particular in physics and its different branches?

I will divide the answer to this question in two parts. First I'll present a sketch of Hilbert's work in physics, focusing on the unpublished part and leaving the published work in the background. Second, I'll discuss some of the objections, which Weyl raised in his obituary of Hilbert against an application of the axiomatic approach in the natural sciences. This gives me the opportunity to correct a subtle misunderstanding in Weyl's apprehension of the axiomatic approach in science. Finally I'll try to identify those aspects and moments in Hilbert's axiomatic foundations of physics, which seem to me *methodologically* appropriate and reasonable, and what should perhaps be given up.

3. HILBERT'S WORK IN PHYSICS BETWEEN 1900 AND 1930

In retrospective it's easy to distinguish Hilbert's work in physics into three relatively sharply 'separated' periods, which can be characterised as follows.

The first period begins in 1898, when Hilbert for the first time held a lecture in physics, entitled "Mechanik", and lasts roughly until 1911, when Hilbert remodeled his lecture on mechanics in order to incorporate Einstein's theory of special relativity. This period I will call Hilbert's "silent period" because there is not a single publication in physics by Hilbert during this period—with a small exception to which I'll come in a moment.

The second period started in 1912 with his first publication in physics (Begründung der kinetischen Gastheorie; Math. Ann. 71) and came to an intermediate halt in 1917/18, when he turned again to the foundations of mathematics. This was Hilbert's most productive period in physics: not only appeared during this period the main bulk of his physical papers, but among them are also the most important ones, the two notes on *The foundations of Physics* of 1915/16.

The third period runs roughly from 1919 to the end of his teaching career in 1933. In this period he presented (beside his teaching duties in mathematics) a large number of more or less 'popular' lectures like that on *Natur und mathematisches Erkennen, Raum und Zeit, Grundgedanken der Relativitätstheorie, Wissen und mathematisches Denken, Über die Einheit der Naturerkenntnis*, and related topics, which were not so much 'scientific' lectures in the proper sense, but were suited for an academic audience, which had an interest in the most recent progresses of science.

It is, however, important to note that there is one exception from this last rule. During a relatively short period from 1922 to 1928 Hilbert became interested in the foundations of quantum theory and worked together with Nordheim and von Neumann on the mathematical methods of the old and the new[!] quantum theory. Out of this collaboration grew the well known *three man paper* "Über die Grundlagen der Quantenmechanik", published in 1928. This paper formed the point of departure for von Neumann's famous book *Mathematische Grundlagen der Quantenmechanik*—a milestone in the axiomatic presentation of the new quantum-theory. (See the end of this paper.)

In this paper I will primarily deal with the first period and its intricacies and put the other two periods more or less aside. This has several simple reasons: The most important one is this: For a proper understanding of Hilbert's axiomatic approach to science the first period is the most significant one. In it he not only made himself familiar with the physics of his time but also developed the idea how to expand or transfer the axiomatic approach from geometry to science, and what this implied, i.e., what had to be done, what changes had to be made in order to execute this program successfully. Another reason lies in the fact that relatively little is known about the first period (due to the circumstance that until now almost nothing had been published from this period) whereas the other two periods are relatively well known, which means, they have been closely studied not only by scientists but also by philosophers and historians of science, of course, with different interests and intensity. Another reason is that the most important theories of the later periods (general relativity and quantum theory) are 'unnecessary complicated' for the purpose of this paper, namely a critical discussion of the axiomatic approach in science.

4. HILBERT'S 'SILENT' PERIOD FROM 1898 TO 1911

As already mentioned, during the 'silent' period Hilbert held quite a number of lectures in physics, but didn't publish anything with one *exception*: there is one published statement that testifies that he had rather early an interest not only in mathematics but also in physics. This is the 'sixth problem' of his famous list of 23 problems presented at the Mathematical Congress in Paris 1900. Let me begin the discussion of the first period with this problem; it bears the title: *Mathematische Behandlung der Axiome der Physik* and opens with a reference to the axiomatic investigations of geometry, published the year before:

> Durch die Untersuchungen über die Grundlagen der Geome-
> trie wird uns die Aufgabe nahe gelegt, *nach diesem Vorbilde*
> *diejenigen physikalischen Disziplinen axiomatisch zu behan-*
> *deln, in denen schon heute die Mathematik eine hervorra-*
> *gende Rolle spielt: dies sind in erster Linie die Wahrschein-*
> *lichkeitsrechnung und die Mechanik.*[12]

One important point should be immediately clear: The 'axiomatic treatment' of physical disciplines is motivated through the success of the axiomatic approach in geometry. This means that the 'axiomatic point of view' in physics has to be roughly the *same* – at least 'in principle' – as that in geometry. In other words, the application of the *axiomatic method* in physics is only legitimate insofar as one takes it as an *expansion* (or transfer) of the 'axiomatic point of view' in geometry. This does not imply that they have to be *exactly* the same; there may be certain differences due to the differences between geometry and physics. On the other hand, Hilbert was deeply convinced (from the very beginning of his occupation with geometry) that geometry is a 'natural science', indeed the most fundamental of all natural sciences, and this is the deeper reason why the axiomatic point of view should be 'transferable' to physics.[13]

Before I explain what the real challenge of the sixth problem is, what kind of research program Hilbert wanted to install by his request to axiomatise physics in accordance with the paradigm-example (*Vorbild*) of geometry, and what the systematic advantage of such an axiomatic investigation could be, let me make a couple of remarks with respect to the *historical context*, in which the request was put forward; this may help to dispel some of the mistrusts, with which the request was taken in by many of Hilbert's scientific contemporaries.[14]

In 1900 the scientific situation in physics was obviously rather different from today. The two "revolutions" in physics had not yet taken place and the dream of a *unified theory* of all physical branches within mechanics was much alive. The task to axiomatise physics in accordance with the example of geometry was put forward by Hilbert at a time, when there was great hope that physics could be completed within the next decade or so. Hilbert himself pointed out in Paris that already significant results had been achieved regarding the foundations of mechanics and refers in this context to the writings of Mach, Hertz, Boltzmann and Volkmann[15] His hope was, of course, that the difficult "limit processes" in mechanics leading from *discrete* systems of bodies to *continuous* media and vice verse would be solved in such a way that the equivalence of both approaches could be recognised.[16]

In this connection it is important to note that Hilbert's list of problems entails a further problem, which has a similar *peculiar* status as the sixth problem. That is the problem entitled "Weiterführung der Methoden der Variationsrechnung". The calculus of variations was not in good shape in 1900 and had not received the attention, which it deserved in Hilbert's view, because it is (beside geometry and the concept of probability) the third pillar on which all physics rest—not only classical mechanics but also general relativity and quantum theory, as Hilbert later was able to show by his axiomatic presentation of these theories. For this reason he attached the development of the calculus of variation to his list of problems—being well aware that this problem (like the sixth problem) was of a far more general nature than the other mathematical problems.[17]

On the other hand, the situation in geometry was much more 'healthy' in 1900. Not only that remarkable clarity had been achieved in recent years regarding the 'logical status' of the different axioms and sentences, but Hilbert had also been able – just in 1900(!) – to solve the 'problem of continuity' by introducing the so-called axiom of 'completeness' in addition to the Axiom of Archimedes. What would be more natural than the request to transfer the axiomatic point of view from geometry to physics?

In spite of the importance of the sixth problem and the historical circumstances just outlined it comes as a surprise that Hilbert did not publish a single paper in physics during the next decade. This nurtures the suspicion that Hilbert himself was not really convinced that the sixth problem was well posed, not to speak of the auspices of a solution in the next years. This suspicion is, however, totally unjustified as we will see if we investigate what Hilbert really did in physics in the first decade of the century.

It is beyond the scope of this essay to present a detailed description of Hilbert's work in physics during his 'silent period' between 1898 and 1912. Instead I will concentrate on two points: (a) Hilbert's physical world view taking mechanics as the foundation of all natural sciences; (b) The axiomatic structure of mechanics and the logical analysis of its principles.

4.1. Mechanics as the foundation of all natural sciences

During his silent period Hilbert held, beside many lectures in pure and applied mathematics, altogether seven extended lectures on mechanics,[18] which covered the whole domain of physics known at the beginning of the 20th century: The lectures included beside mechanics in the proper narrow sense also such disciplines as hydrodynamics, thermodynamics, and also electrodynamics, which we regard as *independent* disciplines (not belonging to mechanics in the narrow sense). In other words, Hilbert regarded these disciplines as *branches* of mechanics in a broad comprehensive sense: He had to show how these disciplines could be *derived from* or *reduced to* mechanics. This was not an easy task, because he had to figure out, which special assumptions were necessary in order to 'embed' the different disciplines into mechanics proper.

A modern reader might think this is an impossible task due to the fact that *space-time-theory* of electrodynamics is fundamentally different from that of classical mechanics.[19] However, although correct, this was not obvious before Minkowski had clarified the *conventionalist* muddle regarding the Einstein-Lorentz-Fitzgerald theory of space-time. Hence, Hilbert could pursue his goal of mechanics as a true universal theory, into which every known physical fact, every phenomenon of [inanimate] nature could be embedded, so long as it was not definitively clear that the classical theory of space and time had to be substituted by a new theory of 'space-time' not distorted by conventionalist misinterpretations.[20]

In order to mediate at least an idea, how Hilbert tried to achieve the goal of a universal theory of physics founded on mechanics in the proper sense, let me explain the schema according to which the lectures on mechanics were set up. Hilbert arranged the whole content of mechanics according to a sequence of criteria into a number of systems, subsystems, sub-subsystems etc. The main distinction is that of the *number* of physical particles (respectively mass points) taken together in a system of (interacting) bodies: (i) systems with exactly *one* particle, (ii) systems with a small *finite* number of particles, and most importantly for the envisaged goal (iii) systems with an *infinite* number of particles.

This division looks quite simple and obvious; the interesting point is, however, that the last category of systems is in Hilbert's view *identical* with what we call *continuum-mechanics*. This means it embraces all those disciplines like hydrodynamics, thermodynamics and electrodynamics which are treated in the usual textbooks under the label 'continuum-mechanics', but whose logical relation to mechanics in the proper sense is often enough left in the dark. I will return to this point in a moment. First let me explain the further criteria for the distinction into subsystems. Next the 'space-time-relations' between the particles are considered: *Constant* (or approximately constant) distances between the particles lead to systems of *rigid bodies* like crystals, metals etc.; their investigation forms the discipline called 'Festkörperphysik' in German. *Smoothly changing* distances between particles lead to *fluids,* whose behaviour is studied in hydrodynamics. Finally, systems with *rapidly changing* distances between (discontinuously moving) particles represent *gases*; they are treated in the 'kinetic theory of gases', which is essentially a *statistical* theory (in distinction to thermodynamics, which merely gives a phenomenological description of gases).

This is only a rough sketch of the first steps. To get a more 'fine grained' division one has to consider further 'boundary conditions' such as the *time-dependence* of the internal forces between, and the external forces on the particles and other *stability* conditions. But I will not do this now. Instead I will return briefly to electrodynamics and the question of its embedding into mechanics.

The problem of a reduction of the phenomena of electromagnetism to those of mechanics is as old as the theory of electromagnetism itself. Already Maxwell, one of the founders of modern electrodynamics tried to construct "mechanical analogies" (models) for the electromagnetic phenomena. When Hilbert entered the topic in 1902 with a lecture on "Mechanik der Continua" already a long list of (unsuccessful) attempts existed including (among others) such famous names as that of Kelvin, Hertz and Thomson, who tried to explain the existence of electrons and electromagnetic phenomena by a kind of circular motion, so-called *vortexes* in an incompressible fluid. Hilbert picked up this idea and generalised it to the notion of *absolute rotations* within an *elastic* and *incompressible ether* in analogy to the Newtonian '*Weltformel*' which unites the equations of motion and the law of gravitation in one formula (a system of four partial differential equations). Hilbert could in fact accomplish an *embedding* of electrodynamics into a peculiar system of classical mechanics exactly because he had a superb understanding of the calculus of variations and furthermore was in the process of developing new theory of integral equations already since 1902. This brings me to the second point: Hilbert's analysis of the principles of mechanics.

4.2. The axiomatic structure of mechanics and the logical analysis of its principles

If one compares Hilbert's lectures on mechanics with contemporary textbooks one can recognise several interesting points.[21] (1) Hilbert's lectures entail a kind of 'logical analysis' and comparison of the different laws and principles of mechanics such as Newton's laws of motion or the principle of least action and so on. (2) The laws and principles do not occur at random, in an *accidental* order so to speak, but in a peculiar, highly specific array. (3) This array is by no means the *historical* order of their discovery but seems to obey a different strategy. Hence, the question arises: *What* was the guiding idea of their arrangement if not the order of their discovery? According to which point of view arranged Hilbert the different principles of mechanics into a determined sequence? The short answer is roughly this: the sequence is ordered according to the "power of explanation" of the respective principles. What do we mean by "explanatory power"? In order to give an answer to this question we have to look to the sequence itself.

In the third extensive lecture on mechanics in winter-term 1905/06 we find the following sequence: it begins with a discussion of Newton's laws of motion for one-body systems with different boundary conditions, then follow the *minimal principles* of Gauß and Hertz, namely the principles of the *least constraint* and the *straightest path* [*kleinster Zwang* and *geradeste Bahn*] and continues with the deduction of the Lagrange-equations from Hertz' principle and closes with a detailed discussion of the *Integral-principles* of Hamilton, Euler and Jacobi.[22]

The guiding idea behind this arrangement seems to be the following: One starts with the most simple physical system, that of a 'free moving' particle on which certain constants, i.e., time-independent forces act. This is mainly what Newton did, when he considered apples *and* planets as falling under the *same* laws of motion. The next more complicated case is a system, in which the particles are *bound to move* on certain lines or surfaces. This is the case of Gauß and Hertz, which consider the constraint that the particles have to move on a given line (or surface) as an additional *condition*, comparable to "external forces" acting on the particle: Under all *possible* motions it is the *real* one, which suffers the least external constraint (Gauß) or takes the straightest path (Hertz). Eventually one considers even more complicated systems, in which the external conditions are not only spatial but also explicitly *time-dependent* like the motion of a ship under the changing influence of its rudder. This and similar cases are best treated by means of the *integral-principles* of Hamilton, Euler and Jacobi. Hence, a first answer to what 'explanatory power' means, is that the principles are arranged such that they can cover more and more *complex* physical cases regarding the internal—and external *constraints*. This raises the expectation that the explanatory more powerful principles *entail* the weaker ones. Whether this is indeed the case has to be studied separately, because the answer depends on a correct understanding of the calculus of variation and is therefore non-trivial.

Now, I have already mentioned that the calculus of variation was not in "good shape" at the turn of the century, when Hilbert began to lecture in physics. For this reason the question of the logical relations between the different principles could not be answered immediately. First the foundations of the calculus of variation had to be clarified, before anybody could answer with certainty a question like that: 'Does Gauß's principle of 'the least constraint' entail Lagrange's equations or not'. For this reason, Hilbert not only included 'the development of the calculus of variation' in his list of problems, but also set himself to work: Already in his first lecture on mechanics he declared that he would continue this lecture in the following summer (1899) under the title *Variationsrechnung*, which he in fact did. And again in the winter-term 1904/ 05, not long before he returned to mechanics the second time, he held a very elaborate and extensive lecture on the calculus of variation.

This lecture was extremely important, not only from the mathematical point of view, but also from a *physical perspective*: it had a significant influence on the next lecture on mechanics because it led to a thoroughgoing *rearrangement* of the principles of mechanics. In the lecture of 1905/06 we find for the first time a sharp *distinction* between *differential-* and *integral-principles* in mechanics. The principles of Gauß and Hertz belong to the first category, all other known principles to the second category (as the collective noun indicates). Now, some reader might think that this is only a *mathematical* difference, possessing little or no significance at all for our *understanding of physics*. But this would be a *mistake*, at least in my view. In order to make clear what the *logical* difference between these two kinds of principles is, and why this difference is also important for our understanding of physics, let me quote a relevant section from the lecture on mechanics in 1905/06, § 11, pp. 121-122:

> The principle, which we have taken so far as the only essential axiom, was Gauß's principle of the least constraint or.. the essentially equivalent principle of Hertz of the straightest path. ... both have in common that they are differential principles, which means: The accelerations are determined at a fixed instant of time t as functions of place and velocity by the request that an explicitly known function (the constraint) is made to a minimum; ..Both are therefore minimal-principles for functions of finite many (3) variables.

We now make an important step forward to a new class of principles of mechanics, which again are minimal-principles; but the essential New is that now certain determined integrals come into consideration, which depend on the course of the function of motion or the path-curve in a finite interval of time or space: Integral-principles:

A certain time interval (t_1, t_2) is considered and an integral of t is given in these limits, whose kernel (Integrandus) contains 3 functions $x(t)$, $y(t)$, $z(t)$ and their derivations; for every determined choice of these functions the integral takes on a numerical determined value, and the task is to determine (under certain additional conditions) 3 such functions that the integral takes on a minimal value.

This is a completely new, essentially more difficult task (than the determination of a differential-function in a single instant of time); we cannot fix now an instant of time t in such a way that as Unknown merely quantities occur – the values of functions in this instant –, but Unknown are functions in a whole interval, on whose complete course in the interval depends the expression, which has to be turned into a minimum. (my translation; for a better understanding see the whole quote in German below).[23]

Whether one agrees with Hilbert's position or not, one point is unquestionable: From the *logical point of view* there is a significant difference between the first and the second case. In the first instance one has to determine the *minimum of a known function* at a certain *instance* of time or space, whereas in the second case one has to determine the minimum of a *function of functions* in a certain time- or space *interval*, depending on the *value-course* of the *unknown* functions in the interval. The *logical* difference between the two cases could be not more impressive. Hence, the only question which remains to be answered is the question, whether the *logical* difference implies any difference with respect to our *physical* understanding of the two types of principles. My tentative answer is: Yes, there is a difference with respect to our understanding of the *physical* situation in both cases. In case of the differential-principles of Gauß and Hertz we are in agreement with the principle of '*causa-efficiens*' that the cause does not come after the effect, or at least we don't contradict it, whereas in the case of the integral-principles this is not so clear, because in this case we have *first* to calculate the integrals over a *finite* interval of time $t_1 - t_2$, and *then* to take the minimum (among all admitted functions) in order to identify the only *real* motion among all

possible motions in the interval. This procedure seems to be in much better agreement with the *teleological* principle of 'final causes' than with the
usual principle of 'causa efficiens'; according to the former a process is determined *now* by the need (or desire) to reach a certain aim in the *future*;
this seems to be precisely what we do if we use the integral-principles to
determine the *real* motion of a system of particles: we look at all *possible*
motions of all particles in a fixed time-interval and then take that motion as
real which makes the integral e.g. of the Hamilton-function to a minimum.

Whether the integral-principles have *ontological* implications or are only
calculating devices, 'mathematical instruments' so to speak, is an interesting
but difficult question, which I will not go into. Let me only mention that
H. Hertz believed that the integral principles violated the principle of *causa
efficiens* and for that (and some other) reasons didn't choose an integral– but
instead the differential-principle of the 'straightest path' as the only principle
of mechanics.[24]

Let me close the chapter on Hilbert's silent period with the remark that
the lectures on mechanics of this period entail many *new* axiomatic representations for old branches of physics, in particular those for continuum mechanics such as the theories of elasticity, hydrodynamics including transport-
equations, electrodynamics and the theory of electrons, and last not least of
thermodynamics. Unfortunately, still very little is known about these comprehensive chapters on almost all branches of physics; this roused the impression, as if Hilbert had no interest in physics before 1912, the year of his
first publication in physics. But the contrary is true; a long period of interest in classical physics came to a halt when Hilbert realised in 1911 that he
had to re-build his lecture on mechanics in spite of Einstein's new theory of
relativity. That he published his first paper in physics just in the next year
is from this perspective nothing but an *accident*. (But it is, of course, no
accident at all if we take the development of the theory of integral-equations
into account). This brings me to the next two periods.

5. HILBERT'S MOST EFFECTIVE PERIODS IN PHYSICS

Regarding the second and third period I can be rather brief because: (1) most
of the work from these periods has been published by Hilbert himself – if we
disregard for the moment the extensive lectures in physics from these periods; (2) most of his publications have been studied intensively (with one
exception, to which I'll come in a moment)—not only by the contemporary physicists but also by historians of science. Therefore I will confine
my remarks to a few points which can help to complete the picture and to

correct some popular misunderstandings with respect of Hilbert's contributions to physics. This will enhance, I hope, our present understanding of the
axiomatic approach to science.

Hilbert's physical papers of the second period (1912–1918) deal with
three rather different topics: (a) the foundations of the kinetic theory of
gases, (b) the theory of radiation, and most important (c) the theory of general relativity including the development of generalised field equations. The
reception of these papers by his fellow physicists was very different. The first
paper on the foundations of the kinetic theory of gases was largely ignored
by the physicists, although it was a very important paper. The second paper
on the theory of radiation involved him in a somewhat 'heated' dispute with
E. Pringsheim[25] about the *sense* and *significance* of an axiomatic presentation of a physical discipline. The third group of papers,[26] the two "notes"
on *The Foundations of Physics* were immediately absorbed by the leading
experts of the time—first and foremost by Einstein, but led simultaneously
to a certain 'tension' between Hilbert and Einstein.

If we ask in retrospect, with the wisdom of hindsight, what have these
papers in common (if they have something in common at all) then my answer would be the following: In every one of the three papers Hilbert tried
to close a certain 'logical gap' with respect to a flawless foundation of the
discipline in question. This is most obvious in case of the kinetic theory
of gases: Boltzmann's 'logical deduction' of the so-called 'master-equation'
from classical mechanics is from a rigorous logical point of view a 'pseudo-
deduction'; an *irreversible* equation like the master-equation cannot be deduced (in a strict logical sense) from a *reversible* theory—also not by incorporating a theory of probability.[27] Hilbert first clearly recognised this gap,
showing which 'presuppositions' were missing for a sound deduction and
then tried to 'close the gap' by using his new theory of integral-equations as
a means of finding a hierarchy of 'approximate solutions'. For this reason I
can't agree with S. Brush[28] if he blames Hilbert to play a mere 'mathematical
game'. In case of the second paper things are quite similar: Hilbert noticed
that the 'proofs' of Kirchhoff's *law of radiation*, given by Pringsheim and
Planck, were 'defective' (from a logical point of view) and tried to rem-
edy the situation by means of his theory of integral-equations. The subsequent debate became rather heated, because Pringsheim had put in question
whether Hilbert's paper was of any *physical* significance at all.

With respect to the third group of papers, the two notes on *The Foundations of Physics*, things are more complicated. At first glance it does not
look as if Hilbert is closing a 'logical gap'; the impression is much more
that he applies his mathematical skills in order to give a kind of 'a priori'

foundations of physics resulting in a certain type of *generalised field equations*. (This is at least a view many physicists shared.) One should, however, be aware that Hilbert had his own 'research program', so to speak, different from Einstein's approach to general relativity regarding a unification of Newton's gravitation and the electromagnetic phenomena. Hilbert's goal – standing in the tradition of G. Mie – was a *generalised field* theory of gravitation and electromagnetism that included a description of the electron as a 'material' object. And with respect to this goal a large gap had to be closed from Mie's original theory to the desired one.

I will not maintain that closing a 'logical gap' is the *single* aspect, under which Hilbert's physical papers of the second period can be seen; there is, of course, also his interest in the foundations of science. But one should be clear that both aspects are closely connected—both are driven by the axiomatic point of view. 'Closing a logical gap' is almost the same action as 'finding a hidden presuppositions', both result in a better *axiomatic presentation* of a field of phenomena.

Before I turn to the third period let me mention that Hilbert held eleven lectures in physics in the period from 1912 to 1918, i.e., ca. one per term, covering all areas of theoretical physics known in his time. Most remarkable are the two lectures on the *Foundation of Physics* in the summer- and winter-term of 1916 and 1916–17; they form the 'backbone' for the assertion that Hilbert pursued a *different* research-program than Einstein.

The third period from 1918 to 1930 is too complex and different in character to be treated closely in this essay, but two points must be mentioned, because they have some relevance for the last chapter, in which I consider Weyl's objection against an axiomatic point of view in science. Hilbert's interest in physics was not confined to the theory of general relativity; he also kept an eye on quantum theory and statistical physics in general. Now it was an 'open secret' that quantum theory was in an unsatisfactory state about 1920. But this was not the only problem according to Hilbert's view. The notion of probability and its application in physics was also rather unclear. Because Hilbert saw a cloth connection between these two problems, he first held in summer 1922 a lecture on statistical-mechanics before he after-wards turned to the mathematical methods of quantum theory in the winter-term 1922/23 and again in 1926/27. All three lectures deserve a distinguished place in the development of physics during the twenties. The first lecture is a jewel of logical clarity and physical distinctness; it entails a detailed *comparison* of Maxwell's, Boltzmann's and Gibbs' foundations of statistical mechanics; it's a paradigm-example for every philosopher of science interested in such questions. The second lecture is a careful description of the state of the art of the old quantum theory before the 'new quantum theory' had been developed by Heisenberg, Born, Jordan, Schrödinger, Dirac and others. The third lecture then gives a profound and comprehensive 'overview' of the new theory and its mathematical methods. It became the point of departure of von Neumann's book *Mathematische Grundlagen der Quantenmechanik* – a milestone in the axiomatic foundation of modern quantum theory.

Let me close this chapter on Hilbert's work in physics with the remark that I hope it has become clear by this review first *what* the axiomatic point of view is, and second what it is *good for* in physics. Nevertheless, the axiomatic point of view has been criticised as 'unsuited' for the aims and purposes of physics hampering more than promoting the progress of physics. Therefore we have to inquire:

6. THE LEGITIMACY OF AN AXIOMATIC POINT OF VIEW IN SCIENCE

Among the different critics of Hilbert's axiomatic approach to science is perhaps H. Weyl (beside Einstein) the most prominent, at least the best informed one with respect to Hilbert's way of thinking and working in science. Let me therefore begin the discussion of the legitimacy of an axiomatic point of view in physics with a quotation from Weyl's obituary at the occasion of Hilbert's death in 1943 published in the yearbook of the American Philosophical society:

> Already before Minkowski's death in 1909, Hilbert had begun a systematic study of theoretical physics, in close collaboration with his friend who has always kept in touch with the neighboring science. Minkowski's work on relativity theory was the first fruit of these joint studies. Hilbert continued them through the years, and between 1910 and 1930 often lectured and conducted seminars on topics of physics. [He greatly enjoyed this widening of his horizon and his contact with physicists, whom he could met on their own ground.] The harvest however can hardly be compared with his achievements in pure mathematics. The maze of experimental facts which the physicist has to take into account is *too manifold*, their expansion *too fast*, and their aspect and relative weight *too changeable* for the axiomatic method to find a firm enough foothold, except in the thoroughly consolidated parts of our physical knowledge.
>
> [Men like Einstein or Niels Bohr grope their way in the dark toward their conceptions of general relativity or atomic

structure by another type of experience and imagination than those of the mathematician, although no doubt mathematics is an essential ingredient. Thus Hilbert's vast plans in physics never matured.] (Weyl, GA, Vol. IV, p. 171 my emphasis)

This is, on the one hand, a nice corroboration of what I said about Hilbert's work in physics by one of the closest and best informed witnesses of the time. But, on the other hand, it also entails a sublime critique of Hilbert's axiomatic approach to physics. As most readers know Weyl has not always been on best terms with Hilbert: There has been a time when the pupil rebelled against his former teacher and tried (in tandem with Brouwer) to instigate a "revolution" with respect to the foundation of *mathematics*. But it's very important to note that Weyl had made his '*peace*' with Hilbert and his axiomatic point of view in mathematics *at the time of the obituary*. All the more astonishing is it that he remained skeptical regarding the fruitfulness and final success of the axiomatic approach with respect to physics.

I think one has to take Weyl's objection seriously and to ask what seems justified and where does it go astray possibly. If I had to judge in this delicate matter I would say the following:

As long as physics is a rapidly growing science, a science in quick progress, so to speak, like in the first decades of the 20th century, the *increase* in new experimental results is too fast and extensive, the *creation* of new theoretical ideas is so diverse and unpredictable, that one cannot expect that the axiomatic point of view can be applied with *success* to science, if one understands by 'success' that it *enhances the process of research*: the discovery of new facts and the invention of new ideas. So far I agree with Weyl's skepticism regarding an adoption of the axiomatic point of view in science: In fact an axiomatic presentation of a physical theory, whose 'life-span' is not more than some month, because it is 'overturned' by new discoveries and inventions, is of little or no value at all.

But one has to ask: Is this really the task for which the axiomatic point of view was invented? Is this the point Hilbert had in mind, when he stated in Paris the sixth problem? Is success (in the just stated sense) what Hilbert achieved in his own work, for example when he investigated the kinetic theory of gases or the general theory of relativity? I think not. Instead I suspect that Weyl's statement entails a sublime and to some extent dangerous misunderstanding.

It was never Hilbert's aim (perhaps with the exception of the first 'note' on the foundations of physics) to be at the *research-front* of physics and, more important, the *axiomatic method* was not designed for that goal. On the contrary, its task was a completely different one: it should promote the *logical analysis* of theories and enhance their axiomatic presentation; this means for example: It should facilitate the recognition of logical gaps in an existing theory such that the gap can be closed in the further development. Or it should improve our understanding of a theory, i.e., our knowledge of the logical dependence and independence of the principles of a certain discipline. Or it should made explicit the logical relations among different theories and their axiomatic presentations. In other words: the aim of [adopting] the axiomatic point of view in physics is not the progress of the research-front nor an extension of our empirical knowledge, but (like in mathematics) a *reflection* and *contemplation* of the logical structure of a given theory (taken from physics or elsewhere). This is to a good degree a *philosophical* task (and should not be condemned for this reason).

Of course, Weyl is correct if he supposes that the axiomatic approach makes *more* sense in cases where a physical theory is already 'thoroughly consolidated' than in cases where the theory is still 'in flux'. But let me stress, the axiomatic point of view does not become *senseless*, as Weyl seems to suggest, if a theory is still in a state of change and reconstruction, because a *real* progress frequently first becomes possible, if the 'hidden' logical gaps and other errors such as inner contradictions have been identified and eliminated. Let me close with a quote from A. S. Wightman's paper on Hilbert's sixth problem in which he makes this explicit by referring to the development of quantum-mechanics in the twenties of the last century. (I could not express my views better than he did 30 years ago.)

> A great physical theory is not mature until is hat been put in a precise mathematical form, and it is often only in such a mature form that it admits clear answers to conceptual problems. In this sense, although quantum-mechanics was discovered in 1925-6, it did not become a mature theory until the appearance of von Neumann's book [1932]. Thus, although von Neumann had nothing to do with the *discovery* of quantum-mechanics, he had a great deal to do with the *creation* of quantum mechanics as the mature theory we know today. In this extent, Hilbert's axiomatic approach showed itself important for physics. (in F. E. Browder *Mathematical Developments Arising from Hilbert Problems* (1976) Providence, p. 158, *Italics* mine).

Nachwort: This essay was written before the book *David Hilbert and the Axiomatization of Physics (1898–1918)* by L. Corry had appeared. Many of the historical points, which I mentioned, are also discussed in the book. However, I saw no reason to change the central claims of my essay.

University of Göttingen
Germany

NOTES

[1] This is the first sentence of H. Hertz's famous introduction to his book *Die Prinzipien der Mechanik* of 1894.

[2] See H. Putnam's essay "What Theories are Not" and F. P. Ramsey's paper "Theories".

[3] There are, of course, also the important contributions of the Italian School. However, these will be ignored in this essay because they did not have (for whatever reasons) the same public efficacy as Hilbert's contributions.

[4] Just to remind the reader: Hermann Weyl was a former disciple of Hilbert. He became the most severe critic of Hilbert's "formalistic" approach in the twenties regarding a *New Foundations of Mathematics* (such the title of Hilbert's programmatic paper). This should, however, not be confused with Weyl's objections against the axiomatic approach in general and the axiomatic presentation of physical theories in particular. Regarding the former he was not dogmatic and confessed later that both approaches, the axiomatic and the constructive one, are equally legitimate (although his heart pulsed further on the constructive side). This confession fits extremely well with the little known fact that Weyl held a two-terms lecture in the beginning of the thirties (when he had become the successor of Hilbert in Göttingen) with the title "Axiomatik" in which he followed Hilbert's trail. In the presentation of Weyl's position I will ignore this intermediate approximation to Hilbert's axiomatic point of view and concentrate primarily on his objections against the axiomatic presentation of physical theories.

[5] Hilbert's first lecture on geometry dealt with projective geometry in a quite traditional manner, without the slightest indication that he would adopt an axiomatic point of view in the years ahead.

[6] 'logical perfection' roughly means the following: the wanted axiom system shall be (deductively) *complete* and logically *simple* in the sense that it entails *no superfluous* assumptions; the first requirement implies that no assumption is missing; the second that all axioms are logically independent of each other.

[7] There are, of course, many different misunderstandings, which I'll not discuss in detail. It begins with Frege's reading of GG as a second-order theory of logic and culminates in Einstein's dictum that "nur das Logisch-Formale gemäß der Axiomatik den Gegenstand der Mathematik [bildet], nicht aber der mit dem Logisch-Formalen verknüpfte anschauliche oder sonstige Inhalt". (Einstein, Mein Weltbild, p. 120) In the following I will only present a kind of synoptic sketch of the common misunderstandings.

[8]"Geometry, like arithmetic, requires only a few and simple principles for its logical development. These principles are called the axioms of geometry. The establishment of the axioms of geometry and the investigation of their relationships is a problem which has been treated in many excellent works of the mathematical literature since the time of *Euclid*. This problem is equivalent to the logical analysis of our *spatial intuition*. The present investigation is a new attempt to establish for geometry a *complete*, and as *simple* as possible set of axioms and to deduce from them the most important geometric theorems in such a way that the meaning of the various groups of axioms, as well as the significance of the conclusions that can be drawn from the individual axioms, come to light".

[9]The core of Frege's logical interpretation is this: He regards words like 'point', 'straight line' and 'between' not as descriptive expressions with a specific sense and reference in his, Frege's sense (i.e., as denoting n-place truth-functions) but as second-order *variables* whose universe of discourse is the set of all 'thinkable' n-place predicates (truth-functions). On this reading Hilbert's axioms become 'logical relations' between second order variables for n-place truth-functions (predicates) of first order.

[10]The question of the 'choice of the axioms' has to be stated a bit more precisely: First, even in the old view of Euclid there is a certain freedom of choice because only the axiom-system as a whole is constrained by the specific content, and second, also on the new point of view the choice of axioms is not *completely* arbitrary, because the axiom-system as a whole has to be logically consistent.

[11]There are several good books about Euclid's *Elements* and his constructive procedure: Ian Mueller *Philosophy of Mathematics and Deductive Structure in Euclid's 'Elements'*, MIT-Press (1981); B. Artmann *Euclid: The Creation of Mathematics*; Springer (1999).

[12]Hilbert (1900) "Through the investigations of the foundations of geometry we become confronted with the task to treat those physical disciplines axiomatically, according to the model of geometry, in which already today mathematics plays a prominent role: these are in the first line the calculus of probability and mechanics". (Hilbert, Gesammelte Werke, Bd. 2, p. 290).

[13]See the lectures on 'Projective Geometry' (1891) and on 'The Foundations of Geometry' (1894), where Hilbert stresses the close connection between geometry and physics. See also my paper "Geometry, Intuition and Experience" where I explain why this view is compatible with the idea that geometry is a mathematical discipline.

[14]It has been noticed that the sixth problem has a peculiar position among the list of 23 problems. It is neither a problem of *pure* mathematics (as most of the other problems), nor of *applied* mathematics (like some of the later problems) but a problem of the *applicability* of a logical (mathematical) method to a natural science, which is fundamentally different from mathematics. (This is at least the standard view around 1900.) Furthermore, it's the only problem (beside the last problem regarding the calculus of variations), which has no 'determined' solution; like the last problem it's more a *research program* than a definitive problem. For these (and

some other) reasons the sixth problem has been taken in with reservations. Most physicists have been – and presumably still are - skeptical regarding the soundness of the problem and the significance of its eventual solution.

[15]Except Volkmann, these physicists are known as the last representatives of the classical, pre-relativistic period.

[16]"So regt uns beispielsweise das Boltzmannsche Buch über die Prinzipe der Mechanik an, die dort angedeuteten Grenzprozesse, die von der atomistischen Auffassung zu den Gesetzen über die Bewegung der Kontinua führen, streng mathematisch zu begründen und durchzuführen. Umgekehrt könnte man die Gesetze über die Bewegung starrer Körper durch Grenzprozesse aus einem System von Axiomen abzuleiten suchen, die auf der Vorstellung von stetig veränderlichen, durch Parameter zu definierenden Zuständen eines den ganzen Raum stetig erfüllenden Stoffes beruhen – ist doch die Frage nach der Gleichberechtigung verschiedener Axiomensysteme stets von hohem prinzipiellen Interesse." (Hilbert, Mathematische Probleme, 1900).

[17]Initially, the problem was only mentioned in the introduction to the Paris-lecture. Eventually Hilbert attached it as the 23rd problem to his list of 10 or 11 problems he had actually talked about in Paris. For more historical details see Rüdiger Thiele: "Über die Variationsrechnung in Hilberts Werken zur Analysis" in N. S. 5 (1997).

[18]The number depends (among other things) on the question whether one counts a lecture extending over two terms as one or two lectures; I have counted a two-term lecture as two lectures

[19]The classical law of the addition of velocities is not, and cannot be valid in electrodynamics in spite of the fact that the velocity of light is a finite, reference-system invariant constant and (at the same time) an upper bound for the velocity of any propagation. This implies the necessity of a generalisation of Galilei's principle of relativity of rest and motion.

[20]See my "Hilbert's Criticism of Poincare's Conventionalism" in *Henry Poincarè: Science and Philosophy*, ed. Gerhard Heinzmann et al., Berlin 1996. A more detailed analysis is given in "The Refutation of Conventionalism" in Majer & Sauer "Intuition and Axiomatic Method in Hilbert's Foundation of Physics", in *Intuition and the Axiomatic Method*, ed. R. Huber & E. Carson, Springer 2006.

[21]Hilbert himself presents a list of textbooks (Lehrbücher) in his first lecture on mechanics (1898/99), which entails among others books by Lagrange Jacobi, Kirchhoff, Helmholtz, Hertz, Boltzmann, Mach, and Poincaré.

[22]The reader should be aware that this is not the historical order. Although there are some points of 'coincidence' with the historical order, for example Newton's laws of motion, Hilbert's array obeys a different point of view.

[23]Das Prinzip, das wir bisher als einziges wesentliches Axiom genommen hatten, war das Gaußsche Prinzip des kleinsten Zwanges, oder das im wesentlichen für eine gewisse Gruppe von Bewegungen gleichwertige Hertz'sche Prinzip der geradesten Bahn. Dem Charakter nach haben beide das gemein, daß sie Differentialprinzipien sind, d.h. zu einem fest gewählten Zeitpunkt t die Beschleunigungen x", y", z"

als Funktionen des Ortes x, y, z und der Geschwindigkeit x', y', z' durch die Forderung bestimmen eine explizit bekannte Funktion von x", y", z" (den Zwang) ... zum Minimum zu machen; beim Hertzschen Prinzip gilt die Betrachtung für einen bestimmten Punkt s der Bahnkurve; an Stelle der zeitlichen Ableitungen treten die nach s, und an Stelle des Zwanges die Krümmung, die bei variablem d^2x/ds^2, d^2y/ds^2, d^2z/ds^2, zum Minimum zu machen ist. Beides sind also Minimalprinzipien für Funktionen von endlich vielen (3) Variablen, deren Theorie bereits in der elementaren Differentialrechnung enthalten ist; die Hauptsache ist, daß der Charakter dieser Variablen als unbekannter Funktion von t bzw. s gar nicht in Betracht kommt, indem man nur einen festen Zeitpunkt t bzw. Stelle s betrachtet.

Wir machen nun einen wichtigen Schritt vorwärts zu einer neuen Klasse von Prinzipien der Mechanik, die bemerkenswerterweise wieder Minimalprinzipien sind; das *wesentlich Neue* aber ist, daß jetzt gewisse bestimmte *Integrale* in Betracht kommen, die von dem Verlauf der Bewegungsfunktion oder Bahnkurve in einem *endlichen Zeit- oder Raumintervall* abhängen:

Integralprinzipe: Es wird etwa ein Zeitintervall (t_1, t_2) betrachtet und es ist ein Integral nach t in diesen Grenzen gegeben, dessen Integrandus 3 Funktionen x(t), y(t), z(t) und deren Ableitungen enthält; für jede bestimmte Wahl dieser Funktionen erhält das Integral einen numerisch bestimmten Wert, und die Aufgabe ist, unter gewissen hinzutretenden Bedingungen solche 3 Funktionen zu bestimmen, daß das Integral einen Minimalwert erhält. *Das ist eine ganz neue wesentlich schwerere Aufgabe; wir können jetzt nicht mehr eine Zeit t so festlegen, daß als Unbekannte lediglich Größen – die Funktionswerte an dieser Stelle – erscheinen, sondern unbekannt sind Funktionen in einem ganzen Intervalle, von deren ganzem Verlauf daselbst der zum Minimum zu machende Ausdruck abhängt.* Diese Prinzipe verschaffen dafür aber auch einen viel tieferen Einblick in alle in Betracht kommenden Fragen und bilden den eigentliche Kern der Mechanik; ...

Die mathematische Disziplin, die die Theorie dieser neuen Aufgabe enthält, ist die Variationsrechnung, die so der Differentialrechnung als Fortführung und Verallgemeinerung an die Seite tritt; was diese für die Lehre von den Größen und den Minimalwerten der *Funktionen von Größen* bietet, das tut jene für die Minima der *Funktionen von Funktionen*, darunter [werden] Ausdrücke verstanden, die von dem Verlaufe unbekannter Funktionen in *ganzen Intervallen*, nicht bloß an *einzelnen Stellen* abhängen. Wir werden uns später zum tieferen Verständnis der Mechanik notwendig eingehender mit den Theorien der Variationsrechnung befassen müssen; (the underlinings are Hilbert's; the italics are mine).

[24] See Hertz "Die Prinzipien der Mechanik" (1894) pp. 27, where Hertz discusses the *admissibility* of Hamilton's principle and considers the 'metaphysical' objection that Hamilton's integral-principle violates the law of 'cause and effect' insofar "it's making the present motions dependent on consequences which can first arise in the future, and therefore (not only) imposes intentions on the lifeless nature, but, what is much worse, it imposes senseless intentions to nature". Hertz then discusses "the usual answer, which the present physics has in stock against such attacks", but,

interestingly enough, Hertz sides more with the metaphysical attack than with the physical defense.

[25]In this debate several physicists were involved, among others C. Caratheodory and Max Planck.

[26]The numbering depends on the art of counting: I take the convolute of papers to the same topic as unit.

[27]See my paper "Lassen sich phänomenologische Gesetze "im Prinzip" auf mikrophysikalische Theorien reduzieren?" in *Phänomenales Bewusstsein*, ed. by M. Pauen & A. Stephan, mentis-verlag 2002.

[28]See his book S.G. Brush *The Kind of Motion called Heat*, North-Holland Publishing House (1976).

REFERENCES

Boltzmann, L. (1897). *Vorlesungen über die Principe der Mechanik*, Vol. 1., Johann Ambrosius Barth, Leipzig.

Boltzmann, L. (1904). *Vorlesungen über die Principe der Mechanik*, Vol. 2., Johann Ambrosius Barth, Leipzig.

Brush, S. G. (1976). *The Kind of Motion called Heat*, North-Holland.

Einstein, A. (1934). *Mein Weltbild*, Ullstein, Berlin. Ullstein Buch Nr. 35024, 1984.

Frege, G. (1903). Über die grundlagen der Geometrie, *Jahresberichte der DMV* **12**.

Hertz, H. (1894). *Die Prinzipien der Mechanik*, Johann Ambrosius Barth, Leipzig. From *Gesammelte Werke*, vol. III.

Hilbert, D. (1891). Projektive Geometrie, in (Hilbert, 2004).

Hilbert, D. (1893/94). Die Grundlagen der Geometrie, in (Hilbert, 2004).

Hilbert, D. (1898). Mechanik der Continua. Unpublished manuscript, Niedersächsische Staats- und Universitätsbibliothek, Cod.Ms. D. Hilbert, 553.

Hilbert, D. (1898/99). Euklidische Geometrie, in (Hilbert, 2004).

Hilbert, D. (1899). Grundlagen der Geometrie, in (Hilbert, 2004). Festschrift zur Feier der Enthüllung des Gauss-Weber-Denkmals in Göttingen.

Hilbert, D. (1900/32). Mathematische Probleme, *Gesammelte Abhandlungen*, Vol. 2., Springer, Berlin. Lecture held at the international Congress of Mathematics in Paris 1900; first published in "Nachrichten von der königl. Ges. der Wissenschaften zu Göttingen", 253-297.

Hilbert, D. (1902). Grundlagen der Geometrie, in (Hilbert, 2004).

Hilbert, D. (1905/06). Mechanik der Continua. Unpublished manuscript, Library of the Mathematical Institute, University of Göttingen, (henceforth: MI).

Hilbert, D. (1911). Mechanik der Continua. Unpublished manuscript, MI.

Hilbert, D. (1911/12). Kinetische Gastheorie. Unpublished manuscript, MI.

Hilbert, D. (1912). *Grundzüge einer allgemeinen Theorie der linearen Integral-gleichungen*, Teubner, Leipzig/Berlin.

Hilbert, D. (1915). Die Grundlagen der Physik, *Nachrichten von der königl. Gesellschaft der Wissenschaften zu Göttingen* pp. 395–407.

Hilbert, D. (1915/16). Grundlagen der Physik. Unpublished manuscript, MI.

Hilbert, D. (1926/27). Mathematische Methoden der Quantentheorie. Unpublished manuscript, MI.

Hilbert, D. (1932). *Gesammelte Abhandlungen*, Vol. 1–3, Springer, Berlin.

Hilbert, D. (1987). *Grundlagen der Geometrie*, 13. edn, Teubner, Stuttgart. Mit Anhängen und Supplementen, ed. by Paul Bernays.

Hilbert, D. (2004). *David Hilbert's Lectures on the Foundations of Geometry: 1891–1902*, Springer, Berlin. Edited by M. Hallett & U. Majer.

Mach, E. (1889). *Die Mechanik in ihrer Entwickelung historisch-kritisch dargestellt*, F.A. Brockhaus, Leipzig.

Majer, U. (1996). Hilbert's Criticism of Poincaré's Conventionalism, *in* G. H. et al. (ed.), *Henry Poincaré: Science and Philosophy*, Berlin.

Majer, U. (1998). Heinrich Hertz's Picture-Conception of Theories, *in* D. B. et al. (ed.), *Heinrich Hertz: Classical Physicist, Modern Philosopher*, Kluwer, Dordrecht.

Majer, U. (2001a). The Axiomatic Method and the Foundations of Science: Historical Roots of Mathematical Physics in Göttingen, *in* M. Redei and M. Stöltzner (eds), *John von Neumann and the Foundations of Quantum Physics*, Kluwer, Dordrecht.

Majer, U. (2001b). Lassen sich phänomenologische Gesetze 'im Prinzip' auf mikrophysikalische Theorien reduzieren?, *in* M. P. &. A. Stephan (ed.), *Phänomenales Bewußtsein: Rückkehr zur Identitätstheorie?*, Mentis-Verlag, Paderborn.

Majer, U. (2003). Husserl and Hilbert on Geometry, *in* R. Feist (ed.), *Husserl and the Sciences*, University of Ottawa Press.

Majer, U. and Sauer, T. (2006). Intuition and Axiomatic Method: Hilbert's Foundation of Physics, *in* R. H. &. E. Carson (ed.), *Intuition and the Axiomatic Method*, Springer, Dordrecht.

Neumann, J. (1932). *Mathematische Grundlagen der Quantenmechanik*, Springer, Berlin.

Putnam, H. (1962). What theories are not, *in* E. Nagel, P. Suppes and A. Tarski (eds), *Logic, Methodology and Philosophy of Science*.

Ramsey, F. (1929/78). Theories, *in* D. H. Mellor (ed.), *Foundations*.

Sauer, T. (1999). The Relativity of Discovery: Hilbert's first note on the foundations of Physics, *Arch. Hist. Exact Sci.* pp. 529–575.

Thiele, R. (1997). Über die Variationsrechnung in Hilberts Werken zur Analysis, published in N.S.5.

Weyl, H. (1933). Axiomatik, unpublished manuscript, MI.

Weyl, H. (1987). *Gesammelte Abhandlungen*, Vol. 1–3, Springer, Berlin.

Wightman (1976). Hilbert's Sixth Problem: Mathematical Treatment of the Axioms of Physics, *in* F. E. Bowder (ed.), *Proceedings of the Symposium in pure Mathematics of the American Mathematical Society.*

HELMUT PULTE

THE SPACE BETWEEN HELMHOLTZ AND EINSTEIN: MORITZ SCHLICK ON SPATIAL INTUITION AND THE FOUNDATIONS OF GEOMETRY[0]

1. BIOGRAPHICAL INTRODUCTION

Comparable to Rudolf Carnap who – in his last letter to Otto Neurath in 1945[1] – named Neurath the "big locomotive" of the "Vienna circle-train", Moritz Schlick can perhaps be called both the *designing engineer* and the *engine-driver* of this train: From 1924 to 1929, the Vienna group around Schlick was the germ of what *later* became known as the Vienna school of logical empiricism or neo-positivism and shaped its philosophical programme considerably. Carnap, in his autobiography, remarked that Schlick's *Allgemeine Erkenntnislehre* (Schlick, 1918) anticipated a number of the circle's later philosophical achievements and formed the nucleus of many of its formal elaborations. On the other hand, Schlick's assassination in 1936 marked the end of the public period, i.e., the organised activities and broad perception of the Vienna circle. Beyond doubt, *he* was the one who kept successfully integrated the divergent philosophical and ideological tendencies of the group, and therefore became the "father figure" of the movement.

This paper will concentrate on *one* period and *one* aspect of Schlick's work that can perhaps best illustrate the "Interaction between Mathematics, Physics and Philosophy" within *early* logical empiricism *before* Vienna: The development of Schlick's understanding of *intuition*, especially of *spatial intuition*, in connection with the scientific development of his *early* career. The main object is not a systematic critique of his concept of intuition, but a historical investigation of its changing meaning and relevance in Schlick's scientific thinking. As the structure of this paper is (by and large) shaped by his different publications at different times from 1910 to 1921, it will start with a few *biographical* and *bibliographical* remarks:

Schlick was educated as a physicist in Berlin. As one of his doctor-father M. Planck's favourite disciples[2] he wrote his thesis *Über die Reflexion des Lichtes in einer inhomogenen Schicht* (Schlick, 1904), which became his first publication. Schlick then turned to philosophy. As a rare exception within the Vienna-circle he did not confine his interests to philosophy of science and epistemology, but also published on ethics and practical philosophy in general (see, for instance, his "Lebensweisheitslehre" (Schlick, 1908) and "Fragen der Ethik" (Schlick, 1930b)). His first publications concerning our subject, however, include his *Habilitationsschrift*, published as "Das Wesen

V.F. Hendricks, K.F. Jørgensen, J. Lützen and S.A. Pedersen (eds.), Interactions: Mathematics, Physics and Philosophy, 1860-1930, pp. 185–206.
© *2006 Springer.*

der Wahrheit nach der modernen Logik" (Schlick, 1910) and his essay "Gibt es intuitive Erkenntnis?" (Schlick, 1913).

The *second* part of this paper presents a discussion of these and other early publications of Schlick. The focus will be on Schlick's understanding of *intuition* as a guarantee of evidence and certainty of our knowledge (a subject which is, of course, of special importance to mathematics). This critical survey will end with Schlick's main philosophical work (Schlick, 1918 [Engl. 1974]) *Allgemeine Erkenntnislehre* ("General Theory of Knowledge").

Next to Philipp Frank, Schlick was the first member of the Vienna Circle who recognized the philosophical relevance of Einstein's two theories of relativity and tried to work out its implications with respect to space and the foundations of geometry. Starting with "The Philosophical Significance of the Principle of Relativity" (Schlick, 1915), he published from 1915 onwards a number of papers on this subject which Einstein himself at several times praised as the best *philosophical* interpretations of his ideas at all.[3] Well known became Schlick's booklet "Space and Time in Contemporary Physics" (Schlick, 1917), which achieved four editions from 1917 to 1922. In this year (1922) he succeeded Mach and Boltzmann on the chair for the Philosophy of the Inductive Sciences at the University of Vienna. Even more than Frank's papers, "Space and Time in Contemporary Physics" made relativity theory popular in early logical empiricism. Without doubt, Schlick very much contributed to appoint Einstein (next to Russell and Wittgenstein) to one of the three "leading representatives of the scientific world view", as they were called in the Vienna circle-programme from 1929.[4] Schlick's interpretation of Einstein's theories – as far as his concept of space and his understanding of geometry is concerned – will be discussed in the *third* part.

Though Schlick must have started to study Helmholtz sometime before 1915 (see (Schlick, 1915, p. 150)), his first and only detailed discussion of Helmholtz's views on geometry is from 1921, the one-hundredth anniversary of his birth. On this occasion, Schlick and Paul Hertz published Helmholtz's *Schriften zur Erkenntnistheorie* (Helmholtz, 1921). Schlick himself added extensive and detailed comments on two of Helmholtz's most important papers concerning our subject: "Über den Ursprung und die Bedeutung der geometrischen Axiome" (Helmholtz, 1870) and "Die Tatsachen in der Wahrnehmung" (Helmholtz, 1878). In 1922, a collection of lectures on Helmholtz, given on the same occasion, appeared; they include Schlick's "Helmholtz als Erkenntnistheoretiker" (Schlick, 1922). His comments and critical discussions will be shortly examined in the *fourth* part, though *intuition* in his *early* period will remain the main subject.

2. SCHLICK'S BASIC EPISTEMOLOGICAL PROBLEM: INTUITION, EVIDENCE AND CERTAINTY OF KNOWLEDGE

From the very beginning, Schlick's discussion of *intuition in general* is linked to what he later calls the "fundamental problem of epistemology" or the "problem of absolutely certain knowledge"[5]: In the history of philosophy, intuition is usually regarded not only as a source of knowledge, but as a source of evident, indubitable, and *certain* knowledge. And this, of course, will be the link to Schlick's discussion of geometry: The claim both of traditional rationalism as well as of Kantian apriorism is that *this* kind of privileged knowledge is realized in *mathematics*, and that its essential features – evidence and certainty – can be transferred to *empirical* knowledge by applying mathematics.

Already Schlick's paper "On the Nature of Truth in Modern Logic" (Schlick, 1910), contains the central argument that he will later use repeatedly in order to demolish what he labels the "theory of evidence"[6], i.e., the view that true knowledge consists entirely in or depends essentially on an immediate and *evident* experience called intuition (*Anschauung*). Schlick does not deny evidence as a psychological entity, and he agrees with Wilhelm Wundt's attitude that "all immediate evidence has intuition as its source" (Schlick, 1910, p. 441). But the claims of the "theory of evidence" in question go further: According to it, evidence is the decisive (if not the only) criterion of truth. Against *this* view Schlick argues that evidence is nothing but a subjective feeling underlying complex and opaque psychological influences not accessible to objective control. As the history of philosophy and the history of science have shown, supposed evidence has frequently led to erroneous (and even absurd) conclusions.

In order to save evidence in the light of such experiences as a meaningful concept, it would have to be disentangled from certainty (*Gewißheit*): Now, one might say that a proposition that seemed to be evident, but later turned out to be wrong was only experienced as (or felt) to be *certain* (and not as truly *evident*). According to Schlick, two possibilities arise from this expected loophole: Either true evidence and deceiving certainty are experienced as basically different. This would mean that a criterion exists that distinguishes evidence and certainty—a criterion, of course, which may not be *post hoc*, in order to avoid circularity of the argument.

But in this case deception by evidence (*Evidenztäuschung*) could not happen at all, and the introduction of an attribute "certainty" *different* from the attribute "evidence" would be entirely meaningless.

Or, on the other hand, the experience of evidence and the experience of certainty are *not* basically different. In this case only *later* investigations may reveal whether certainty *with* evidence or certainty *without* evidence was experienced. But this possibility would admit that the immediate experience of evidence is no criterion of truth *at all*, and that such a criterion has to be found in *later* experiences. But *these* can not be experiences of evidence in order to avoid circularity. This means that the separation of evidence and certainty fails in this case, too. To sum up Schlick's argument: Evidence can be *no* criterion of truth, and there is no immediate experience or intuition as a source of truth available to us, either.[7]

Evidence is also unnecessary as an *additional* characteristic of some propositions next to truth. Schlick asks: "What point is there in establishing self-evidence if we can verify the truth of a judgement directly by the presence of its essential features"? (Schlick, 1918, p. 131; [Engl. 1974], p. 150). This is a problem of his theory of verification, where he strongly parallels factual truths (*Tatsachenwahrheiten*) and conceptual truths (*Begriffswahrheiten*) (Schlick, 1910, pp. 435–458, esp. p. 445). Of special importance in this context is his criterion of truth for logical and mathematical propositions or, as Schlick puts it, for conceptual truths. Here we find that evidence still plays a more prominent role than he is willing to admit in his former criticism of the so-called 'theory of evidence'.

If we desist from the question of truth-transfer to propositions by deduction and restrict our attention to the truth of first premises or *axioms*, Schlick's view seems to be roughly this:

Axioms can not be characterized by a conceptual necessity (*begriffliche* [. . .] *Notwendigkeit*) (Schlick, 1910, p. 441). (Euclid's parallel postulate, for example, is not necessary in this sense, as is shown by the existence of non-Euclidean geometries.) Axioms can neither be characterized by evidence in the sense of an immediate experience of their truth. They *need to be* exemplified, but they also *can* be exemplified, and this is the point where *intuition* becomes important. The exemplification consists in the translation of general and abstract concepts into concrete and intuitive ideas (*Vorstellungen*). Only at *this* level intuition can gain evidence and clarity (Schlick, 1910). The truth of an axiom, however, is not comprehended by an immediate experience of evidence of this kind, but by a more complicated process of *identification*:

The general proposition is applied to an intuitive example, which gives occasion for an immediate experience of a certain fact (*unmittelbar erlebter Tatbestand*) which is expressed in a second proposition. And it is the identity of both propositions that brings about evidence and verifies the axiom. Schlick indeed draws a parallel between verification of factual propositions by outward experience and verification of logical and mathematical propositions by intuition, which he also labels "inward experience", "immediate experience" or "perfect experience" (Schlick, 1910, pp. 447f.). But this parallel stresses their *process character* rather than their *outcome*: Verification of axioms does not mean a *foundation* of their truth in inner experience (comparable to verification of a factual proposition), but rather a *recognition* (*Konstatierung*) of their truth in inner experience. Schlick *sharply separates* the epistemological status of the results in both cases:

> [...] full certainty [...] is not possible in the case of factual propositions, [...] because their verification depends on perception, in short on the outward world, [whereby] its laws are never known to us completely and with certainty. In the case of propositions about the relations of concepts, however, no perceptions are needed; they are verified by immanent processes, so to speak by tools which always accompany our mind; they are known to it more perfect than the relations of real things, and that is why they can be absolutely true to [our mind]. (Schlick, 1910, p. 447)

Though Schlick appeals to the works of Hilbert, Russell and Couturat and indicates that Kant's theory of intuition and mathematical certainty will no longer be acceptable in the light of their logical foundation of mathematics, his own understanding of so-called 'conceptual propositions' shows some Kantian reminiscences . In agreement with this seems to be, for example, his support of Poincaré's view that mathematical *induction* will never be reducible to logic, but is rather a necessity of our thinking, rooted in the nature of our understanding: Here, too, "we find the ‚eternal verities‘ founded in inner or perfect experience" (Schlick, 1910, p. 453).[8]

But over the next few years, Schlick's assessment of intuition and his philosophy of mathematics in general will change, and the more or less implicit 'Kantian connection' of intuition and the truth of conceptual propositions will vanish. Within this process two different phases can be separated: The first one (from about 1913 to 1917), is a more 'destructive‘, the second one (from 1918 onwards) is a more 'constructive‘ period. Schlick's reception of Einstein's doctrine seems to me of interest for both periods, though his departure from intuition obviously has internal epistemological reasons, too.[9] Relativity theory, he remarks in his paper "Die philosophische Bedeutung des Relativitätsprinzips" with respect to Kant's understanding of spatial and temporal intuition, "forces us to wake up from a little dogmatic slumber" (Schlick, 1915, p. 153).

The more 'destructive' period: Already in his paper "Gibt es intuitive Erkenntnis?" from 1913 Schlick makes clear "that the deepest insights of the present time, especially in theoretical physics have shown that now and again the intuitive representation [of knowledge] has to be abandoned, just in order to preserve knowledge in its whole purity" (Schlick, 1913, p. 485). He has not only spatial intuition in mind here, which becomes his main subject with Einstein's general theory of relativity from 1915 onwards. His general point is that Einstein's special theory of relativity violates certain presumingly intuitive and evident assumptions of Newtonian absolute space and time, and he now sharply criticises Kant for trying to give a philosophical foundation for these assumptions by mixing up conceptual knowledge of physical objects with the pure intuition of space and time. (Schlick, 1913, p. 485) It seems to be a plausible conjecture that *via* his reception of Einstein's special theory of relativity *Kant* became the main target of his reflexions on intuition from 1913 onwards, while other philosophers mentioned (as Husserl and Bergson, for example) were of secondary importance.

Schlick's destruction of intuition as an epistemologically relevant concept is linked to his earlier definition of knowledge and his sharp *demarcation* of knowledge and intuition: Knowledge is always conceptual; I *know* an object if I designate it by concepts in a univocal manner, and then integrate these concepts into the whole net of concepts and judgements already established. Hence, knowledge *always* demands two elements: something that is recognized (*etwas, das erkannt wird*) and the thing as *what* it is recognized (*dasjenige, als was es erkannt wird*) (Schlick, 1913, p. 485). Knowledge is neither an immediate experience nor a mental representation of reality, but consists entirely in the unambigious relation between objects and concepts.[10] These concepts are no images or pictures but mere *signs* – an understanding of concepts which is strongly influenced by Helmholtz's theory of signs.[11] Though Schlick's reception and modification of this theory is beyond the scope of this paper, some implications concerning his understanding of *intuition* should be considered:

First, it would contradict the essentially *symbolic* character of knowledge[12] to assume something like *intuitive knowledge*. Intuition "survives" only as a (more or less immediate) experience (*Erleben*) of a single object, and has nothing to do with the comparing, ordering and relating activity that gains proper knowledge. Therefore, Schlick's (short and concise) conclusion runs as follows: "*Intuitive* knowledge is *a contradictio in adiecto*." (Schlick, 1913, p. 481)

Schlick does *not* dissociate himself explicitly from his *former* use of intuition, which was meant to secure the truth of the propositions of logic and mathematics. He just mentions casually that the talk of "intuitive knowledge" might be perhaps justified in those propositions where (in the process of thinking) concepts can be represented by intuitive ideas (*anschauliche*

Vorstellungen). (Schlick, 1913, p. 484f.) But there is no allusion to (or revival of) his earlier view that truth and evidence of logical or mathematical axioms might be gained by the pretty weird process of identification sketched earlier. Quite contrary, Schlick insists on the purely *psychological* character of this kind of intuition, which is "comparable to the colouring in order to display the details of microscopical objects" (Schlick, 1913, p. 485). And he emphasizes that modern developments, especially in physics, have shown that fundamental new insights sometimes require the *renunciation* of intuitive support.

I call this first period *destructive* because intuition is deprived of any epistemological relevance. It is, as he says, "quite the opposite of knowledge" (Schlick, 1913, p. 486). But on the other hand, Schlick obviously sticks to the conviction that the propositions of pure mathematics and the mathematical sciences must be somehow epistemologically privileged, i.e., characterized by truth and (at least in the first area) by certainty. With intuition lost, they can no longer be regarded as synthetic a priori in Kant's sense. But what else can be their distinctive characteristic? With respect to this problem, Schlick's papers from 1913 onwards are rather disappointing, and it is not before his publication of *Allgemeine Erkenntnislehre* in 1918 that he treats it in a more *constructive* manner. But even during the period of destruction, Schlick's rejection of Kant's theory of intuition remains *ambiguous*:

In his discussion of Richard Hönigswald's interpretation of Einstein's theory of special relativity[13], given in his *Zum Streit über die Grundlagen der Mathematik* (Hönigswald, 1921) for example, Schlick comes to the conclusion that Kant's doctrine of space and time as forms of pure intuition has *not necessarily* to be abolished, but should be somehow *modified*:

He wants to deprive Kant's forms of intuition "from all quantitative, all mathematical, all metrical attributes"; as defining elements of a subjective, necessary and a priori form of intuition remain "only qualitative attributes of space and time"—or "in short the genuinely *temporal of time*, the specifically *spatial of space*" (Schlick, 1915, p. 163). But Schlick's discussion leaves *totally unclear* what these elements might be. Moreover, in the context of his whole epistemological framework purely *qualitative* forms of space and time can be nothing but psychological constructs without 'foundational output', and this is obviously *more* than a mere modification of Kant. All in all, Schlick's analysis of intuition during this period seems half-hearted and inconclusive.

The more 'constructive' period: Schlick's attempt in his *Allgemeine Erkenntnislehre* to save an area of *exact* and *certain* knowledge *without* recourse to intuition depends predominantly on his interpretation of what he (and others) describe as *'Hilbert's* method of implicit definition', i.e., the definition of the basic concepts of geometry by their axioms.[14] This approach is characterized by Schlick as a "path that is of the greatest significance for epistemology" (Schlick, 1918, p. 31; [Engl. 1974], p. 33).[15] His argument, which differs from Hilbert's original intention, can be summed up as follows: Knowledge needs objective and exact concepts instead of subjective and vague ideas. Explicit definitions of a concept rely on its attributes and in the end – in order to avoid a *regressus at infinitum* – on concrete or ostensive definitions of attributes. Therefore, they inevitably end in subjective and inexact immediate experiences (*Erleben*) (Schlick, 1918, p. 28; [Engl. 1974], p. 29). Schlick turns to the method of implicit definition in order to save (as he says) "the absolute certainty" and "rigor" of knowledge without any endangering appeal to intuition (Schlick, 1918, p. 29; [Engl. 1974], p. 30). Thus axiomatics, allegedly developed along Hilbert's line, becomes Schlick's model of scientific conceptualization – not only for geometry, but also for other branches of mathematics (as number theory) and even for the empirical sciences (especially theoretical physics).

Schlick's leading idea obviously is to relate the dichotomy of implicit and explicit definition to his older dichotomy of knowledge and intuition (where intuition includes both empirical intuition, *Erlebnis*, and Kant's pure intuition). Implicit definitions, according to *his* understanding, avoid any kind of intuition: They determine concepts completely and precisely through their logical relations which are fixed by a system of axioms (Schlick, 1918, p. 30ff.; [Engl. 1974], p. 31ff.). Schlick uses a nice picture in order to contrast explicit and implicit definitions:

> In the case of ordinary definitions, the defining process terminates when the ultimate indefinable concepts are in some way exhibited *(Aufzeigen)* in intuition [...] This involves pointing to something real, something that has individual existence. [...] In short, it is through concrete definitions that we set up the connection between concepts and reality. Concrete definitions exhibit in intuitive or experienced reality that which henceforth is to be designated by a concept. On the other hand, implicit definitions have no association or connection with reality at all; specifically and in principle they reject such association; they remain in the domain of concepts. A system of truths created with the aid of implicit definitions does not

at any point rest on the ground of reality. On the contrary, it floats freely, so to speak, and like the solar system bears within itself the guarantee of its own stability. None of the concepts that occur in the theory designate anything real; rather, they designate one another in such fashion that the meaning of one concept consists in a particular constellation of a number of the remaining concepts. (Schlick, 1918, p. 35; [Engl. 1974], p. 37)

Schlick's *Allgemeine Erkenntnislehre* does not say very much about the question how to come from the 'solar system' down to the 'ground' of reality, or, to be less metaphorical, how to connect precise concepts and relations of the axiomatic system with concrete empirical objects or intuitive mathematical models and examples. This problem will soon be discussed in the context of *spatial* intuition. But first it is important to note the implications of Schlick's new approach to what he called the "fundamental problem of epistemology", i.e., the certainty and evidence of our knowledge—or at least of some parts of it. Schlick sums up his position in these words:

[...] it is to this point that our consideration returns again and again – the moment we carry over a conceptual relation to intuitive examples, we are not longer assured of complete rigor. When real objects are given us, how can we know with absolute certainty that they stand in just the relations to one another that are laid down in the postulates through which we are able to define the concepts?

Kant believed that immediate self-evidence assures us that in geometry and natural science we can make apodictically certain judgements about intuitive and real objects. For him the only problem was to explain how such judgements come about, not to prove that they exist. But we who have come to doubt this belief find ourselves in an altogether different situation. All that we are justified in saying is that Kantian explanation might indeed be suited to rendering intelligible an *existing* apodictic knowledge of reality; but *that* it exists is not something that we may assert, at least not at this stage of our inquiry. Nor can we even see at this point how a proof of its existence might be obtained. (Schlick, 1918, p. 36; [Engl. 1974], p. 38)

Two aspects of Schlick's conclusion seem to me especially important: *First*, Schlick claims that Kant's theory of mathematical intuition and evidence in its 'historical form' is no longer tenable; therefore his explanation of the applicability of mathematics to 'real world problems' is *obsolete*. He does not claim, however, that any new attempt to establish synthetic principles a priori will necessarily fail. This marks a strong contrast to his later and neither proven nor provable claim to have shown that synthetic principles a priori are "a logical impossibility" (*eine logische Unmöglichkeit*). (Schlick, 1930a [repr. 1938], p. 25)

Secondly, and more general, Schlick's sharp demarcation of concept and intuition, of logico-mathematical thinking and empirical reality 'shuts the door' for mathematical certitude and evidence when mathematics is applied to the realm of empirical phenomena. Exact and certain knowledge is restricted to formal properties, defined by the axiomatic system. Empirical content is always infected by the subjectiveness and uncertainty of immediate experience and intuition. That is the point of view soon adopted by Einstein in his lecture "Geometrie und Erfahrung" (Einstein, 1921) and summed up in his famous *dictum*: "As far as the propositions of mathematics refer to reality, they are not certain, and as far as they are certain, they do not refer to reality" (Einstein, 1921, pp. 385f.). Schlick's final solution of his so-called 'fundamental problem of epistemology' is – with respect to intuition – definitely a *negative* one. Paraphrasing Einstein, one might say: As far as we can gain absolute certainty of knowledge, it does not depend on intuition, and as far as we have intuition at our disposal, it can not found knowledge at all.[16]

3. INTUITIVE SPACES AND CONCEPTUAL SPACE: EINSTEIN'S GENERAL THEORY AND SCHLICK'S RECEPTION

In the second edition of his *Allgemeine Erkenntnislehre* (Schlick, 1918, [²1925]), Schlick takes into account Einstein's "Geometrie und Erfahrung".[17] He sums up his affirmative discussion as follows: "Geometrical space is a conceptual tool (*Hilfsmittel*) for designating the ordering of the real. There is no such thing as pure intuition of space, and there are no *a priori* propositions about space" (Schlick, 1918, [²1925], p. 326; [Engl. 1974], p. 255). This statement draws attention to a fundamental problem in Schlick's concept of geometry: If the basic concepts of geometry are defined by an abstract and uninterpreted system of axioms, it is by no means clear how they should somehow contribute to a designation of – in his terms – 'the ordering of the real'. Schlick's attempt to solve this problem depends strongly on a reinterpretation of intuition:

For Schlick, subjective and private experience is *real*, and the question is how the subjective and private experience of *spatiality* and *temporality* of these data can be linked to objective concepts of space and time. This happens, as he says in his *Allgemeine Erkenntnislehre*, "always [...] in accordance with the same method, which we may call the *method of coincidences*. It is of the greatest significance epistemologically" (Schlick, 1918, p. 234; [Engl.], p. 272).[18] His leading idea concerning this method can be described as follows: We achieve intuitive experience of spatiality through our different senses, the visual sense, the tactile sense, etc. There are as many intuitive spaces as there are different senses, though these spaces are totally different in *qualitative* respect.[19] We can, however, experience at the same time a *singularity* in two different intuitive spaces. For example, I can look at the tip of my pencil and, at the same time, I can touch it with my finger. The result is a *coincidence* of two different singularities with different qualities. These two singularities are now coordinated to the same point in objective space by abstracting from the qualitative properties of the different spaces.[20] Other coincidences will yield other points in objective space, and the system of points gained can be extended to a continuous manifold by our thinking.

Though, generally speaking, Schlick perceives Einstein's general theory of relativity rightly as a scientific theory that fits quite well to the epistemological framework of the *Allgemeine Erkenntnislehre* and not as a theory that shaped its epistemological content, it seems that his understanding of the method of coincidences owes to it at least one insight: Einstein's general theory worked as a *mediator* in order to come from this *topological* concept of objective space to a *metrical* concept. Schlick insists that "all measurements, from the most primitive to the most advanced, rest on the observation of spatio-temporal coincidences" (Schlick, 1918, p. 236; [Engl.], p. 275). At the end, all measurements of physical magnitudes, not only distances or time intervals, are based on these spatio-temporal coincidences and thereby on coincidences of singularities in different intuitive spaces. At first sight, this seems to be nothing but an allusion to Einstein's early and frequently repeated claim that his special theory of relativity is a theory about coincidences of events (i.e., coincidences of world lines).[21] This could, in a way, even be said for Newtonian physics. But in his "Space and Time in Contemporary Physics" (Schlick, 1917). Schlick makes clear *what* the peculiarity of Einstein's *general* theory is: *Both* classical physics *and* the special theory of relativity sticked to the idea of an "Euclidean structure" of space – the first one by assuming the existence of a rigid rod for measurements of length, the second one by assuming that all measurements in a system can be performed by a rod which is at rest with respect to this system. The metrical determination was expected to be entirely independent of other physical conditions (of gravitational fields, for example). This preference of Euclidean metric, however, was removed by Einstein's general theory.[22] The elimination of a certain and fixed background geometry matches Schlick's idea how objective space (or rather space-time) is constructed when we proceed from our different intuitive spaces: The experienced coincidences of this *intuitive level* are first brought into a *topological* order. The spatio-temporal manifold thus reached is nothing but the embodiment or essence (*Inbegriff*) of objective elements defined by the method of coincidences. (Schlick, 1917 [³1920], p. 83) Measurement and metric are, in a way, secondary: Measurement always *depends* on coincidences in space and time, and the metrical properties built into the laws of physics always *serve* the attempt to yield a general representation of the coincidences in space-time.[23] According to Schlick, this is the procedure by which we can construct mathematical models for a system of axioms of mathematical physics.[24] Two remarks may be appropriate in order to make this point more explicit:

First. It must be said that Schlick is not very clear or explicit about the question how the abstract level of implicit definitions at the top and the model-building at the ground are related in detail. In so far, his theory of coincidences seems to be unsatisfactory.

Second. As the restricting conditions of Riemannian metric are neither accepted as evident or even necessary by Schlick, *and* as the transition from coincidences to measurement (and metric) always is in need of physical specifications and, indeed, of arbitrary physical assumptions, (Schlick, 1917 [³1920], pp. 55f.) Schlick rejects Cassirer's thesis that the very concept of the Riemannian line element includes synthetic a priori-knowledge of space. (Schlick, 1921, p. 101)[25]

Without going into the details of Schlick's criticism of Cassirer from 1921, his main point here as well as in other papers on relativity from this period can be summed like this: 'Space' is a medal with *two* sides, a *conceptual* one *and* an *intuitive* one. The *method of implicit definition* allows to define space by precise concepts. The *method of coincidences* gains an understanding of its intuitive basis. Kant was right in stressing both the conceptual and the intuitive side of space. But without the method of implicit definitions on the one hand and the method of coincidence on the other hand, he could not but mixing up conceptual and intuitive elements of space (in his fiction of space as a form of *pure intuition*).[26] Einstein, however, opened a new way of bringing back spatial intuition *without* mixing it up with conceptual knowledge. His general theory of relativity has shown that the same method of coincidence by which we proceed from empirical or psychological intuition to objective space underlies the physical construction of space.

According to Schlick, this is probably the most important epistemological outcome of Einstein's general theory.[27]

To refer back to Schlick's rather opaque 'modification-claim' discussing Einstein's theory of special relativity in 1915: Kant's approach should not be given up, but should be essentially modified; space should be freed from all quantitative attributes ('the specifically spatial of space'). His coincidence-argument from 1917 onwards can be understood as an elaboration of this idea: The empirical or psychological intuition of space became his *substitute* for Kant's pure intuition of space, though this substitute had not to carry on the foundational burdens of Kant's original conception.

4. SPATIAL INTUITION AND THE PROCESS OF CONCEPTUALIZATION: SCHLICK'S CRITICISM OF HELMHOLTZ'S APPROACH

1921 was a very fertile year in Schlick's career: He did not only discuss Cassirer's[28] and Reichenbach's[29] analysis of Einstein's theories, but also edited (with Paul Hertz) Helmholtz's *Schriften zur Erkenntnistheorie* (Helmholtz, 1921). In this context, Schlick commented two of Helmholtz's most important papers in some detail: "On the Origin and Significance of the Axioms of Geometry" (Helmholtz, 1870) and "The Facts in Perception" (Helmholtz, 1878). Helmholtz's investigations of the concepts of space and the foundations of geometry are closely linked to his research in the field of physiology, especially on visual and tactual perception. In short, it is *our own* mobility in space, our ability to occupy different positions with respect to perceived objects, that makes spatial localization of objects possible and is, indeed, decisive for our *concept of space* itself.[30]

Starting from the existence and free mobility of rigid bodies as a pre-condition of congruence and measurement, his principal aim (in the first in-stance) was to establish the Euclidean structure of physical space as an intu-itively necessary concept. *In so far* he defended a Kantian-like understand-ing of spatial intuition: Though intuition is brought in by visual and tactual perception, it serves as a guarantee of the Euclidean character of physical geometry.

It is pretty clear that Schlick must have been sympathetic with the empir-ical origin of Helmholtz's intuition of space, but it is also clear that he could not accept its 'Euclidean services'. In the two papers commented by Schlick, however, Helmholtz changes his position considerably: The axioms of Eu-clidean geometry are no longer considered as being necessary by their foun-dation in a transcendental form of intuition. We can imagine other spaces with other axioms and we can – what is most important for Helmholtz's empiricist approach – imagine lawful sense-experiences in those spaces without contradictions. Intuition of space therefore can *not* yield the necessity of *any* system of geometrical axioms, and in so far Euclid's geometry becomes *contingent*. (Helmholtz, 1870 [repr. 1921], p. 22.)

This new consequence was, of course, most welcome to Schlick. To his mind there remains, however, an important inconclusiveness in Helmholtz' argument: Though spatial intuition can not yield any necessity of geometrical axioms, Helmholtz sticks to Kant's idea that there is something like space *as a subjective and transcendental form of intuition*. This seems to be Schlick's basic problem with Helmholtz's foundation of geometry. Three remarks about Schlick's criticism may be sufficient here:

First. According to Helmholtz, the really interesting features of spatial intuition are those which can *not* be grasped by axiomatic systems like Euclid's. After all, that is why he states that "space can be transcendental without the axioms being it", as he puts it in his famous *dictum*.[31] In this context, Helmholtz describes several sense experiences which come close to those of the *Allgemeine Erkenntnislehre*, and can therefore be integrated into Schlick's psychological intuition analysed by the method of coincidences. Consequently, they pose no problem for Schlick and are commented affirmatively.

Second, and perhaps more important: Helmholtz's intuition of space is *richer* than Schlick's, because it operates with the free mobility of rigid bodies and therefore includes the idea of *constant curvature* of space. Helmholtz makes quite clear that he does *not* misunderstand perfect rigidity as a meaningful *empirical* concept; for him it is rather a concept *a priori*, introduced in order to make measurement possible.[32] *And* at the same time we have to presuppose that "the measuring instruments which we take to be fixed, actually (*wirklich*) are bodies of unchanging form. Or that they at least undergo no kinds of distortion other than those which we know [...]" (Helmholtz, 1870 [repr. 1921], p. 18; [Engl. 1977], p. 19)

According to Schlick, Helmholtz's argument results from an inadmissible *extension* of spatial intuition or, to put it otherwise, from a confusion of intuition and conceptual knowledge. This is his indignant comment on Helmholtz's last sentence:

> In the little word 'actually' there lurks the most essential philosophical problem of the whole lecture [of Helmholtz]. What kind of sense is there in saying of a body that it is *actually* rigid? According to Helmholtz's definition of a fixed body

[...], this would presuppose that one could speak of the distance between points 'of space' without having regard to bodies; but it is beyond doubt that without such bodies one can not ascertain and measure the distance in any way. [...] If the content of the concept 'actually' is to be such that it can be empirically tested and ascertained, then there remains only the expedient [...] to declare those bodies to be 'rigid' which, when used as measuring rods, lead to the *simplest* physics. Those are precisely the bodies which satisfy the condition [of coincidences] adduced by Einstein. Thus what has to count as 'actually' rigid is then not determined by a logical necessity of thought or by intuition, but by a convention, a definition. (Helmholtz, 1870 [repr. 1921], p. 33, n. 40; [Engl. 1977], p. 34).[33]

In short, Helmholtz's idea of constructing space on the basis of the free mobility of rigid bodies is *totally* rejected by Schlick. Helmholtz's most important 'intuitive link', i.e., the relation of a priori-rigidity and of empirical measurement – was obviously 'cut off' for Schlick by *general relativity*. And, considered the other way round: Schlick applied the philosophical lesson he drew from Einstein's theory to Helmholtz's approach, i.e., sharply to separate abstract concepts and intuitive experience – thereby, however, he failed to appreciate Helmholtz's mathematical contribution to the foundations of geometry.

Third, and probably most important, there is a profound difference in the epistemological perspective of Helmholtz and Schlick that has to be considered: Schlick's residual of spatial intuition rests, as mentioned earlier, on immediate sense experience. Though it is linked to objective, physical space by his method of coincidence, the whole conceptual framework of objective space is built up *deductively* or 'top down', starting with axioms, because only the method of implicit definitions can guarantee *precise* basic concepts.

Helmholtz's approach is quite contrary. Gregor Schiemann describes it aptly as an *inductive* or 'bottom up'-conceptualization:[34] Axioms are of minor importance. We proceed from the perception of qualities, whereby the lawlikeness of spatial relations is brought in by our own free mobility. On *this* basis we build up spatial concepts by association or unconscious induction. Without going into the details of Helmholtz's argument, it can be said that according to his approach there can be no sharp separation of intuition and conceptual knowledge, as it is claimed by Schlick. Intuition can change by learning—we can learn, for example, how it would be to

move in a space with negative curvature. Helmholtz points out the difference between his idea of intuition and the traditional one, which supposes
a "flash-like evidence" of spatial intuition. (Helmholtz, 1878 [repr. 1921],
p. 161) And against Kant he claims that "the most essential progress of science", especially physiology, was the decomposition of traditional intuition
into "elementary processes of thinking" which, as he says later, "can not yet
be expressed by words" (Helmholtz, 1878 [repr. 1921], p. 172). This is, of
course, not compatible with Schlick's tidy separation of intuition and knowledge, and his comments on Helmholtz's writings on geometry reflect this
'dualism'.

5. CONCLUDING REMARK

The purpose of this paper was to light up Schlick's changing understanding of intuition as a result of an 'interaction between mathematics, physics
and philosophy': Schlick's 'turning off' from a form of intuition, which is
Kantian-like in so far as intuition guarantees the truth of the propositions
of geometry, seems to be influenced by his reception of Einstein's theory
of relativity, especially the method of coincidences. In addition, his *later*
argument for a sharp separation of knowledge and intuition makes use of
the method of implicit definition which he attributes to Hilbert's *Grundlagen der Geometrie*. Schlick's 'new' separation is directed both against Kant
and Helmholtz. It may be asked, however, if he does justice to Helmholtz's
approach, according to which spatial intuition can be dissolved into 'elementary processes of thinking'. This approach seems not to be so far away from
Einstein's method of coincidences, by which Schlick wants to do justice to
intuition: The 'Space between Helmholtz and Einstein' seems to be more
restricted than he is willing to admit.

Ruhr-Universität Bochum
Germany

NOTES

[0]Revised version of my talk given at the conference on "The Interaction between Mathematics, Physics and Philosophy from 1850 to 1940" in September
2002. Many thanks to the participants for critical and fruitful discussions and especially to Jesper Lützen for organisation and hospitality. A former version of this
paper was red by Carsten Seck and Ralf Kuklik, whom I would like to thank for
their critical discussion, too.

   In the following titles are abbreviated by names and the first date of appearance. If quotations are *not* from the first edition, the year of publication is added

in square brackets. If an English translation is quoted, the bracket contains, in addition, the abbreviation "Engl.". In cases where the English short title is added in round brackets, the note (or reference) refers to the first (German) title or quotation.

[1] Letter from Carnap to Neurath from the 23.8.1945; the quotation is drawn from (Hegselmann, 1992, p. 23).

[2] See, for instance, (Heilbron, 1988, pp. 188f.).

[3] Cf. (Stadler 1997, pp. 201f.; Hentschel 1986, pp. 476ff.; Hentschel 1990, pp. 377f.; Howard 1984, pp. 618f.; Howard 1988, pp. 204ff.).

[4] Cf. (Carnap/Hahn/Neurath, 1929 [repr. 1999], pp. 166f.).

[5] Cf. (Schlick, 1934, pp. 79-80, 94-95).

[6] Cf. (Schlick, 1910, pp. 389ff.).

[7] Cf. (Schlick, 1910, pp. 390–392) – on p. 392 evidence is, strange enough, reintroduced not as a sufficient, but as a necessary feature of truth; (Schlick, 1918, pp. 130–131, 68–69).

[8] Cf. (Goldfarb, 1996, p. 214).

[9] Cf. (Schlick, 1918, p. 148f.) for his dissociation of his former theory of truth.

[10] Cf. (Friedman, 1999, p. 20).

[11] Cf. (Friedman, 1999, esp. p. 20).

[12] Cf. (Schlick, 1913, p. 481).

[13] Cf. (Howard, 1994, p. 51).

[14] It is well known that the notion 'implicit definition' was already introduced by J. D. Gergonne in 1818, and also, that this notion was, at the end of the 19th century, applied by the school of Peano for the *different* approach of definitions by postulates; see (Otero, 1969/70). F. Enriques and others soon applied it from 1904 onwards to Hilbert's axiomatics, because Hilbert explicitly stated that the axioms are *also definitions* of the basic concepts, without using the term 'implicit definition' (see (Hilbert, 1902, pp. 71–72; Gabriel, 1978) where the confusion of the older and newer meaning of the term by the Peano school and its impact on its later use is analysed). That Schlick's interpretation and adaptation of Hilbert's approach are problematic, was already pointed out by Majer (2001, pp. 214–216).

[15] Cf. (Goldfarb 1996, p. 214).

[16] Cf. (Schlick, 1918, p. 130; [Engl. 1974], p. 149).

[17] Cf. (Ferrari, 1994, p. 436, n. 64).

[18] Cf. (Ferrari, 1994, pp. 435f.).

[19] Cf. (Friedman, 1999, pp. 37f.).

[20] See (Schlick, 1918, p. 235f.; [Engl.], pp. 274f.); cf. (Schlick, 1917, p. 96f.).

[21] Cf. (Coffa, 1991, p. 198, 399, n. 6).

[22] Cf. (Schlick, 1917 [^{3}1920], pp. 46f.).

[23] (Schlick, 1917 [^{3}1920], pp. 50f.) "Alle Weltbilder, die hinsichtlich der Gesetze jener Punktkoinzidenzen übereinstimmen, sind physikalisch absolut gleichwertig." (p. 51).

[24] Cf. (Friedman, 1999, p. 38).

[25] Cf. (Cassirer, 1921, p. 101; [repr.1957], p. 93).

[26] Cf. (Schlick, 1921, pp. 108–109).

[27]Cf. (Ryckman, 1992, pp. 494f.).

[28]Cf. (Ryckman, 1991; Ferrari, 1994).

[29]Cf. (Friedman 1994).

[30]Cf. (Helmholtz, 1868 [repr. 1921], pp. 38-40 (introductory part); 1866 [repr. 1883], pp. 610-612).

[31]Helmholtz's title of the "2. Beilage" from "Facts in Perception"; cf. (Helmholtz, 1878 [repr. 1921], p. 140).

[32]Helmholtz (1870 [repr. 1921], p. 18) discusses influences of temperature and other minor factors.

[33]Cf. (Helmholtz, 1870 [repr. 1921], p. 30, n. 31; [Engl. 1977], p. 31).

[34]See (Schiemann, 1997, p. 350) with respect to (Helmholtz, 1878 [repr. 1921], pp. 123f.).

REFERENCES

Carnap, R., Hahn, H. and Neurath, O. (1929). Wissenschaftliche Weltauffassung – Der Wiener Kreis. Publication of the 'Verein Ernst Mach', Berlin, repr.: Fischer, 1995 [repr. 1999], 125–171.

Cassirer, E. (1921). *Zur Einsteinschen Relativitätstheorie. Erkenntnistheoretische Betrachtungen*, Bruno Cassirer, Berlin. Repr.: (Cassirer, 1957, 1–125).

Cassirer, E. (1957). *Zur Modernen Physik*, Wissenschaftliche Buchgesellschaft, Darmstadt.

Coffa, J. A. (1991). *The Semantic Tradition from Kant to Carnap: To the Vienna Station*, Cambridge University Press, Cambridge.

Einstein, A. (1921). Geometrie und Erfahrung[expanded version of an address, given at the 'Preussische Akademie der Wissenschaften zu Berlin'; 27.01.1921], *in The Collected Papers of Albert Einstein, vol. 7: The Berlin years: writings, 1918-1921*, Princeton University Press, Princeton, 2002, pp. 383–386. Engl.: *Geometry and Experience, in* (Einstein, 1922, 25–56).

Einstein, A. (1922). *Sidelights on Relativity*, Dover, London. Repr.: New York 1983.

Ferrari, M. (1994). Cassirer, Schlick und die Relativitätstheorie. Ein Beitrag zur Analyse des Verhältnisses von Neukantianismus und Neopositivismus, *in* E. W. Orth and H. Holzhey (eds), *Neukantianismus. Perspektiven und Probleme*, Königshausen & Neumann, Würzburg, pp. 418–441.

Fischer, K. R. (1995). *Das Goldene Zeitalter der Österreichischen Philosophie*, WUV Univ.-Press, Wien. Repr.: *Österreichische Philosophie von Brentano bis Wittgenstein*, UTB, Vienna, 1999.

Friedman, M. (1983). Critical Notice: Moritz Schlick, Philosophical Papers, *Philosophy of Science* **50**: 498–514.

Friedman, M. (1994). Geometry, Convention, and the Relativized A priori: Reichenbach, Schlick, and Carnap, *in* W. Salmon and G. Wolters (eds), *Logic, Language, and the Structure of Scientific Theories*, University of Pittsburgh/Universitätsverlag Konstanz, Pittsburgh, Konstanz.

Friedman, M. (1997). Helmholtz's Zeichentheorie and Schlick's Allgemeine Erkenntnislehre: Early Logical Empiricism and Its Nineteenth-Century Background, *Philosophical Topics* **25**(2): 19–50.

Friedman, M. (1999). *Reconsidering Logical Positivism*, Cambridge University Press, Cambridge.

Gabriel, G. (1978). Implizite Definitionen - Eine Verwechselungsgeschichte, *Annals of Science* **35**: 419–423.

Giere, R. N. and Richardson, A. W. (eds) (1996). *Origins of Logical Empiricism*, University of Minnesota Press, Minneapolis.

Goldfarb, W. (1996). The Philosophy of Mathematics in Early Positivism, *in* R. N. Giere and A. W. Richardson (eds), *Origins of Logical Empiricism*, University of Minnesota Press, Minneapolis.

Hegselmann, R. (1992). Einleitung: Einheitswissenschaft - das positive Paradigma des Logischen Empirismus, *in* (Schulte and McGuinness, 1992, 7–23).

Heidelberger, M. and Stadler, F. (eds) (2001). *History of Philosophy of Science. New Trends and Perspectives*, Springer, Dordrecht.

Heilbron, J. L. (1988). *Max Planck. Ein Leben für die Wissenschaft. 1858-1947*, Hirzel, Stuttgart.

Helmholtz, H. von (1866). Über die thatsächlichen Grundlagen der Geometrie, *Verhandlungen des naturhistorisch-medizinischen Vereins zu Heidelberg* **IV**: 197–202. Repr.: (Helmholtz, 1883, 610–617).

Helmholtz, H. von (1868). Über die Thatsachen, die der Geometrie zugrundeliegen [03.06.1868], *Nachrichten von der Königlichen Gesellschaft der Wissenschaften und Georg-August Universität zu Göttingen* **9.**: 193–221. Repr.: (Helmholtz, 1921, 38–69). Engl.: *On the Facts Underlying Geometry, in* (Helmholtz, 1921[1977], 39–71).

Helmholtz, H. von (1870). Über den Ursprung und die Bedeutung der geometrischen Axiome, *in* (Helmholtz, 1921, 1–37). Lecture, held at the 'Docentenverein zu Heidelberg'; a comprehensive elaboration of (von Helmholtz, 1868). Engl.: On the Origin and Significance of the Axioms of Geometry, *in* (Helmholtz, 1921[1977], 1–38).

Helmholtz, H. von (1878). Die Tatsachen in der Wahrnehmung [address given during the 'Stiftungsfeier der Friedrich-Wilhelm-Universität zu Berlin]', *in* (Helmholtz, 1921, 109–175). Engl.: The Facts in Perception, *in* (Helmholtz, 1921[1977], 115–185).

Helmholtz, H. von (1883). *Wissenschaftliche Abhandlungen*, Leipzig.

Helmholtz, H. von (1921). *Schriften zur Erkenntnistheorie*, Springer, Berlin. Ed. and explained by Paul Hertz and Moritz Schlick. Engl.: *Epistemological Writings*, ed. [with introduction and bibliography] by R.S. Cohen and Y. Elkana; translated by M. F. Lowe, Boston, 1977.

Hentschel, K. (1986). Die Korrespondenz Einstein-Schlick: Zum Verhältnis der Physik zur Philosophie, *Annals of Science* **43**: 475–488.

Hentschel, K. (1990). *Interpretationen und Fehlinterpretationen der speziellen und allgemeinen Relativitätstheorie durch Zeitgenossen Albert Einsteins*, Birkhäuser, Berlin.

Hilbert, D. (1899). Grundlagen der Geometrie, *Festschrift zur Feier der Enthüllung des Gauss-Weber-Denkmals in Göttingen* pp. 1–92.

Hilbert, D. (1902). Sur les problèmes futurs des mathématiques, *Compte rendu du 2me Congrès International des Mathematiciens. Paris,* pp. 58–114.

Hönigswald, R. (1921). *Zum Streit über die Grundlagen der Mathematik*, Heidelberg.

Howard, D. (1984). Realism and Conventionalism in Einstein's Philosophy of Science: The Einstein Schlick Correspondence, *Philosophia Naturalis* **21**: 616–629.

Howard, D. (1988). Einstein and Eindeutigkeit: A neglected Theme in the Philosophical Background to General Relativity, *in* J. Eisenstaedt and A. J. Kox (eds), *Studies in the History of General Relativity*, Vol. 3. of *Einstein Studies*, Birkhäuser, Boston, 1992, pp. 154–243.

Howard, D. (1994). Einstein, Kant, and the Origins of Logical Empiricism, *in* W. Salmon and G. Wolters (eds), *Logic, Language, and the Structure of Scientific Theories*, University of Pittsburgh/Universitätsverlag Konstanz, Pittsburgh, Konstanz.

Krüger, L. (ed.) (1970). *Erkenntnisprobleme der Naturwissenschaften. Texte zur Einführung in die Philosophie der Wissenschaft*, Cologne.

Majer, U. (2001). Hilbert's Program to Axiomatize Physics (in Analogy to Geometry) and its Impact on Schlick, Carnap and other Members of the Vienna Circle, *in* (Heidelberger and Stadler, 2001, 213–224).

Orth, E. W. and Holzhey, H. (eds) (1994). *Neukantianismus. Perspektiven und Probleme*, Königshausen & Neumann, Würzburg.

Otero, M. H. (1969/70). Gergonne on Implicit Definition, *Philosophy and Phenomenological Research* **30**: 596–599.

Ryckman, T. A. (1991). Conditio sine qua non? Zuordnung in the early Epistemologies of Cassirer and Schlick, *Synthese* **88**: 57–95.

Ryckman, T. A. (1992). '(P)oint-(C)oincidence Thinking': The Ironical Attachment of Logical Empiricism to General Relativity, *Studies in the History and Philosophy of Science* **23**(1): 471–497.

Schiemann, G. (1997). *Wahrheitsgewissheitsverlust: Hermann von Helmholtz' Mechanismus im Anbruch der Moderne. Eine Studie zum Übergang von klassischer zu moderner Naturphilosophie*, WBG, Darmstadt.

Schlick, M. (1904). *Über die Reflexion des Lichtes in einer inhomogenen Schicht*, PhD thesis, Berlin.

Schlick, M. (1908). *Lebensweisheit. Versuch einer Glückseligkeitslehre*, Beck, Munich.

Schlick, M. (1910). Das Wesen der Wahrheit nach der modernen Logik, *Vierteljahrsschrift für wissenschaftliche Philosophie und Soziologie* **34**: [new series 9, no. 4], 386–477. Repr.: Schlick 1986, 31–109.

Schlick, M. (1913). Gibt es intuitive Erkenntnis?, *Vierteljahrsschrift für wissenschaftliche Philosophie und Soziologie* **37**: [new series 12, no. 3], 472–488.

Schlick, M. (1915). Die philosophische Bedeutung des Relativitätsprinzips, *Zeitschrift für Philosophie und Philosophische Kritik* **159**(2): 129–175. Engl.: The Philosophical Significance of the Principle of Relativity, translated by P. Heath, *in* (Schlick, 1978/1979, vol. 1, 153–189).

Schlick, M. (1917). Raum und Zeit in der gegenwärtigen Physik. Zur Einführung in das Verständnis der allgemeinen Relativitätstheorie, *Die Naturwissenschaften. Wochenschrift für die Fortschritte der Naturwissenschaft, der Medizin und der Technik* **5**(11): 162–186. Repr. [as a monograph]: Berlin [1]1917, [2]1919 [enlarged and improved edition with the changed subtitle: *Zur Einführung in das Verständnis der Relativitäts- und Gravitationstheorie*], [3]1920 [enlarged and improved edition], [4]1922 [enlarged and improved edition]. Engl.: Space and Time in Contemporary Physics, translation of the 4th edition by H. Brose and P. Heath *in* (Schlick, 1978/1979, vol. 1, 207–269).

Schlick, M. (1918). *Allgemeine Erkenntnislehre*, Berlin [2]1925. Engl., ed. by M. Bunge, *General Theory of Knowledge* [translated from the 2nd edition by A. E. Blumenberg], Vienna, 1974.

Schlick, M. (1921). Kritizistische oder empiristische Deutung der neuen Physik? Bemerkungen zu Ernst Cassirers Buch, "Zur Einsteinschen Relativitätstheorie", *Kant-Studien* **26**: 96–111. Engl.: Critical or Empiricist Interpretation of Modern Physics? Translated by P. Heath, *in* (Schlick, 1978/1979, vol. 1, 322–334).

Schlick, M. (1922). Helmholtz als Erkenntnistheoretiker [lecture, given at the celebration of the one-hundreth anniversary of his birth for the 'physikalische, physiologische und philosophische gesellschaft zu Berlin'], *in Helmholtz als Physiker, Physiologe und Philosoph. Drei Vorträge*, Müllersche Hofbuchhandlung, Karlsruhe, pp. 29–39.

Schlick, M. (1930a). Gibt es ein materiales Apriori?, *Wissenschaftlicher Jahres-bericht der Philosophischen Gesellschaft an der Universität zu Wien, [1930/31]* pp. 55–65. Repr.: (Schlick, 1938, 19–30). Engl.: Is There a Factual a Priori?, *in* Readings in Philosophical Analysis, ed. by H. Feigl and W.S. Sellars. New York, 1949, pp. 277–285.

Schlick, M. (1930b). *Fragen der Ethik*, Vienna. Repr. ed. and introduced by Rainer Hegselmann, Frankfurt a.M. 1984, ²2002.

Schlick, M. (1934). Über das Fundament der Erkenntnis, *Erkenntnis* **4**: 79–99. Repr.: Krüger 1970, 41–56. Engl.: On the Foundation of Knowledge, translated by P. Heath, *in* (Schlick, 1978/1979, vol. 2, 370–387).

Schlick, M. (1938). *Gesammelte Aufsätze. 1926-1936*, Gerold & Co., Vienna. Intro-duced by Friedrich Waismann. Repr.: Hildesheim 1969.

Schlick, M. (1978/1979). *Moritz Schlick: Philosophical Papers*, ed. by H. L. Mulder and B. F. van de Velde-Schlick, 2 vols. Kluwer, Dordrecht.

Schlick, M. (1986). *Philosophische Logik*, ed. with introduction by Bernd Philippi, Suhrkamp, Frankfurt a.M.

Schulte, J. and McGuinness, B. (eds) (1992). *Einheitswissenschaft*, Suhrkamp, Frankfurt a.M.

Stadler, F. (1997). *Studien zum Wiener Kreis: Ursprung, Entwicklung und Wirkung des logischen Empirismus im Kontext*, Suhrkamp, Frankfurt a.M.

Turner, J. L. (1996). Conceptual Knowledge and Intuitive Experience: Schlick's Dilemma, *in* (Giere and Richardson, 1996, 292–308).

ROBERT DISALLE

MATHEMATICAL STRUCTURE, "WORLD STRUCTURE," AND THE PHILOSOPHICAL TURNING-POINT IN MODERN PHYSICS[0]

1. INTRODUCTION

Logical empiricism was intended, at least, to embody the philosophical perspective that seemed to be implicit in, and an indispensable motivation for, the revolutionary physical theories of the early 20th century. Beyond offering a philosophical reflection on the state of contemporary physics, however, the logical empiricists sought to connect the revolution in physics with late 19th-century transformations in the philosophy of logic and mathematics– above all, a transformed view of the nature of axiomatic structures and their physical interpretation. Ironically enough, its account of the nature and interpretation of scientific theories came to be seen, in the later 20th century, as a central reason for the failure of logical empiricism as a whole. My aim is not so much to excuse the logical empiricists of such failings, as to understand better the insight that they sought unsuccessfully to capture, and its relevance to the philosophy of physics. They had reason to think that insights concerning structure and interpretation had had sweeping consequences for the foundations of mathematical physics, as well as for traditional philosophical questions regarding the nature of a priori knowledge – and that these insights had played essential roles in the development of 20th-century physics, especially general relativity. But their own account of scientific theories as formal structures presented a seriously distorted view of the situation, and therefore a misleading account of the philosophical principles that had inspired the new theories. I would like to examine an alternative account that emerged more or less contemporaneously with theirs, chiefly through the work of Hermann Minkowski, Hermann Weyl, and Arthur Eddington. Possibly because it was articulated mainly by non-philosophers, and then not in very explicit detail, this account never had the sort of impact on philosophical discussions of general relativity that logical positivism had. But, as we will see, it offered a much clearer understanding, not only of Einstein's theories, but, more generally, of the connections between mathematical structure and the physical world – a deeper understanding of what Weyl referred to as "world-structure".

V.F. Hendricks, K.F. Jørgensen, J. Lützen and S.A. Pedersen (eds.), *Interactions: Mathematics, Physics and Philosophy, 1860-1930*, pp. 207–230.
© 2006 Springer.

2. THE PROBLEM OF INTERPRETATION IN HISTORICAL PERSPECTIVE

Toward the end of the 20th century, logical positivism, having dominated philosophical discussions of science for some decades, was widely rejected, and especially criticized for its seeming detachment from ordinary scientific practice. Even if it is true, however, criticism of this sort is somewhat beside the point. For the logical empiricists' discussions of scientific theories were never meant to describe the general practice of scientific inquiry; nor were they meant to prescribe a method for doing science. In fact the notion that philosophers could prescribe for scientists was quite opposed to their general way of thinking. For the logical empiricists, science had itself undergone a kind of philosophical revolution, and their aim was to understand its implications for the practice of philosophy. More precisely, theoretical physics, especially in the work of Einstein, had been transformed by the clarification of problems that philosophy had never adequately addressed—in particular, the way in which physics connects formal mathematical structure with the world of our experience, and in general, the ways in which abstract concepts acquire an empirical interpretation.

It is unfortunate that, in the post-positivist period, these questions of interpretation have been largely set aside. For the difficulty with the postivists' view was not that these questions are not important, even central to any genuine philosophical understanding of the nature of scientific theories. Rather, it is that the positivists' attempts to answer them were generally not very convincing. They articulated the problem of interpretation in the setting of what is now called a the "syntactic" or "statement" view of theories: a scientific theory is an axiomatic structure or calculus for generating theoretical statements; it is essentially uninterpreted, and its application to the physical world therefore must be fixed by some interpretive rules, "correspondence rules" or "coordinative definitions" that associate principles of the formal structure with empirical statements about observable states of affairs. (Cf. Carnap 1966). The first and most influential example was that of ordinary spatial geometry: Euclidean geometry, for instance, can be understood as a purely formal structure based on certain axioms, and it first becomes interpreted as a set of claims about the world when concepts that occur in the axioms, such as congruence, are coordinated with physical processes, such as the displacement and comparison of rigid bodies. Only by means of such a coordination can it be meaningful to claim that "space is Euclidean." (Reichenbach, 1957, ch. 3). But coordinating principles are necessarily arbitrary; they cannot be considered empirical scientific claims, because they lay down the conditions for making empirical claims in the first place. Like Kant's synthetic a priori, they are conditions of the possibility of experience; but they are by nature analytic rather than synthetic claims, since they say nothing about the world; they state only how a particular concept is to be used in our investigation of the world, or, more precisely, how a concept functions within a particular linguistic framework through which we conceptualize the world. They can only be determined on the basis of convenience – say, on the basis of how simple and useful a system of physical geometry can be built upon them – rather than arrived at by empirical inquiry. So it is quite difficult, on this view, to represent conceptual change in the physics of space and time as a progressive development on any epistemic grounds; the best that can be said is that 20th century physics developed simpler and more convenient formal structures, and that its founders had a better appreciation than their predecessors of the arbitrariness involved. In fact, just this recognition of the "relativity of geometry" – relativity to conventions about measurement – was what the logical positivists regarded as the philosophical turning-point in modern physics.[1]

In recent decades the syntactic view of theories has been largely displaced by the "semantic" or model-theoretic view, on which the axiomatization of physical theories is dispensed with. Instead, theories are characterized directly as model-theoretic structures; their application to the world is said to be embodied in the "empirical hypothesis" that the world is a model of the relevant structure.[2] So, instead of an axiomatization of Euclid's geometry and a collection of coordinating principles, the semantic view offers a direct characterization of the structure, e.g. as a set of points R^3 with the Pythagorean distance relation defined between pairs of points; the relevant empirical hypothesis is that the world, or more precisely physical space, is a model of this structure. This representation claims the advantage that it better reflects the actual practice of science, in which the axiomatization of theories rarely seems useful, and therefore only involves philosophers in pointless metamathematical difficulties; it certainly reflects the normal practice of modern texts on general relativity (e.g. Misner et al. 1973, Hawking and Ellis 1972, Wald 1984) far better than, say, Reichenbach's *Axiomatik der Raum-Zeit Lehre* (1924). It is true that one could hardly characterize the structure without stating what is taken to be true in all models – that is, the axioms of the theory. But at least the task of giving a rigorous axiomatization is set aside. As far as the problem of interpretation goes, however, it should be noted that the semantic view leaves us no better off than we were before. For the model-theoretic presentation of the theory alone does nothing to address the question: what does it mean to say that space is a model of Euclidean geometry? In order to answer it, we would have to consider again the very sort of question that the logical positivists were preoccupied with, namely how do we determine what in our experience is supposed to represent any given element of the geometrical structure? It can be said, in defense of the syntactic view, that it placed this problem in the foreground, whereas the semantic view merely overlooks it; a structure is assumed to have an "intended" interpretation, and it is not deemed necessary to explain how this comes about. In any case, how we are to regard the physical interpretation of formal structures – as empirical, conventional, or something else – is clearly independent of which of these views of theories we might prefer (cf. Demopoulos 2003).

This problem of interpretation is a peculiar product of the 19th century. It could not have arisen for Kant, for example; on his view, there was no such thing as "uninterpreted" mathematics. Every mathematical theory is a theory of some particular domain of objects, and those objects are cognizable for us only to the extent that they are constructed in sensible intuition. Only "general logic" is truly empty, in the sense that it is truly general and therefore about relations that may hold among any objects whatever, considered abstractly only as possible terms in those relations. Mathematics, however, is essentially characterized by the fact that its objects are defined by some constructive procedures that are intuitively evident. Indeed, as Kant emphasized on many occasions, in both his "pre-Critical" and "Critical" phases, this fact was the central distinction between the exact sciences and metaphysics. Metaphysics could attempt to emulate mathematics by being deductively rigorous, developing philosophical arguments "*in more geometrico*" after the manner of, say, Leibniz or Spinoza. But this method could never secure universal assent, or indeed any degree of objectivity, for a metaphysical system; it was never merely the deductive structure of mathematics that guaranteed its objectivity, but its capacity to *define* the objects of mathematical knowledge, the starting points for mathematical reasoning, by constructive procedures that left no room for doubt or ambiguity. Mathematics first comes into being, as something distinct from general logic, by the synthesis of its fundamental objects in intuition. And thus it comes into being as a structure that is already, and completely, interpreted. Its very formal rigour depended on this fact, since (as far as Kant knew) mathematics depended on constructive proofs for reasoning about concepts, particularly infinity and continuity, for which logic was (then) inadequate.[3] The view of rationalist philosophers such as Leibniz, that mathematics was a kind of purely intellectual knowledge, free of any empirical taint and ultimately reducible to logical identities, was revealed to be illusory; it reflected only those philosophers' unawareness of just how pervasive the appeals to intuition were in the mathematical reasoning of the time.

The Kantian solution to the problem of interpretation was abandoned for various obvious reasons, at least three of which are immediately relevant to our theme. First, the rigorization of analysis in the 19th century, through the work of Bolzano, Weierstrass, and others, achieved much of what Kant had considered impossible: analytic definitions of concepts such as continuity and infinity, and rigorous analytic proofs of propositions regarding them, that did not appeal to intuitive constructive procedures. It would appear that the need for intuitive constructions revealed, not an essential property of mathematical reasoning, but the limited power of "general logic" in Kant's time. The possibility of a purely formal mathematics no longer seemed so remote.

Second, the development of non-Euclidean geometry undermined the Kantian connection between Euclidean geometry and the form of spatial intuition. This would appear obvious, now, but it is worth recalling that it was not merely the development of consistent geometries other than Euclid's, or even their eventual use in the natural sciences, that most seriously challenged Kant's view. Rather, it was the recognition, especially by Helmholtz (1870b) and Poincaré (1913), that certain non-Euclidean geometries could be as evident as Euclidean geometry to intuition itself. Helmholtz demonstrated that precisely those intuitive constructive procedures that convince us of the truth of Euclid's postulates could, in different circumstances, convince us that the world is non-Euclidean. In a sense this was a predictable result of Kant's analysis. For Kant had argued persuasively that our only grounds for geometrical knowledge are those synthetic procedures by which we construct geometrical objects and prove geometrical propositions; given this, it was only necessary to imagine how such synthetic operations might unfold according to different laws, revealing a world with a different geometrical structure. Having accepted that the content of geometry is irreducibly empirical, Kantianism had no grounds on which to resist Helmholtz's argument. The very experiences that made Euclidean geometry "intuitable" ("anschaulich") could apply to any geometry of constant curvature.

Third, and consequent upon the first two points, was the eventual recognition that the structure of geometry could be separated from its content altogether (cf. Nagel 1939). This was exhibited by the discovery of the inter-translatability of Euclidean and non-Euclidean geometries. It became

possible to exhibit Euclidean models of non-Euclidean spaces, as in Bel-
trami's model of pseudospherical space as the interior of a disc with a peculiar distance-function; it also became possible to give relative consistency
proofs for non-Euclidean geometries, since the sameness of logical structure,
for the two intertranslatable systems, implied that non-Euclidean geometries
were as consistent as Euclid's. Moreover, the consistency of geometry could
be proven relative to that of arithmetic, and so could be shown to be independent of any particularly geometrical content. This general development may
be said to have culminated in Hilbert's axiomatization of geometry (1899);
here the primitive terms are not alleged to be known intuitively, or defined independently, but are only implicitly defined by the geometrical axioms themselves. Thus the geometry could not be said to have (e.g.) points, lines, and
planes as its subject-matter, as if these were assumed to be definable in some
other terms prior to the axioms. Instead, the subject-matter of geometry is
whatever might satisfy the axioms—perhaps tables, chairs, and beer-mugs,
as in Hilbert's famous dictum. By the turn of the 20th century, then, it was
evident that the question of interpreting geometry had been entirely divided
from every important question about its formal structure, and so it had to be
answered on entirely independent grounds.

This was the context in which the logical positivists understood the origins and the significance of special relativity and, even more, general relativity. Spacetime was no longer to be understood as having a direct physical interpretation, through the displacement of rigid measuring-instruments
and the progress of ideal clocks. Instead, it was to be understood as an
abstract formalism lacking in empirical content. In Riemann's theory of differentiable manifolds, general relativity had a mathematical framework that
represented spacetime as something amorphous, or lacking in any particular
physical properties. This was to correspond to the fact that in our experience, the only objectively knowable properties of spacetime are supposed to
be "the meeting of material points of our measuring-instruments with other
material points" (Einstein, 1916, p. 123); therefore any arbitrary coordinate-
transformation that preserves "point-coincidences" – a diffeomorphism of a
Riemannian manifold – is assumed to be an allowable transformation that
preserves what is empirically meaningful about spacetime (cf. Schlick 1917,
ch. 3). For the logical positivists, spacetime is therefore an empty framework, acquiring its interpretation from decisions that we make, on pragmatic
grounds, about how to connect its features with observable objects.

In some sense this view was exaggerated. The fact that general relativity requires spacetime to be locally Minkowskian (i.e. that special relativity

holds in the infinitely small) already imposes a degree of physical interpretation on the structure of spacetime that goes well beyond mere point coincidences. The local structure is, on this assumption, the conformal structure of spacetime in the tangent space to any point; it is, in other words, a geometrical structure that is already interpreted as representing the local causal structure of a spacetime in which light propagation is (locally) an invariant limiting velocity for causal propagation, and the local metrical structure of a spacetime in which the invariance of the speed of light defines the metric interval. From a certain kind of empiricist standpoint, it might be said that the observable behavior of light rays – for example, the null result of the Michelson-Morley experiment – provides direct empirical evidence for the local structure of spacetime; general relativity does not undermine that evidence, but only reveals its purely local character. For the logical empiricists, however, the locally Minkowskian structure of spacetime is only a convention that is adopted in order to connect the general framework of Riemannian manifolds with the physical world. Indeed, since the "emptiness" of spacetime geometry of any empirical content is supposed to be a fundamental lesson of general relativity, it follows as a matter of course that its local structure must be imposed as a matter of arbitrary stipulation.

3. RELATIVITY AND "WORLD-STRUCTURE"

Contemporaneously with the work of the logical positivists, there emerged a competing interpretation of general relativity that represented the theory from a completely different philosophical perspective: not as a theory of the relativity of motion, but as a theory of the spatiotemporal structure that determines states of motion; not as a theory that takes away "the last vestige of physical objectivity" from space and time, but as a theory of the objective structure underlying our limited local perspectives on space and time; not as an uninterpreted formalism, but as itself a deep and revealing interpretation of the spatiotemporal aspects of our experience. This was the view of spacetime geometry as "world-structure."

To see spacetime theory in this light was to pick up a thread from 19th century philosophy of geometry, one which was absolutely crucial to the development of physical geometry beyond Kantianism, but for which the logical empiricist approach had no particular use. The logical empiricists certainly saw that a great transformation had taken place in the 19th century, and they saw this as central not only to the emergence of relativity, but also to their entire conception of science. But from their perspective, the seminal principle in this transformation was that geometry is conventional: once it is recognized as a purely formal structure, it follows that its formal propositions must be translated into claims about the physical world by means of stipulations about the physical objects that are to represent geometrical objects. But, carefully understood, Poincaré's conventionalism was not really based on the idea that geometry has no empirical content at all. More precisely, Poincaré certainly acknowledged that the structure of geometry may be separated from any empirical content, and contributed much to the general understanding of that separation, and what it means for our understanding of mathematics in general. And he thought of the general theory of Riemannian manifolds as something separate from empirical geometry altogether, because it represented a class of geometries that are not synthetically constructible; only the geometries of constant curvature are truly synthetic geometries, on his view, because they are susceptible of classical constructive proofs, and the rest are merely formal systems. But in the case of synthetic geometry – physical geometry – the empirical interpretation of the formal structure, according to Poincaré, was given from the start by the way in which we become acquainted with it. For space is nothing more than a conceptual representation of a central and primitive element of human experience: the existence of a group of displacements, or changes of perspective, that are definitive of spatial displacements, and thereby distinguished from all other changes that a human being might observe. In other words, every human being's ability to shift positions, and to reverse such a shift at will, acquaints her directly with the group of rigid motions, and thereby with the isometries of space. So the association between metrical structure and rigid motions is not assigned by convention, but discovered in the course of every human being's adaptation to the immediate environment. Again, convention enters only because the principle of rigid motion is insufficient to establish the global structure of space, beyond imposing the condition of constant curvature upon it.

In other words, even though the late 19th century learned to separate the formal structure of geometry from its empirical content, and on that basis to understand the possibility of an infinite variety of abstract geometries, it did not really separate physical geometry from its basis in intuition. To the extent that geometry concerned the nature of space, its structure still had a privileged interpretation, given by the intuitive operations that first define our conception of space. The essential advance beyond Kant was not to dislodge intuition from its central role, but to show that it was susceptible of a further analysis; to show that the intuitive conception of space was not based in an irreducible "form" of intuition, but in the simple experiences with physical objects and their movements that constitute the real content of our intuitions. (Cf. DiSalle 2006).

It is not difficult, in retrospect, to understand why this aspect of the 19th century view should have been difficult to translate into the setting of early 20th century physics. In the context of special relativity, what had seemed to be an objective way of constructing the geometry of space – measurement by the diplacement of rigid bodies – now was seen as constructing only a particular local perspective, dependent on a particular state of motion. Similarly, the determination of simultaneity by light-signalling, and the measurement of time by the progress of ordinary clocks, only determines simultaneous events and time-intervals for a given inertial frame. These are elementary facts about special relativity, of course, but it is important to appreciate their bearing on the problem of interpretation: they seem – at least, they seemed to the most prominent philosophical interpreters of general relativity – to create a conceptual gulf between the structure of spacetime and the constructive procedures of spatial and temporal measurement. The latter therefore could no longer constitute the basis for a natural interpretation of physical geometry; they are conventions that define arbitrary local perspectives on geometry.

If this is a fair summary of the situation of general relativity, and the philosophical confusion surrounding its interpretation and application, one might well ask, how is it possible that a useful physical theory resulted from all of this? Why did Einstein produce a theory of the structure of space-time, seemingly in spite of himself? Several possible answers come to mind. First, we might conclude from this that the actual construction of scientific theories need not be affected by the philosophical attitudes of scientists. In that case the philosophical motives to which Einstein attached so much importance, and which seemed so convincing to his philosophical followers, would simply be irrelevant to the construction or the evaluation of the theory that seemed to develop in spite of them. Or, instead, we could infer, as Kuhn often suggested, that philosophical convictions belong to the subjective and psychological motives for theory choice, helping to explain a conceptual transformation for which no rational scientific argument can be given. Both of these hypotheses grant Einstein's theory a mysterious life of its own, evolving inexplicably along the right mathematical lines without regard for its author's conscious intentions. But there is at least a third possibility, one that acknowledges the intimate connection between the theory and its philosophical context, but without falling into the confusions that grew up around it. On this view, the modern geometrical interpretation of relativity is not merely our retrospective interpretation of an old theory. Rather, it evolved from a conception of physical geometry that was intimately connected with the development of the theory, and indeed that was indispensable to its construction as a mathematical theory. It was not a revolutionary idea of 20th century philosophy, but the evolution of an older idea in the foundations of geometry. And it did indeed originate in a philosophical development, that is, in philosophical reflections on the relations between mathematics and experience: how mathematical structures acquire a physical intepretation, and how physical experience acquires a mathematical interpretation; how geometrical principles are presupposed in empirical inquiry, and how unexpected experience can sometimes challenge the presupposed framework. In short, it arose from a series of reflections on how physical geometry can be, at the same time, both a framework for and an object of empirical inquiry.

Kant could only understand one aspect of this relation, since on his view, mathematics had an interpretation that we would never have the occasion, or even the possibility, to revise. The logical positivists learned from developments in the 19th century that geometry certainly is open to revision. But since they viewed interpretation as essentially an arbitrary assignment of meaning, they could not articulate a meaningful sense in which the revision of basic interpretive principles could arise directly from empirical inquiry. Indeed, this difficulty was entrenched in Carnap's distinction between internal and external questions (1956). Empirical inquiry is only well defined within a framework for interpreting phenomena as representing formal structures; therefore empirical questions are only well posed as questions internal to a particular framework. Questions about the appropriateness of any interpretive principles are by hypothesis external to the relevant framework, and so are matters for purely pragmatic discussion. Evidently this is inherent in the very nature of linguistic frameworks as Carnap (1956) conceived them. Such a view makes it difficult, if not impossible, to understand the empirical origins of interpretive principles, or how re-interpretation may forced by new and better empirical knowledge. Yet that is precisely what led to the 20th century transformations in the physics of space and time, and those who understood this point understood those transformations – and so the philosophical significance of general relativity – much more clearly than the logical positivists ever did.

To speak of a better understanding of general relativity might seem to make the entire analysis somewhat unhistorical. If we identify a certain contemporary interpretation of general relativity as the most natural and appropriate one for ourselves, and therefore prefer it to the one articulated by Einstein and his followers, it might seem anachronistic to represent those who agree with us as the most important philosophical tradition. But this is only an apparent problem. What distinguishes the people I am considering is not that they anticipated our contemporary interpretation, but that, in their own time, they made it possible to understand relativity as an empirical theory of the structure of spacetime. And so the modern geometrical view of relativity is a direct descendant of their view, because it was they who gave the theory a form in which it could serve as the basis for the empirical study of the properties of space and time. In recognizing this we separate the genuine physical-geometrical content of the theory from Einstein's various philosophical motives – not simply in order to separate what now seems good about the theory from what seems philosophically misguided, but in order to answer a genuine historical question: how was it possible for Einstein's work, in spite of the seeming confusion of philosophical motives that brought it into being, to give rise to a physical theory of spacetime curvature? While Einstein was arguing that space and time had "lost the last vestiges of physical objectivity," how did others manage to show that general relativity somehow captures whatever objective knowledge we have of spacetime structure?

The problem was precisely analogous to that faced by Helmholtz and Poincaré in the 19th century, when it was necessary to reveal the physical principles that are the basis for our objective knowledge of space, in order to reveal the sense in which the structure of space is open to empirical investigation. But in their case, it was sufficient to provide a conceptual analysis of a familiar concept, that of space, and to show that our means of coming to know its approximately Euclidean structure could be used to determine any number of other structures, provided only that they are of constant curvature. In other words, the intimate connection between the geometry of space and the structure of the group of rigid motions was revealed by a straightforward, though subtle, analysis of how we come to distinguish spatial changes from other changes in our environment. In the case of spacetime, there was the preliminary difficulty of introducing the concept itself for the first time, and showing for the first time that certain parts of our physical knowledge constitute knowledge of spacetime. Helmholtz and Poincaré offered an analysis of our established knowledge of space; Minkowski, Weyl, and Eddington had to show that we actually do have knowledge of spacetime, if only we can bring to light what is implicit in the known laws of physics. Then, following the example of Helmholtz and Poincaré, they could argue that this characterization of our knowledge is definitive of our knowledge of physical geometry—that there is no other source of insight into the nature of space and time to which we could appeal from the results of physics.

The first and most familiar step from relativity to "world-structure" is Minkowski's formulation of special relativity as a four-dimensional geometry. Perhaps because it is so familiar, its philosophical significance is easy to overlook. We know that Minkowski represented the invariance of the speed of light as a geometrical invariance, that is, as defining an indefinite metric on the four-dimensional set of physical events or "world-points"; thus he replaced the Newtonian decomposition of spacetime into hyperplanes of simultaneity – which Einstein had shown to be dependent on the state of motion of the observer – with the invariant light-cone defined at each spacetime point. So the group of Lorentz transformations between different inertial coordinate systems, as analyzed by Einstein, can be seen as the group of isometries of spacetime, preserving not spatial or temporal intervals but the spacetime interval. And, in a well-known remark, Minkowski proposed that "the relativity postulate" is a misleading name for the theory characterized by this structure:

> Since the postulate comes to mean that only the four-dimensional world in space and time is given by phenomena, but that the projection in time and space may still be undertaken with a certain degree of freedom, I prefer to call it the *postulate of the absolute world* (or, briefly, the world-postulate). (1909, p. 80)

It is common to read this as claiming that the relations expressed by special relativity are *explained* by the existence of an underlying spacetime.[4] But the essential mathematical-physical content of this idea is only that the invariance properties of electrodynamics have a four-dimensional group structure analogous to that of a four-dimensional pseudo-Euclidean space, i.e. with an imaginary time-coordinate, where the Lorentz transformations are analogous to rotations in Euclidean space. And this idea had already been expressed by Poincaré (1905), simply as a fact about the invariance properties of the laws of electrodynamics, and entirely within the setting of a Newtonian spacetime with an electromagnetic ether. So the mere existence of the structure does not suffice to identify it as the fundamental structure of spacetime, or the underlying reason for the truth of special relativity as opposed to the theory of Lorentz (1895). Analogously, the mere possibility of deriving the Lorentz transformations from the invariance of the velocity of light and the relativity principle does not by itself show that the latter explain the former, for it is only a logical relation that does nothing to explain why one is a more plausible starting-point than the other. It is not self-evident, then, that the four-dimensional geometry that Minkowski formulated must be thought of as the structure of the "absolute world."

To understand the philosophical import of Minkowski's analysis, we need to look closely at his interpretation of the structure. Neither of the prominent philosophical views of interpretation is very illuminating here. On the positivist view, special relativity is based on the convention that light-signals define simultaneity; the structure of spacetime has to be thought of as a formalism that we interpret by the arbitrary coordination of isometries with the invariance of the velocity of light. So the only possible argument for Minkowski's view would be that these conventions define a more pragmatically useful structure than Newtonian spacetime and Lorentzian electrodynamics. On the semantic view, we have the "empirical hypothesis" that the world is a model of the Minkowski structure, and that this is the "structural explanation" for the invariance of the velocity of light (cf. Hughes 1987, p. 221). But the empirical evidence for this hypothesis is equivalent to the evidence for the Lorentz theory, and so again the only possible argument is a pragmatic one. This is not to say that such pragmatic arguments are useless or even insufficient. Whatever their merits, however, such arguments simply do not capture the actual reasoning of Minkowski. His argument was that the structure he identified is not at all an explanatory hypothesis for special relativity. Rather, it is the structure *implicit in* special relativity; Einstein's arguments for special relativity are in themselves arguments that "the world" has that structure. The spacetime structure does not explain why the velocity of light is invariant by appealing to some deeper level of reality; rather, a world in which the velocity of light is the fundamental invariant simply *is* a world with a particular spacetime structure. So the problem of interpretion is not to assign a meaning to a formalism by designating physical phenomena to represent it. Instead, it is to learn to interpret the geometrical significance of what physics tells us about space, time, and electromagnetism.

This, at least, is the way Minkowski presents the matter. Einstein has shown, he writes, that the relativity postulate "is not an artificial hypothesis, but rather a novel understanding of the time-concept that is forced upon us by the appearances" (1908, p. 56). It is Lorentz's theory that involves a hypothesis to explain the differences of local time for electrons in relative motion; Einstein's theory merely acknowledges this empirical fact as revealing something about the nature of time.

> Lorentz called the t' combination of x and t the local time of the electron in uniform motion, and applied a physical construction of this concept, for the better understanding of the hypothesis of contraction. But the credit of first recognizing clearly that the time of the one electron is just as good as that

of the other, that is to say, that t and t' are to be treated identically, belongs to A. Einstein. (1905, p. 81).

Minkowski, then, at least by his own account, is not introducing a deeper theory or a theoretical entity that is responsible for the relations that Einstein had articulated.

> [We] are compelled to admit that it is in four dimensions that the relations here taken under consideration first reveal their inner being in full simplicity, but on a three dimensional space previously imposed upon us they cast only a very complicated projection. (1909, p. 83).

The "postulate of the absolute world" is not the explanation for what Einstein had regarded as merely relative; the world-postulate is simply a better name than "theory of relativity" for what Einstein's theory *actually says*. In short, in spite of its evident break with spatial and temporal intuition, special relativity does implicitly contain a physical geometry which, *ipso facto*, has a direct physical interpretation. As surely as in the case of intuitive spatial geometry, invariant geometrical relations directly express objective physical relations. Thus the link between Minkowski's spacetime and Klein's classification of geometries is more than a mere structural analogy, or the application to physics of a useful formalism. It is in fact a demonstration that that the insights won in the 19th century into the foundations of physical geometry, by Helmholtz, Poincaré, Klein, and others, could be made independent of spatial intuition – not just in the logical sense, because the structure could be separated from the content or theorems proved analytically, but because the structure could be understood as the spatio-temporal structure that our physical knowledge directly expresses.

In the setting of general relativity, evidently, there could be no such interpretation of spacetime geometry through its group of isometries, since spacetime could no longer be assumed to have isometries, or any non-trivial symmetries. Thus the idea of an equivalence-class of privileged reference-frames could no longer be assumed to apply, though in certain peculiar physical circumstances (in specific solutions of Einstein's equation), symmetries in the distribution of matter would allow for spacetime symmetries. This is not the place to consider the collection of philosophical issues raised by the move from Lorentz invariance to general covariance (see instead Earman 1989), but one aspect of the matter is immediately relevant. From the point of view of Einstein and the positivists, the restricted character of Lorentz invariance made it desireable, on general epistemological grounds, to seek a wider covariance group with no privileged coordinate systems and, presumably, no preferred states of motion. Whatever philosophical significance we attribute to general covariance, it is clear by now that it does not have that connection with the relativity of motion that it was initially purported to have; in spite of Einstein's conviction that an "extension of the principle of relativity" to all states of motion was epistemological necessary (1916), general relativity maintained the notion of a privileged trajectory. This suggests something of the great philosophical significance of Minkowski's world-postulate for general relativity. Its most obvious significance is that general relativity, as an identification of gravitation with spacetime curvature, could not have been formulated except from the starting-point of Minkowski's four-dimensional geometry.[5] Less obviously, Minkowski's viewpoint made it possible to think of the extension of the principle of relativity in a completely different light. It emerged that the identity of inertia and gravitation, as derived from Einstein's equivalence principle, was not primarily a way of making motion completely relative (which, in any case, it failed to accomplish). Rather, it led immediately to the idea that gravitational free-fall is a privileged state of motion, and in fact identical with geodesic motion in spacetime. And this made the equivalence principle into a constructive basis for spacetime geometry, just as the light-postulate had been for Minkowski's geometry. But since these geodesics would be determined, as gravitational trajectories, by the varying distribution of matter and energy, spacetime would be generally non-homogeneous, and Minkowski's geometry approximately correct only on very small scales. In short, instead of the conclusion that a new epistemology is required, and that what we thought was absolute is really relative, the world-postulate suggested a different sort of conclusion from the generalization of relativity: what we thought was global turns out to be really local; what we thought was static turns out to be dynamical. In the hands of Weyl and Eddington, the legacy of Minkowski's view was an understanding of general relativity, not as a theory of the relativity of motion, but as the theory of a dynamical "world-structure" implicit in the behavior of falling bodies and light-rays.

Weyl's general approach to physical geometry, and the contrast between his realistic view of spacetime structure as opposed to Einstein's relativitism, have been the subjects of sufficient commentary already. His major account of general relativity, *Raum-Zeit-Materie* (1918), expressed this contrast clearly, if briefly. For him the true significance of general relativity rested in "the assumption that the World-metric is not given a priori, but the quadratic groundform is to be determined by matter through generally invariant laws." (1918, pp. 180–181); as for the physical content of the theory, the "essential kernel," he saw it less in the requirement of general invariance than in this principle [that gravitation is a mode of expression of the metric field]" (1918, p. 181). While he apparently shared some of Einstein's enthusiasm for Mach's principle, he was much more critical of it than the logical positivists,[6] who considered it merely the physical application of an undeniable epistemological principle. Weyl pointed out that the phenomena associated with "absolute rotation" are "in part an effect of the fixed stars, relative to which the rotation takes place"—adding in a footnote, "In part, because the mass-distribution in the world does not uniquely determine the metric field..." (1918, pp. 175–176). Later, in the more philosophical setting of *Philosophie der Mathematik und der Naturwissensschaften* (1927), he broadly criticized the Machian emphasis on the relativity of motion:

> Incidentally, according to the general relativity-postulate, without any basis in a world structure, the concept of relative motion of several bodies is left hanging in the air just as much as the concept of absolute motion of a single body.... Thus a solution of the problem along the lines of Huyghens and Mach, eliminating the world-structure, is impossible. (1927, p. 66).

Thus he saw the kind of relativism advocated by the positivists as reflecting a poor understanding of general relativity, and, more broadly, as a failure to understand the role of geometrical structure in physics.

In order to understand his kinship with Minkowski more completely, we need to consider his remarks on our *knowledge* of geometrical structure, and on the geometrical interpretion of our physical knowledge. These make it clear that the kinship between the two goes beyond the belief in an objective world-structure, and encompasses the same kind of insight into the ways in which a structure is revealed by physical phenomena. This begins with the most elementary assumptions about the events that we experience:

> A definite *structure* is already ascribed to the four-dimensional extensive medium of the external world if one believes in a division of the universe in the sense that it is objectively meaningful to say of any two different events, localized in space-time, that they are happening at the same place (at different times) or at the same time (at different places).... One attributes to the world a metrical structure by assuming that the equality of time-intervals and congruence of spatial configurations have an objective meaning.... (1927, p. 87).

When we formulate more complicated physical laws, we implicitly introduce more complicated geometrical structures; as the invariance of the speed of light introduced the local Minkowski structure, our conception of inertial motion introduces another kind of structure, since "the experiences which prove the dynamical inequivalence of different states of motion teach us that the world bears a structure" (1927, p. 88), namely the inertial structure or affine structure of spacetime. The philosophical objection to spacetime theories before general relativity, then, has nothing to do with their treatment of spacetime structure as something physically objective; it has to do with their failure to recognize what is revealed by the nature of free-fall, namely that such a structure cannot be a static and uniform background for behavior of matter, but must stand in genuine physical relations with matter (1927, p. 89). Minkowski's and Newton's spacetimes – flat spacetimes, thus allowing for a privileged class of global inertial frames – leave out this dynamical relation that is revealed by free-fall, and so they attempt to impose on all of spacetime a framework that can only be applied to the smallest regions. In order to understand the significance of the identity of inertia and gravitation, in sum, we need to see that it commands us to interpret free-fall trajectories as revealing the world-structure on a larger scale.

Weyl's work was complemented by that of A.S. Eddington, who offered both an alternative geometrical presentation of general relativity, and also an alternative presentation of its fundamental philosophical point of view. But the essential affinity between the two views is not difficult to see. Eddington wrote a great deal on philosophical issues connected with general relativity, especially in *Space, Time, and Gravitation* (1920). But here we are not concerned with Eddington writing "as a philosopher" or for a broad public; it is much more illuminating, I think, to consider the philosophical reasoning that appears in what he wrote directly for other physicists and mathematicians – the philosophical reasons that he deemed inseparable from the scientific case for general relativity. His text *The Mathematical Theory of Relativity* (1923) was a significant influence on the assimilation and evolution of the theory, but he first presented the case in lectures to the Royal Astrophysical Society in late 1917, his "Report on the Relativity Theory of Gravitation" (1918). And it is clear from this work that, beyond an account of the mathematics of the theory and its evidential basis, Eddington appreciated the need for a philosophical defense of the theory's basic principles. This was not, in the case of Einstein's logical positivist followers, a critique of the epistemological defects of earlier theories; rather, it was a defense of the identification of gravity with spacetime curvature, through a philosophical analysis of the means by which we can come to know anything at all about spacetime geometry. Thus it was not the sort of radical epistemology of geometry that Einstein and the positivists proposed, but an argument that general relativity, despite its radical features, did not offend against the physicist's sense of physical geometry as an object of scientific investigation.

Making this argument required a clearer articulation than was available, at least from Einstein or his philosophical followers, of the relation between general relativity and earlier theories of "absolute" spacetime structure. From this would emerge a clearer understanding of the principle that there are no privileged coordinate systems or frames of reference.

> Although we deny absolute space, in the sense that we regard all space-time frameworks in which we can locate natural phenomena as on the same footing, yet we admit that space – the whole group of possible spaces – may have some absolute properties. It may, for instance, be homaloidal or non-homaloidal.... You cannot use the same co-ordinates for describing both kinds of space, any more than you can use rectangular coordinates on the surface of a sphere; that is, in fact, the geometrical interpretation of the difference. (Eddington, 1918, p. 23).

In other words, while coordinate systems are not themselves physically meaningful, it is a fact about spacetime (an "absolute property") that it does or does not admit certain kinds of coordinatization. A famous remark of Einstein's, about the need to "free oneself of the notion that coordinates have a direct physical meaning," thus requires a fairly careful interpretation. If coordinate systems have no meaning, it is because the very idea of a classical coordinate system implies a global imposition of structure on spacetime, whereas general relativity asserts that spacetime is in general non-uniform. So the coordinatizability of a given spacetime in a given way does have a direct physical, or physical-geometrical, meaning. In fact it is just an expression of the identity between gravity and spacetime curvature.

Eddington emphasized that in speaking of curvature as an "absolute property," general relativity is not merely introducing a strange metaphysical hypothesis, and therefore raising the question whether spacetime really does have the structure that the theory says it has. His account is worth quoting at length:

> The reader may not unnaturally suspect that there is an admixture of metaphysics in a theory which thus reduces the gravitational field to a modification of the metrical properties of space and time. This suspicion, however, is a complete misapprehension, due to the confusion of space, as we have defined it, with some transcendental and philosophical space. There is nothing metaphysical in the statement that under certain circumstances the measured circumference of a circle is

less than π times the measured diameter; it is purely a matter for experiment. We have simply been studying the way in which physical measures of length and time fit together– just as Maxwell's equations describe how electrical and magnetic forces fit together. The trouble is that we have inherited a preconceived idea of the way in which measures, if "true," ought to fit. But the relativity standpoint is that we do not know, and do not care, whether the measures under discussion are "true" or not; and we certainly ought not to be accused of metaphysical speculation, since we confine ourselves to the geometry of measures which are strictly practical, if not strictly practicable. (1918, 29).

This discussion of "the relativity standpoint" calls to mind, at least initially, the attitude of the logical positivists: it suggests that general relativity is not merely a theory of space, time, and gravitation, but also a philosophical stand against metaphysics, and for the reduction of suspect metaphysical ideas to their observational content. But from what we have seen so far, it should be clear that Eddington's standpoint subtly differs from the anti-metaphysical view of the positivitists. For them, space and time were arbitrary constructions from material consisting only of point-coincidences. Eddington, however, does not advocate this sort of "eliminative" reduction; rather, he is trying to portray the structure of spacetime as accessible to empirical measurement—to show that the "absolute" structure revealed to us by such measurements is no empty metaphysical notion, but a legitimate object of empirical inquiry. Its "absoluteness" consists not in transcending the empirical evidence, in the sense criticized by Mach and the logical positivists, but in being, so to speak, empirically recalcitrant: far from being "amorphous," it has recalcitrant properties that severly limit the sort of geometrical structure that we can impose upon it. Not even the general equivalence of coordinate systems can render spacetime amorphous, for its properties determine the sort of coordinates that we can possibly construct. (It is, so to speak, amorphous to us, in the sense that it cannot be assumed in advance to have any global structure whatever, but it nonetheless does have the structure that is determined, in accord with Einstein's equation, by the distribution of matter and energy, and we come up against this structure in our efforts to impose inertial frames.)

Eddington's analysis thus recapitulates, in the context of relativistic space-time, what Helmholtz had tried to articulate in the context of space. There is a geometrical structure implicit in, and revealed by, certain characteristic physical criteria—criteria not purely phenomenal (like point-coincidences) but conceptualized in the formulation of some elementary theoretical principles. And the criteria are not arbitrarily chosen, but revealed by some conceptual analysis to be fundamental to our spatial or spatio-temporal knowledge. In other words, they are essential to our capacity to form any notion of physical geometry at all; as Helmholtz argues that the properties of space are known to us through the principle of free mobility, Eddington argues that the large-scale structure of spacetime is revealed by the properties of inertial motions. Therefore the preference for some other picture of physical geometry – as Euclidean over non-Euclidean space, or flat over curved spacetime – to the extent that it defies what our physical criteria reveal, must be seen as an instance of transcendent metaphysical prejudice. The retreat from the consequences of Einstein's theory, then, is a retreat from physical geometry altogether.

4. CONCLUSION

We have seen that the problem of the interpretation of structure, as posed in typical philosophical accounts, has little bearing on the evolution of spacetime theory as an interpretation of physical phenomena. The reason for this, it now appears, is that the usual accounts reflect an artificial separation of mathematical formalism from the physical principles that give rise to them. To think of a spacetime geometry as an abstract formalism, awaiting some convention or hypothesis that will fix its empirical content, is to forget that the formalism emerges from the attempt to understand the geometrical content of our empirical knowledge. In the 19th century, Helmholtz and Poincaré explicated the concept of space by showing that our knowledge of certain motions is, implicitly, knowledge of the structure of space; spatial geometry is the formal interpretation of this elementary knowledge. The common theme of Minkowski, Weyl, and Eddington is that our physical theories, analogously, implicitly embody our knowledge of "world-structure." The connection is not intuitively obvious, as in the case of spatial geometry and rigid displacements, but it is empirically just as direct, once the spatio-temporal content of dynamical principles is clearly understood. Understanding this connection, rather than seeing a need for conventional interpretation, was the philosophical turning-point in modern physics – the beginning of understanding that the most abstruse and counter-intuitive mathematical structures could promise insight into the deep structure of the empirical world.

University of Western Ontario
Canada

NOTES

[0] I would like to thank William Dempoulos for many discussions on the topics of this paper. I also thank David Hyder, Michel Janssen, and Ulrich Majer for their insightful comments.

[1] The first clear application of this idea to general relativity is Schlick (1917). For further comment on the motivations and eventual fate of conventionalism see DiSalle (2002).

[2] See, for example, van Fraassen (1989), Hughes (1987). My discussion is deeply indebted to that of Demopoulos (2003), to which the reader is referred for an extended analysis of this problem in a different philosophical context.

[3] See Friedman (1992, 1999b) for an extended discussion of this problem, and also DiSalle (2002).

[4] For a concise and compelling presentation of this view see Hughes (1987, 221–23). For a contrasting view see DiSalle (1995).

[5] For recent and illuminating discussion of this matter, and its connection with 19th-century work in the foundations of geometry, see Friedman 2002.

[6] In the literature of the philosophy of science, in fact, the serious re-assessment of Mach's view only began in the late 1960's, with works such as Stein (1967). See also DiSalle (2002).

REFERENCES

Carnap, R. (1956). Empiricism, semantics, and ontology, *Meaning and Necessity*, University of Chicago Press, Chicago, pp. 205–221.

Carnap, R. (1966). *An Introduction to the Philosophy of Science*, Dover Publications reprint 1995, New York.

Demopoulos, W. (2003). On the rational reconstruction of our theoretical knowledge, *British Journal for the Philosophy of Science* **54**: 371–403.

DiSalle, R. (1995). Spacetime theory as physical geometry, *Erkenntnis* **42**: 317–337.

DiSalle, R. (2002). Conventionalism and modern physics: a re-assessment, *Noûs* **36**: 169–200.

DiSalle, R. (2006). Kant, Helmholtz, and the meaning of empiricism, *in* M. Friedman and A. Nordmann (eds), *Kant's Legacy for the Exact Sciences*, MIT Press, Cambridge, Massachusetts. In press.

Earman, J. (1989). *World Enough and Spacetime: Absolute and Relational Theories of Motion*, MIT Press, Cambridge, Massachussetts.

Eddington, A. S. (1918). *Report on the Relativity Theory of Gravitation*, Fleetwood Press, London.

Eddington, A. S. (1920). *Space, Time, and Gravitation. An Outline of General Relativity Theory*, Cambridge University Press, Cambridge.

Eddington, A. S. (1923). *The Mathematical Theory of Relativity*, Cambridge University Press, Cambridge.

Einstein, A. (1905). On the electrodynamics of moving bodies, *in* A. Einstein, H. A. Lorentz, H. Minkowski and H. Weyl (eds), *The Principle of Relativity*, Dover Books, New York, pp. 35–65.

Einstein, A. (1916). The foundation of the general theory of relativity, *in* A. Einstein, H. A. Lorentz, H. Minkowski and H. Weyl (eds), *The Principle of Relativity*, Dover Books, New York, pp. 109–164.

Einstein, A., Lorentz, H. A., Minkowski, H. and Weyl, H. (eds) (1952). *The Principle of Relativity*, Dover Books, New York.

Friedman, M. (1992). *Kant and the exact sciences*, Harvard University Press, Cambridge, Massachusetts.

Friedman, M. (1999a). *Reconsidering logical positivism*, Cambridge University Press, Cambridge.

Friedman, M. (1999b). Geometry, construction, and intuition in Kant and his successors, *in* G. Scher and R. Tieszen (eds), *Between Logic and Intuition: Essays in Honor of Charles Parsons*, Cambridge University Press, Cambridge.

Friedman, M. (1999c). Poincaré's conventionalism and the logical positivists, *Reconsidering logical positivism*, Cambridge University Press, Cambridge.

Friedman, M. (2002). Geometry as a branch of physics: background and context for Einstein's 'Geometry and Experience', *in* D. Malament (ed.), *Reading Natural Philosophy: Essays in the History and Philosophy of Science and Mathematics to Honor Howard Stein on his 70th Birthday*, Open Court Press, Chicago.

Hawking, S. and Ellis, G.F.R. (1972). *The Large-Scale Structure of Spacetime*, Cambridge University Press, Cambridge.

Helmholtz, H. (1870a). Die Thatsachen in der Wahrnehmung, *Vorträge und Reden*, Vol. 2., Friedrich Vieweg und Sohn, Braunschweig, pp. 215–247.

Helmholtz, H. (1870b). Ueber den Ursprung und die Bedeutung der geometrischen Axiome, *Vorträge und Reden*, Vol. 2., Friedrich Vieweg und Sohn, Braunschweig, pp. 1–31.

Helmholtz, H. (1884). *Vorträge und Reden*, Vieweg und Sohn, Braunschweig.

Hilbert, D. (1899). *Die Grundlagen der Geometrie*, Teubner, Leipzig.

Hughes, R.I.G. (1987). *The Structure and Interpretation of Quantum Mechanics*, Cambridge University Press, Cambridge.

Kant, I. (1787). *Kritik der reinen Vernunft*, Felix Meiner Verlag, Berlin. Reprint 1956.

Klein, F. (1872). *Vergleichende Betrachtungen über neuere geometrische Forschungen*, A. Duchert, Erlangen.

Lorentz, H. (1895). Michelson's interference experiment, *in* A. Einstein, H. A. Lorentz, H. Minkowski and H. Weyl (eds), *The Principle of Relativity*, Dover Books, New York, pp. 3–7.

Mach, E. (1901). *Die Mechanik in ihrer Entwickelung, historisch-kritisch dargestellt*, Brockhaus, Leipzig.

Minkowski, H. (1908). Die Grundgleichungen für die elektromagnetischen Vorgünge in bewegten Körper, *Nachrichten der königlichen Gesellschaft der Wissenschaften zu Göttingen, mathematisch-physische Klasse* pp. 53–111.

Minkowski, H. (1909). Raum und Zeit, *Physikalische Zeitschrift* **10**: 104–111.

Misner, C., Thorne, K. and Wheeler, J. (1973). *Gravitation*, W.H. Freeman, New York.

Nagel, E. (1939). The formation of modern conceptions of formal logic in the development of geometry, *Osiris* **7**: 142–224.

Poincaré, H. (1905). Sur la dynamique de l'èlectron, *Comptes rendues de l'Acadèmie des Sciences* **140**: 1504–08.

Poincaré, H. (1913). *The Foundations of Science: Science and Hypothesis; The Value of Science; Science and Method*, The Science Press, Lancaster, PA. Translated by G.B. Halsted.

Reichenbach, H. (1924). *Axiomatik der relativistischen Raum-Zeit Lehre*, Springer-Verlag, Berlin.

Reichenbach, H. (1949). The philosophical significance of relativity, *in* P. Schilpp (ed.), *Albert Einstein: Philosopher-Scientist*, Open Court, pp. 289–311.

Reichenbach, H. (1957). *The Philosophy of Space and Time*, Dover Publications, New York. Originally published as *Philosophie der Raum-Zeit-Lehre*, Berlin, 1927. Translated by Maria Reichenbach.

Reichenbach, H. (1965). *The Theory of Relativity and A Priori Knowledge*, University of California Press, Berkeley and Los Angeles. Originally published as *Relativitätstheorie und Erkenntnis Apriori*, Berlin 1920. Translated by Maria Reichenbach.

Schlick, M. (1917). *Raum und Zeit in der gegenwrtigen Physik. Zur Einführung in das Verständnis der Relativitäts- und Gravitationstheorie*, Springer, Berlin.

Stein, H. (1967). Newtonian space-time, *Texas Quarterly* **10**: 174–200.

van Fraassen, B. (1989). *Laws and Symmetries*, Oxford University Press, Oxford.

Wald, R. (1984). *General Relativity*, University of Chicago Press, Chicago.

Weyl, H. (1918). *Raum-Zeit-Materie. Vorlesungen über allgemeine Relativitätstheorie*, Springer-Verlag, Berlin.

Weyl, H. (1927). Philosophie der Mathematik und der Naturwissenschaften, *Oldenburg's Handbuch der Philosophie*, Oldenburg Verlag, Munich.

DAVID E. ROWE

EINSTEIN'S ALLIES AND ENEMIES: DEBATING RELATIVITY IN GERMANY, 1916-1920

In recent years historians of mathematics have been increasingly inclined to study developments beyond the traditional disciplinary boundaries of mathematical knowledge.[1] Even so, most of us are accustomed to thinking of the relativity revolution as belonging to the history of physics, not the history of mathematics. In my opinion, both of these categories are too narrow to capture the full scope of the phenomena involved, though interactions between the disciplines of physics and mathematics played an enormously important part in this story.[2] A truly contextualized history of the relativity revolution surely must take due account of all three scientific aspects of the story – physical, philosophical, and mathematical – recognizing that from the beginning Einstein's theory cut across these disciplinary boundaries. Still, one cannot overlook the deep impact of the First World War and its aftermath as a cauldron for the events connected with his revolutionary approach to gravitation, the general theory of relativity.[3]

Historians of physics have, however, largely ignored the reception and development of general relativity, which had remarkably little impact on the physics community as a whole. Hermann Weyl raised a similar issue in 1949 when he wrote: "There is hardly any doubt that for physics special relativity theory is of much greater consequence than the general theory. The reverse situation prevails with respect to mathematics: there special relativity theory had comparatively little, general relativity theory very considerable, influence, above all upon the development of a general scheme for differential geometry" (Weyl, 1949, 536–537). This was the mature Weyl speaking, which is perhaps why he said nothing about the wider implications of relativity for the history of human thought. Back in 1918, when he wrote *Raum-Zeit-Materie* (Weyl, 1918), the philosophical implications of relativity were at the forefront of his mind. Weyl was younger then, but his former mentor, Hilbert, was just as excited about the import of Einstein's ideas for the post-war era (Hilbert, 1992). In this specific cultural context relativity had a deeply polarizing effect that is easy to document but difficult to explain.

A central assumption of this essay is that the highly politicized debates on relativity that took place in post-war Germany cannot be understood in terms of disciplinary developments alone. A variety of other lenses are thus required to study the nexus of political and scientific issues surrounding relativity, one being the semi-popular scientific literature on relativity as well as

V.F. Hendricks, K.F. Jørgensen, J. Lützen and S.A. Pedersen (eds.), Interactions: Mathematics, Physics and Philosophy, 1860-1930, pp. 231–280.
© 2006 Springer.

the popular articles in the press.[4] Another is gained by looking at the activities of leading pro- and anti-relativists during these years.[5] By closely examining these sources and events, new insights emerge that can help us gauge the ideological import of the relativity movement and the countermovement it spawned.

1. PUBLICITY AND FAME

Both movements were, of course, reactions not only to Einstein's ideas but also to his personality and fame (Rowe, 2006). Indeed, a significant factor that shaped the relativity debates in Germany stemmed from the circumstance that Einstein's fame was launched by British and American newspapers. In banner headlines they announced the confirmation of his general theory of relativity in November 1919 (Fölsing, 1993, 513–533). The German press, by contrast, reacted far more soberly. A month later Einstein's achievement made a comparable splash in his native land; but this time the new wave of interest came from an image rather than enlarged print. On 14 December 1919 a brooding face appeared on the cover of the *Berliner Illustriter Zeitung* above the caption: "A New Giant of World History: Albert Einstein, whose research signifies a complete overturning of our view of nature comparable to the insights of Copernicus, Kepler, and Newton." The text that followed offered a compact history of cosmology from Copernicus to Einstein, two great revolutionary thinkers:

> In the year 1543 a new chapter began in human thought and understanding, indeed in the entire development of humanity. This was brought about by the publication of one simple book, *De revolutionibus orbium coelestium*, the work of Nicolaus Copernicus (Grundmann, 1998, 116–117).

Readers of *BIZ* learned that the Canon of Frauenburg "dethroned the earth from its place in the cosmos; man was no longer in the center of the creation." Then came Kepler, Galileo, and Newton, whose laws governed the motions of celestial bodies for two centuries... till various anomalies began to creep in, but even such a "luminary physicist as Heinrich Hertz, the spiritual creator of the wireless telegraph and telephone, sought in vain to resolve the contradictions" in Newton's mechanics. Finally, along came Einstein, who

> ... probed our conventional conceptions of space and time with daring and determination.... By showing that we cannot determine any absolute motion and that all of our physical measurements must involve the concept of time he overcame the

gaping contradictions between electrodynamics and Newtonian mechanics.... The solar eclipse of 29 May 1919 became the crucial test for Einstein's ideas, and the English researchers who worked on obtaining these results had to concede that Einstein had emerged victorious over Newton. Newton's law of gravitation became only a special case of Einstein's theory of relativity. Our notions of space and time must be changed to accord with Einstein's theories, and just as before in Copernicus's time, a change in our image of the world has once again come about. A new epoch in human history has now arisen and it is indissolubly bound with the name of Albert Einstein (Grundmann, 1998, 117).

The *BIZ* stood at the forefront of a new trend in journalism in which photography rather than the printed word dominated the page. Few remembered the stories, but the *BIZ*'s images left a lasting impression on many of its million-plus subscribers, nearly three times that of the popular *Berliner Tageblatt*.[6] Editor-in-chief Kurt Korff exploited the camera's potential for conveying the dramatic events of the day, and its impact on popular culture in Germany was enormous. On top of this, Einstein was unusually photogenic, which helps to account for why people were constantly taking snapshots of him. His son-in-law, Rudolf Kayser, found the famous title-page portrait for *BIZ* at once authentic and powerful, noting how it evoked a sense of awe in and reverence for the man heralded as the greatest scientific genius of the day.[7]

Although Einstein has often been seen as an otherworldly sage totally indifferent to his own worldly fame, he was nevertheless deeply puzzled by the psychological roots of the relativity revolution. Sometime during the 1940s he wrote that:

> For me it was always incomprehensible why the theory of relativity, whose concepts and problems are so far removed from practical life, should have found such a lively, even passionate resonance in the widest circles of the population for such a long time. Since the time of Galileo nothing quite like that has happened. Yet then the church's officially sanctioned view of man's place in the cosmos was shaken – an event of patent significance for cultural and political history – whereas the theory of relativity is concerned with the attempt to refine physical concepts and to develop a logically complete system of hypotheses for physics. How could this have occasioned such a gigantic and long-lasting psychological reaction?[8]

This suggests that, on the one hand, Einstein found the parallels between "his" scientific revolution and the one linked with the names of Copernicus and Galileo far-fetched. On the other hand, he was convinced that the reception of relativity in Germany after 1919 was deeply influenced by political factors, especially anti-Semitism (Stachel, 2002, 63–64).

Like Einstein, Philipp Frank was puzzled and intrigued by the relativity revolution. Frank had succeeded Einstein as professor for theoretical physics in Prague. There, back in November 1919, he delivered a semi-popular lecture on Einstein's theory of relativity that led to a bitter debate with the philosopher Oskar Kraus, a vehement anti-relativist (Hentschel, 1990, 173–176). Thus Frank was an early participant in and witness to the storm that began brewing over Einstein's theory in the tense, ethnically divided atmosphere of Prague. After both Frank and Einstein fled the Third Reich for the United States, they had ample time to contemplate how the theory of relativity quickly became entangled with concurrent political ideologies and cultural slogans. Einstein thereafter encouraged his friend to explore this theme by grappling with the social, political, and psychological roots of the relativity debates. The result was Frank's biographical study, *Einstein, sein Leben und seine Zeit* (Frank, 1949), written during the early war years, but first published in complete form and in the original German in 1949. The undated quote cited above comes from a preface Einstein wrote for this book, probably the abridged English version (Frank, 1947). But this preface never appeared, thereby depriving the world of any insight into the role Einstein had played in promoting Frank's study.

Philipp Frank related an anecdote that nicely captures how the politicized atmosphere in France affected attitudes toward Einstein and relativity. During Einstein's trip to Paris in 1922 a distinguished historian at the Sorbonne commented: "I don't understand Einstein's equations. All I know is that the Dreyfus adherents claim that he is a genius, whereas the Dreyfus opponents say he is an ass. And the remarkable thing is that although the Dreyfus affair has long been forgotten, the same groups line up and face each other at the slightest provocation" (Frank, 1949, 314).

Berlin was a very different place than Paris, yet nevertheless its highly charged post-war milieu had a strikingly similar effect on the reception of Einstein's theory (Levenson, 2003). Nor did Londoners know whether to rejoice or throw up their arms in dismay. Einstein's career as a cultural celebrity began in London with the dramatic scientific meeting of 6 November 1919 which announced the results from two British expeditions that measured the deflection of light passing through the sun's gravitational field (Pais, 1982, 303–312). Their findings, so it seemed, confirmed Einstein's quantitative prediction based on general relativity. This event unleashed strong scientific as well as political currents that brought Einstein far more publicity, but also notoriety, than he had ever experienced before, despite longstanding controversies over the theory of relativity.

In England, where every schoolboy had been taught to revere the immortal genius of Isaac Newton, advocates of general relativity like Arthur Eddington had to step forward carefully. Einstein was forewarned about this state of affairs by Frederick A. Lindemann, head of the Clarendon Laboratory, who informed him that "national feeling was wounded and the world moved to a state of alarm" by the reports on relativity in *The Times*.[9] Knowing this, Einstein wrote a tactful article for the London paper in which he addressed the "downfall of Newton" with these soothing words:

> No one must think that Newton's great creation can be over-thrown in any real sense by this or any other theory. His clear and wide ideas will forever retain their significance as the foundation on which our modern conceptions of physics have been built (*CPAE*, vol. 7, 214).

Einstein and British pro-relativists were quite successful in their efforts to allay fears about the revolutionary character of the new theory of gravitation, but not entirely. The old guard, especially Oliver Lodge and Joseph Larmor, still tried to salvage ether physics in the grand Cambridge tradition of George Gabriel Stokes and co. (Warwick, 2003) The controversies surrounding relativity in England were, however, relatively mild compared with those that took place in Germany, where opposition to Einstein's theory had been mounting since 1916.

2. ARNOLD BERLINER'S *DIE NATURWISSENSCHAFTEN*

A pivotal figure in this story was Arnold Berliner (1862-1942), editor of the weekly scientific newspaper *Die Naturwissenschaften* published by Springer Verlag.[10] This widely read publication brought relativity theory to the scientifically educated public in an accessible and sometimes provocative form.[11] Berliner was the author of *Lehrbuch der Physik*, a highly acclaimed textbook that went through five editions. He worked for many years as a technical expert for Emil Rathenau, founder of the *AEG* (*Allgemeine Elektrizitätsgesellschaft*). Rathenau had purchased the rights to several of Edison's patents in the 1880's and soon thereafter he appointed Berliner to head his filament lamp factory. During his twenty-five-year career with the *AEG* he helped develop incandescent carbon lighting, X-ray bulbs, and early gramophones. By all accounts Berliner's years with Rathenau's firm were successful ones, but his boss was not an easy man to please, and so he chose to resign in 1912.

Just one year later, the first issue of *Die Naturwissenschaften* was already in print, and Berliner's new career as an editor had begun. His success was part of the larger story of how Springer-Verlag came to dominate scientific publishing in Weimar-era Germany.

During and immediately after the war, the Berlin firm of Julius Springer emerged as a dynamic new force in this field. Founded in the mid 19th century, Springer gradually moved from the publication of political pamphlets written by progressive figures from the 1848 era to weightier tracts showcasing German work in the fields of engineering and medicine.[12]   Julius Springer and his two sons, Ferdinand and Fritz, were also great aficionados of the game of chess, which accounts for the Springer logo (the knight piece is called a Springer in German). By 1907, the grandsons, Ferdinand and Julius, still in their twenties, were running the firm which had since grown from four employees in 1877 to 65 in just thirty years. The two cousins established a fairly clear division of labor: Julius was responsible for the applied side – engineering and pharmacy – whereas Ferdinand took on the natural sciences and medicine. In 1911 they moved their company into new quarters on the Linkstrasse near Potsdamer Platz, where it continued to flourish. By 1913 Springer stood at the front ranks among German scientific publishers, producing 379 titles, second only to Teubner in Leipzig, which published a large number of cheap school books (by total list price, the leading publisher was Gustav Fischer in Jena, followed by Springer and then Teubner). The Leipzig giant had a virtual lock on mathematics publications, however, whereas Springer dominated engineering and had strong programs in medicine and the natural sciences as well as in the fields of law, political science, commerce, and trade.

In the wake of World War I, Ferdinand Springer made a strong move to enter the publishing market in mathematics and physics. Teubner, wary of the economic risks of doing business during the early years of the Weimar Republic, had begun to pull out, and Springer quickly seized the opportunity to fill the void created by his competitor's departure (Sarkowski, 1996, 261). Thus, in 1920 the Berlin firm assumed responsibility for publishing *Mathematische Annalen*, the journal made famous by Klein and Hilbert in Göttingen. When the first issue appeared, its new cover revealed that Einstein had now joined Klein, Hilbert, and Otto Blumenthal as the fourth editor, an opportunistic move that reflects Göttingen's long-standing efforts to strengthen ties between mathematics and physics. Soon thereafter, Springer began fostering especially close relationships with two leading Göttingen figures, Richard Courant and Max Born, who became his principal advisers for publications in mathematics and physics, respectively. Courant even received a generous salary from Springer for his ongoing assistance, certainly an unusual arrangement for the time (Sarkowski 264).[13]

It was during the years of expansion right before the war that Arnold Berliner approached Ferdinand Springer about his ideas for launching the semi-popular weekly, *Die Naturwissenschaften* (Sarkowski, 1996, 192–196). Berliner's brainchild was originally modeled on the British journal *Nature*, though his subsequent influence as both editor and entrepreneur was quite singular. In fact, this undertaking amounted to a novel kind of joint-venture, since Berliner was not an official employee of the Springer Verlag, even though he worked in an office in the company's building on the Linkstrasse. He received a monthly honorarium of 400 marks plus 10% of the net profits. Authors were also handsomely paid, anywhere from 160 to 250 marks per sheet of 16 pages; each weekly issue contained about 24 pages and a quarterly subscription cost 6 marks.

Arnold Berliner brought enormous energy and enthusiasm to the task at hand. As a regular participant at the bi-weekly meetings of the Berlin Physics Colloquium, he developed friendly contacts with a number of leading physicists, including Einstein. Berliner also quickly sought out a number of leading experts on relativity. Thus, Erwin Freundlich and Moritz Schlick wrote lengthy articles for *Die Naturwissenschaften* on the astronomical and philosophical significance of relativity, respectively.[14] Berliner later issued these as separate brochures with Springer-Verlag, an arrangement that reflects the special relationship he established with the publishing house that produced his journal. Indeed, personal contacts played a key role in solidifying the ties between Springer and the Göttingen community. Courant, Born, and the slightly older Berliner all came from Jewish families in Breslau, a milieu that produced many of the young talents who gravitated into Hilbert's circle in Göttingen. Berliner met Born through their mutual friendship with Berliner's cousin, the Breslau physician Alfred Neisser (Born, 1978, 79). Already in 1913 Berliner introduced Born to Ferdinand Springer (Sarkowski, 1996, 266). These contacts and circumstances helped Berliner and Springer to promote the mutual interests of mathematicians and physicists who stood at the cutting edge of research on Einstein's new gravitational theory, the general theory of relativity.

At the same time, *Die Naturwissenschaften* served as a forum not only for showcasing this work but also for airing ongoing debates on the foundations and philosophical import of relativity theory (see fig. 1 below and the literature cited after the bibliography). During the war years, Berliner tried to keep the German scientific community well abreast of British interests in testing Einstein's theory. Sir Frank Dyson, then Royal Astronomer at Greenwich, had already begun laying out plans for the two eclipse expeditions in 1917. Despite the wartime blockade and increasingly hostile scientific relations, Berliner was able to obtain copies of Dyson's publications. That same year *Die Naturwissenschaften* informed the German public about this British project in a detailed report written by the Potsdam astronomer Otto Birck (Birck, 1917).

Berliner was also acutely aware of the mounting opposition of certain German physicists to Einstein's general theory of relativity. As the crucial date of 29 May 1919 approached, he informed Einstein about his latest plans to publicize British scientific opinion on the eve of this long-awaited event: "To the joy of Anglophobes like Lenard, Stark, and others, I'm publishing the following communication from *Nature* in the first May issue" (Berliner to Einstein, 9 April 1919). Berliner invited Einstein to check his translation of a notice in *Nature* on Eddington's *Report on the Relativity Theory of Gravitation*, which contained the following reflections:

> Einstein's 1905 paper on the principle of relativity gave the physicists of the world a new subject of controversy. For ten long years the argument went on between those who clung to the ether as the firm foundation of the universe and those more mathematically oriented physicists who found Einstein's elegant abstraction offered a solid stronghold and who were untroubled that ether, space, and time began to totter. And while both sides argued, the originator of all this excitement quietly prepared something still greater: a general relativistic theory of gravitation. Eddington noted that Einstein's theory explained the famous anomaly in the movement of Mercury's perihelion without recourse to a new constant or any trace of a contrived agreement.[15]

Still, the editors of *Nature* were quite ambivalent about what this all meant:

> What can one say about a theory that surpasses Newton's wonderful accomplishments by assuming the complete relativity of time and space? While we are amazed by this magisterial theory and by its grasp of a till now unrecognized conceptual unity, some difficult questions still remain. If this dream of complete relativity is really true, then we approach a point of such generality that we lose touch with ordinary experience. The new law of gravitation lacks the astonishing simplicity that characterizes Newton's law of gravitation. The old problem of rotation is thrown back further, but it remains true that

there are reference systems with respect to which the dynamical phenomena can be represented with the greatest simplicity.
We ask why our first naïve choice of a self-evident system of
measure is such that the material bodies within this system
maintain a nearly constant form and that light has a nearly
constant velocity? Generalization is the highest intellectual
accomplishment, but perhaps it leaves us thirsting for particu-
larity and simplicity. Eddington's report on what is certainly
the most remarkable publication during the war leaves us reflecting about the direction in which the greatest satisfaction
lies (*ibid.*).

Berliner also informed Einstein of another article in *Nature* from 6 February 1919 in which Eddington described the three-pronged character of the
British undertaking to determine whether light has weight: a null result, the
half-value deflection (later called the Newtonian value, but first presented by
Einstein in 1911 on the basis of the equivalence principle); or the full Einsteinian deflection.[16] As it turned out, Eddington and the pro-relativists had
to construct a rather delicate web of arguments about the quality of their data
in order to salvage an outcome consistent with general relativity.[17] But these
technical difficulties were all but ignored amid the hullabaloo that followed
the initial announcement. German anti-relativists afterward objected that the
experts in their country were not given access to the British data analysis
until much later.[18] By the time experts in Germany could scrutinize the results, Einstein had already been crowned the "new Copernicus" by leading
pro-relativists and the popular press.[19] These developments set the stage for
the highly volatile events that took place in Berlin in 1920.

Even before the British announcement, proponents of general relativity had lionized Einstein's achievement, anticipating that empirical support
would vindicate his theory of gravitation. Two early advocates were Hermann Weyl and Max Born. For Weyl, writing in 1918, relativity represented
a revolution in human thought, a conceptual breakthrough that overturned
conventional understanding of the nature of time and space:

With Einstein's theory of relativity human thought regarding
the cosmos has climbed to a higher level. It is as if a wall
that separated us from the truth has suddenly collapsed: now
lie before our searching eye of knowledge wide expanses and
new depths of which we earlier had not even an inkling. We
have taken a giant step closer to grasping the Reason (*Vernunft*) underlying the processes of physical world.... in our
time, a revolutionary storm has broken out, toppling all those

Arnold Berliner and the "GRT Offensive," 1916–1922

Reviews of Literature

(Einstein, Born, Thirring, Kopff, Reichenbach)

Counterattacks on Critics

(Einstein, Laue, Born, Thirring)

Philosophy of Space-Time

(Schlick, Reichenbach)

Die Naturwissenschaften

(Arnold Berliner)

British Expedition of May 1919

(Berliner, Birck, Freundlich)

Empirical Support for GRT

(Freundlich)

FIGURE 1. For the defense belt surrounding Einstein and his theory, see Hentschel 1990, 165.

conceptions of space, time, and matter that have hitherto been regarded as the firmest pillars of natural science; though only to open the way for a freer and deeper view of things. This revolution was essentially realized through the conceptual work of one individual man, Albert Einstein (Weyl, 1918, pp. 1–2).

Born went even further in personalizing Einstein's scientific accomplishments by writing a short biographical sketch as an appendix to his popular book on relativity theory (Born, 1920). Therein he described Einstein as "an unusual person; not a researcher caught up in his abstract thoughts, but rather a full-blooded living individual who participates fully in all things and events of the world with complete and everlasting love for his fellow humans" (*ibid.*). Born speculated that Einstein's breakdown in 1917 came about not merely through overwork but also as a result of the emotional stress he felt due to the ghastly consequences of the war. He also reported that Einstein had expressed total confidence regarding the confirmation of his theory by Eddington and his colleagues: "He believed in his theory because it seemed to him almost self-evident, and it turned out he was right. Today he is a very famous man, perhaps the best-known German scholar; but he has remained all along a simple, unassuming person. We physicists honor him as the leader in a new era of research" (*ibid.*).

When the second edition of Born's book came out two years later, this appendix was missing as was the photo of Einstein that adorned the first edition.[20] As Max Born later related, the decision to omit these came after Max von Laue insisted that such extra-scientific features provided anti-relativists with ammunition for one of their central arguments, namely that the success of relativity theory was not based on hard empirical evidence but rather on the ability of its advocates to stir up public fascination for ideas that were as bizarre as they were incomprehensible (Born, 1978, 198). Thus by late 1920, advocates of relativity theory in Germany found themselves suddenly in an awkward and unfamiliar position.

A major motivation behind the attacks on Einstein and relativity stemmed from envy over his sudden rise to fame. Some of his opponents saw him as the initiator of a massive propaganda campaign aimed at promoting relativity. Even those who were deeply interested in the theory became dismayed by the buzzing excitement after November 1919.[21] Less than a year later, Arnold Berliner became so incensed with the glut of bogus popular literature on relativity that he sent the following alarming message to Hans Reichenbach, Einstein's self-appointed bulldog in the field of philosophy, calling for decisive action:

> I think you must next time write an article [for *Die Naturwissenschaften*] about the inadequacy and incompetence of the people who are now busy popularizing relativity theory, whether by lectures, articles, or brochures. Most of these popularizers naturally acquired their wisdom from some kind of popularized brochures, as it is clear that they cannot read Einstein's original works, and one should really put a stop to what these people are doing. Having learned the bare essentials from one of the little brochures, they write one of their own in order to make public that which they've not even digested themselves. These are the kind of people for whom Schiller's verse applies: "That which they learned yesterday, they already want to teach tomorrow. Alas, these gentlemen have truly short intestines.[22]

3. ERNST GEHRCKE AND THE ROOTS OF THE ANTI-RELATIVIST MOVEMENT

Most accounts of the anti-relativist movement in Germany begin with the highly visible events associated with Einstein's public fame and the backlash that followed in 1920. This, however, overlooks the whole dynamic that led up to the explosion. In tracing the roots of this movement, a single key figure emerges: Ernst Gehrcke.[23] A well-known experimental physicist, Gehrcke worked at the Physikalisch-Technische Reichsanstalt; he also regularly attended Heinrich Rubens's bi-weekly colloquium along with Einstein and Berliner. An old-fashioned ether theorist, Gehrcke thought that Stokes's theory, which assumed that the ether was dragged along by the earth, could still be salvaged. In optics, Gehrcke's main area of expertise, Einstein introduced a radically new treatment of time, one which Minkowski then took as the basis for a new space-time geometry. As a traditional Kantian, Gehrcke claimed that this geometry could not be visualized, but he also tried to undermine relativity by suggesting that the work of Einstein and Minkowski was less original than many thought. He thus claimed that an obscure Hungarian philosopher named Melchior Palagyi (1859–1924) had earlier concocted a similar theory for combining space and time. This assertion, as Klaus Hentschel has pointed out (Hentschel, 1990, 153–155), was patently ridiculous, but Gehrcke kept pushing his claim for Palagyi's priority for over ten years. The tactic of casting doubt on the intellectual property rights of others was thus part of Gehrcke's arsenal of weapons from the beginning; it would later play a major part in his ongoing campaign to expose Einstein's theory as a hoax. Gehrcke leveled this charge against relativity theory already in 1912, when he asserted that its success was the result of a massive propaganda effort.[24]

In 1911 Gehrcke wrote about the "limitations" of the principle of relativity, but in 1913 he published his "objections" to the theory in Berliner's new journal (Gehrcke, 1913a). This article presented four "substantial" arguments against relativity theory alongside the claim that its success was an effect of mass psychology. The four main points were, first, that the principle of special relativity failed to extend beyond inertial frames of motion, making it, strictly speaking, worthless for terrestrial experiments. Second, Gehrcke claimed Einstein's treatment of time dilation led to absurdities such as revealed by the clock paradox. He further accused Einstein of mystifying time by presenting thought experiments involving living organisms rather than just clocks. Third, Gehrcke cited the objection that special relativity dispenses with the ether, and fourth, that it had not shown a way to incorporate gravitation into the theory. He concluded that, in view of these objections, the electron theory of Lorentz should be reinstated, since "the only question that remains is whether one can find any useful pieces among the rubble of the old relativity theory" (Gehrcke, 1913a, 66). He then called "classical relativity theory" a mixture of mutually contradictory premises, and an interesting case of "mass suggestion" in physics. Here he likened the fuss over relativity theory in German-speaking countries with the furor caused in France around the turn of the century by the "discovery" of N-rays, a bogus form of radiation announced by researchers in Nancy.

Such charges could not very well go unanswered, and Berliner thus turned to Max Born, who countered Gehrcke's attack in the very next issue of *Die Naturwissenschaften* (Born, 1913). Berliner was no doubt concerned that the readership he hoped to cultivate might be unduly influenced by such vehement arguments against relativity. At any rate, Born did his best to undermine Gehrcke's authority. Regarding the issue of who was competent to judge the merits of the new theory, Born noted the strong support it had received from mathematicians, who tended to be far more exacting critics than the typical experimental physicist. "Relativity theory," he went on, "not only had withstood mathematical criticism it had received its actual formal clothing from one of the foremost German mathematicians of our time, Minkowski."[25] This was more than mere name dropping, since Born had studied under the since deceased Minkowski in Göttingen. In this case, he appealed to the latter's geometric representation of space-time to counter Gehrcke's claim that Einstein's treatment of time in SRT led to a contradiction.[26]

Gehrcke replied in the very next issue with counterarguments that seemed to evince gross ignorance of the fundamental assertions of the special theory of relativity (Gehrcke, 1913b). Probably he had made no serious effort to understand the theory, but he still wanted to make sure that he got his main message across, namely that he and other physicists found Einstein's arguments untenable. The clock paradox, he insisted, remained unresolved, and this topic formed the substance of his first direct encounter with Einstein one year later.

Gehrcke and Einstein presumably met each other for the first time in May 1914, shortly after Einstein settled in Berlin. Both regularly attended the Berlin physics colloquium, which convened on Wednesdays at the university. As it turned out, the sessions on May 20 and 27 were devoted to the foundations of relativity theory. Einstein reported to Otto Stern that he was very pleased with the atmosphere in Berlin, and that "gravitation elicits just as much respect among my colleagues as skepticism" and so he planned "to lecture on it in the near future in the colloquium."[27] In the course of these colloquium discussions, the phenomenon of time dilation and the clock paradox were taken up. Gehrcke continued to pester Einstein about this afterward, as the latter mentioned to Stern: "yesterday I spoke with Gehrcke. If he had as much intelligence as self-esteem, it would be pleasant to discuss things with him" (*ibid.*). Gehrcke clearly hoped to capitalize on this opportunity in order to obtain Einstein's own account of the clock paradox. Einstein claimed that a clock B set in motion with respect to a synchronized clock A at rest will slow down. Thus, if B should set out on a journey that departs from and returns to the location of A the readings on the two clocks will not coincide. Gehrcke believed this led to a direct contradiction within special relativity. He reasoned that one can view B as at rest while A moves relative to it. He thus posed this very problem to Einstein, who responded as follows: "the clock B, which was moved, falls behind because, in contrast to the clock A, it underwent accelerations. These accelerations are, in fact, irrelevant for the size of the time difference between the two clocks, but their presence nevertheless conditions the slowing of clock B and not that of clock A. Accelerating motions are absolute in the theory of relativity."[28]

Gehrcke recorded this statement and incorporated it into a short announcement that he submitted to *Die Naturwissenschaften*, but which never appeared. According to Gehrcke, this note was initially accepted for publication and was even sent to the printer (Gehrcke, 1924a, 34–35). Afterward, however, Berliner decided to send the page proofs to Einstein who pleaded for him to suppress publication. The explanation Gehrcke received, presumably from Berliner, who in all likelihood was repeating Einstein's words *verbatim*, read: "I am altogether opposed to the publication of this note because – out of context – it can create confusion, despite being to a certain degree correct."[29] Einstein's position no doubt seemed reasonable to Berliner, who must have suspected the motives of the pesky Gehrcke anyway. The outbreak of the war soon afterward diverted Gehrcke's attention and energy, but he never forgot this incident which he regarded as an object lesson in how the Einstein clique tried to stifle criticism of relativity theory.

Within the Berlin physics community, the new theory of gravitation stirred even more interest by early 1916. In November 1915, Einstein had published four brief announcements in the proceedings of the Prussian Academy containing the now famous generally covariant field equations as well as an argument that claimed to account for the discrepancy in the shift of Mercury's perihelion (Einstein 1915a, 1915b, 1915c, 1915d). These results were then elaborated in a lengthy article that sketched the mathematical tools and physical concepts in this new theory of gravitation (Einstein, 1916). This paper was published in March 1916 in *Annalen der Physik*, the journal coedited by Max Planck and Willy Wien, and a leading outlet for contributions to theoretical physics in Germany. Within almost no time, Planck and Wien were staring at another manuscript bearing the title "On the Critique and History of the Recent Gravitational Theories." Its author (no surprise) was Ernst Gehrcke, but many surely must have been surprised by the thrust of his criticisms.

Gehrcke repeated his claim that Einstein's new theory was in no sense a generalization of the so-called special theory (Gehrcke, 1916). In making this point, it appears that Gehrcke had deluded himself into thinking that his earlier criticisms of the special relativity had caused Einstein to rethink his position, and he surmised that "Einstein was no longer satisfied with the old relativity theory." A superficial reading of the introduction to Einstein's 1916 paper might well have led to this faulty conclusion, particularly if the reader was someone who, like Gehrcke, thought that special relativity constituted a flawed theoretical approach. But, predictably, he also found Einstein's treatment of gravitation every bit as problematic as his theory of electrodynamics in inertial frames. The very idea that one could replace the idea of gravitational force, as expressed precisely in Newton's law of universal attraction, by the notion of bodies constrained to move along geodesics in a spacetime manifold was a daring leap that many refused to make, even well after 1916. Gehrcke rejected the legitimacy of Einstein's equivalence principle, and pointed out that the admissibility of arbitrary coordinate systems leads to physically absurd conclusions. This latter theme would soon be taken up by Philipp Lenard, for whom it served as the central issue in his subsequent attack on Einstein's general theory of relativity. It should be pointed out that many open-minded experts shared a similar sense of dizziness when it came to the physical interpretation of gravitational fields induced by the inertial effects of arbitrary motions in general coordinate systems. One finds lengthy discussions of these very issues in Einstein's friendly and cordial correspondence with Lorentz, Gustav Mie, and others during this period (see *CPAE*, vols. 8A, 8B).

If Gehrcke's earlier writings against relativity could be dismissed as weak and ineffectual, here he nevertheless hit on something new and surprising that enabled the antirelativists to mount a brief counterattack. Indeed, the issue he raised proved an important factor in the campaign to discredit not only Einstein's theory but his personal integrity. It concerned an obscure publication from 1898 by a Gymnasium teacher named Paul Gerber, who sought to account for the slight deviation in the movement of Mercury's perihelion by treating gravitation as a force acting between bodies that was transmitted at the speed of light (Gerber, 1898). If we bear in mind that in

1916 the shift in Mercury's perihelion was the only empirical support Einstein could offer in support of his new theory, then it is easy to see why this portion of Gehrcke's criticism was potentially devastating. As Gehrcke himself emphasized, Gerber's theory of gravitation was obviously far simpler than Einstein's. Yet, with it Gerber had obtained precisely the same formula Einstein had derived using general relativity. Gehrcke backed up this claim by writing down the two formulas and showing that they were simply different expressions for the same result. He then added: "one could think that this was a great coincidence, and that Einstein might have arrived at the same result without knowledge of Gerber's work. Such an assumption, however, is weakened by the circumstance that Gerber's publication is discussed in Mach's *Mechanik* (Mach, 1904, 201), and Einstein demonstrated his precise knowledge of the contents of this well-known book in his recent obituary of Mach" (Gehrcke, 1916, 124). With that, Gehrcke clearly expressed his opinion that Einstein was not above stealing other people's work to keep his relativity show going. Up until this time, Gehrcke had maintained his view that relativity theory was a hoax; now he called the leader of the relativity movement a plagiarist.

Einstein let Willy Wien know that he found Gehrcke's attack "tasteless and superficial" and that he had no intention of answering his charges.[30] Nor did anyone else make a move to refute publicly what Gehrcke had written in Germany's most prestigious physics journal—at least not right away. Perhaps with a growing sense of confidence that he had finally called the relativists' bluff, Gehrcke played his next trump card: in 1917 he republished Paul Gerber's lengthier account of his theory (Gerber, 1902) in *Annalen der Physik*.

By this time, the Nobel laureate Philipp Lenard decided that the time was ripe to jump on the anti-relativists' bandwagon.[31] He therefore prepared his own independent version of an ether-based gravitational theory (Lenard, 1918), which appeared in Johannes Stark's *Jahrbuch für Radioaktivität und Elektronik*. Compared with Gehrcke's numerous writings, this text was both more substantive and far less polemical. Lenard made no hint of possible plagiarism, but he did enter a strong plea that Gerber's achievement should receive due credit. Moreover, by coupling this claim with a lengthy critique of general relativity, he clearly sought to throw his weight behind Gehrcke's campaign to discredit Einstein's whole approach. What, after all, was the point of all that mumbo-jumbo about curved Riemannian space-time geometry if the very same formula for the precession of planetary perihelia had already been derived by Gerber in a straightforward fashion?

Lenard's motives can be gauged more precisely from an earlier letter he wrote to Stark, dated 16 July 1917, around the time he composed a first draft of Lenard 1918. In this letter Lenard describes the fourfold purpose of his effort: 1) to support the claims for an ether theory, 2) to do the same for Gerber's results, 3) to locate the Achilles's heel of the principle of general relativity, and 4) to give an account of gravitation based on the ether.[32] A few months later, still before his paper had gone to press, Lenard's equanimity was shaken after he read the criticisms leveled against Gerber's work by both Max von Laue and Hugo von Seeliger. This clearly had a sobering effect, and he immediately telegraphed Stark asking him to withhold publication so that he could revise the manuscript. This Stark did, enabling Lenard to rework his paper, which he resubmitted in February 1918 along with a letter in which he noted that his praise for Gerber had been toned down in the new version.[33] These circumstances help to account for why Lenard adopted a cautious approach in his critique of general relativity.

In Lenard's published essay he aimed to demonstrate that the principle of general relativity could not be upheld, and that a gravitational theory based on ether physics offered a more plausible alternative.[34] For this purpose he took his cue from Einstein's discussion of the relativistic effects inside a moving train car, a thought experiment he had used to illustrate the equivalence principle (Einstein 1917, *CPAE*, vol. 6, 464-466). One imagines that the train car suddenly brakes causing objects within it to be thrown about. As Lenard correctly observed, according to Einstein's theory the passengers inside the car were not entitled to conclude that these effects were necessarily due to a change in the train's state of motion. Such an assertion would be tantamount to regarding the train's motion as absolute when the same effects might have been produced by a suitable shift in the surrounding frame of reference. Lenard found this typically Einsteinian admixture of everyday experience and abstract reasoning far-fetched, and argued that the second possibility could be ruled out as counterintuitive, or as he called it a violation of "sound reason" ("gesunder Verstand"): "If as a result [of the non-uniform motion] everything in the train were wrecked due to the effects of inertia, while everything outside remained undamaged, then I believe no sound mind would draw any other conclusion than that it was the train that altered its motion with a jolt and not the surroundings."[35] These words and images proved to be of immense significance for the debates on relativity during the early years of the Weimar Republic.

4. EINSTEIN'S "DIALOGUE ON ARGUMENTS AGAINST RELATIVITY THEORY"

Einstein had managed to ignore Gehrcke's provocative attacks, but Lenard was someone he knew and respected as a physicist, whatever he thought of him as a person. He saw no need for pointless polemics now that his theory of gravitation had won the support of leading figures like Lorentz, Schwarzschild, Hilbert, and Weyl. By the same token, his semi-popular account of relativity theory (Einstein, 1917) was now available for readers who wished to gain an impression of the fundamental underlying ideas.[36] Still, Berliner had been pestering him for some time to answer the critics of relativity theory in print, and so by late 1918 he decided that the readers of *Die Naturwissenschaften* had waited long enough for a response from his pen. Einstein's "Dialogue on Arguments against Relativity Theory" (Einstein, 1918a) brings to mind Galileo's far more ambitious *Dialogue Concerning the Two Chief World Systems*, and presumably Einstein hoped to convey a similar sense of lively popular scientific discourse. Only two interlocutors enter this mini-debate: a persistent, but open-minded Kritikus who queries Relativist, clearly a thinly-disguised pseudonym for the author. The tone Einstein struck at the outset was at once playful and sarcastic. Whether intended or not, it was also undoubtedly offensive to the two principals at which it was aimed, Ernst Gehrcke and Philipp Lenard, neither of whom shared Einstein's penchant for irony.

Only Lenard was addressed by name, and the Relativist's more substantive remarks addressed specific points the Heidelberg physicist had raised. That Einstein also had Gehrcke in mind, however, can easily be surmised from the opening remarks of Kritikus about his mission and motives:

> Unlike many of my colleagues, I am not so enamored by the honor of my guild that I act like a superior being with supernatural insight and assurance (like a reviewer of scientific literature or even a theater critic). I speak instead like a mortal being, knowing all too well that criticism often goes hand in hand with a lack of creative thought. Nor will I treat you – as one of my colleagues did recently – like a prosecuting attorney, and accuse you of stealing intellectual property or other dishonest actions. My visit is merely occasioned by the need to contribute to the clarification of some points about which opinions still diverge widely (Einstein, 1918a, 115).

From this passage, it seems likely that Einstein's decision to invent Kritikus and then reply to him, rather than to his real critics, was simply his way of avoiding the fanatical Gehrcke. But fanatics are not easily discouraged, and Gehrcke was only just warming up to the challenge. Einstein had published a short rebuttal of Gehrcke's claims regarding Stokes's theory (Einstein, 1918b), but Gehrcke made sure that he had the last word: "Herr Einstein stands on the ground of relativistic theories and has already for years disavowed the ether. It has been regarded as a principal advantage of relativity theory that it dispenses with the ether. Yet the ether, if it exists, cannot be extinguished by any amount of theorizing" (Gehrcke, 1919).

In his "Dialogue" Einstein again took up the clock paradox, noting that the clocks A and B cannot be treated symmetrically, as Gehrcke had done in his earlier critique from 1913. The reasoning behind this, maintained Einstein, was simple enough to grasp: clock A remains throughout in an inertial frame, whereas clock B undergoes accelerations. Thus, when the observer in B's frame rejoins the other observer, both agree that B's clock has fallen behind A's. Probably this was the very same explanation Einstein gave Gehrcke when they discussed the issue back in June of 1914, but the latter remained unconvinced. In his response to Einstein's "Dialogue," Gehrcke claimed that by breaking the symmetry Einstein had, in effect, abandoned the principle of relativity by denying that it was possible to view clock B as at rest and clock A as in motion relative to it (Gehrcke, 1919). This argument was patently false, as any informed reader could have seen, yet Gehrcke insinuated that Einstein was just up to his old tricks, inventing a Kritikus who caved in just when he should have gone on the counteroffensive.

Regarding general relativity, Gehrcke mainly tried to leap to Lenard's defense by lambasting Einstein's more general pronouncements. Thus he characterized Einstein's Relativist as a thinker with his head in the clouds: someone who understands the world in terms of "formulas and mathematical concepts, but lacks the philosophical vein that would enable him to go beyond formulas and concepts" (Gehrcke, 1919, 148). Lenard had tried to undercut Einstein's equivalence principle by means of his thought experiment involving a train that suddenly accelerated so that the objects inside it were thrown about (Lenard, 1918). As Lenard saw it, if Einstein's theory were correct then the passengers inside the train were not entitled to conclude that this effect was due to the train's motion since a suitable shift in the surrounding landscape would have produced the same reaction.

Einstein's Relativist tried to address this criticism, but the issues involved were deep and continued to preoccupy leading theoreticians for some time, including Lorentz, Weyl, and Gustav Mie. Gehrcke, however, contended that Einstein simply tried to duck all the main problems while claiming that Lenard's distinction between real and fictive physical explanations resolved nothing: "When to defend his standpoint the Relativist introduces gravitational fields that are not generated by gravitational masses, when in order to account for a train crash he sets backwards the whole landscape, earth, planets and fixed stars in an induced gravitational field within which only the train's mass is somehow mysteriously shielded and which again instantly and just as mysteriously vanishes, so can one well observe: *this* interpretation, freely accepting the complications, is *a posteriori* incorrect" (*ibid.*). Gehrcke clearly implied that Einstein was nothing but a charlatan, and he offered this assessment of his hocus-pocus with relativity theory:

> The matter amounts to this: in order to get around the inner contradiction in the old, special relativity theory, the new, general relativity theory was created, and this leads, as can be seen in the most varied ways, to untenable physical consequences. Herr Einstein's Kritikus thus has an easy situation compared with that of the Relativist: he merely needs to point out to him that in order to carry on a discussion over physical matters one must take into account along with the *a priori* the *a posteriori* as well (*ibid.*).

This statement nicely captures the essence of Gehrcke's position. After years of prodding, he had finally (with Lenard's help) drawn Einstein into a debate over the "scientific issues" raised by relativity theory. These issues, however, once Einstein addressed them, quickly gave way to *weltanschauliche* concerns, including categories like "gesunder Menschenverstand" that would soon take on sinister connotations. In Gehrcke's case, the physical and philosophical issues he claimed were at stake seemed to dissipate into thin air, revealing nothing but a passion to expose what he thought was a scientific fraud.

Sensing that he was dealing with an elusive moving target, Gehrcke apparently felt Einstein had to keep changing his mind in order to escape the insuperable difficulties that plagued his theory. The status of the ether offered a major case in point.[37] Einstein's original theory of 1905 – later called special relativity (SRT) – had supposedly shown that the notion of an ether was altogether superfluous for electrodynamics. By 1920, however, Einstein had modified his original position: in his Leyden inaugural lecture, he argued that the notion of an ether was, after all, an important constituent of his new general theory of relativity. Lorentz had stripped the ether of virtually all mechanical properties, but he held fast to the notion of an ether that filled all space. Einstein's shift in viewpoint undoubtedly pleased him a great deal (Kostro, 2000, 63–74). Not so with the anti-relativists, however, who saw this move as just another mathematical trick. For them, gravitational fields could only be generated by ponderable masses; thus, the so-called fictional forces that arose in accelerated frames had an altogether different origin for them than the one Einstein attributed to such phenomena, namely they were regarded as inertial effects due to the presence of an ether at rest in space. At the close of the "Dialogue," Kritikus asks about the health of the ether, "that sick man of theoretical physics," who had earlier been pronounced dead by so many. Relativist replies that, on the contrary, the gravitational field might be thought of as an ether within the context of his general theory of relativity. He emphasized, however, that this conception differed radically from the earlier conception of a space-filling ether that had served as the foundation for Lorentz's theory. These were slippery arguments; little wonder that they simply maddened traditional ether-theorists like Gehrcke and Lenard.

In *Night Thoughts of a Classical Physicist* (McCormmach, 1982), Russell McCormmach sketches a portrait of the mental world of one Victor Jacob, a fictionalized senior member of the German physics community whose work centered on ether physics. The year is 1918 and we find Jacob reflecting back on his career, now in its twilight phase. His reminiscences reveal much about the joys, sorrows, and the whole value system of an elderly German physicist who was burnt out and whose life now appeared like a dream that passed before him. Still, classical physics did not suddenly come to an end with Jacob's generation; there were plenty of younger physicists who were intent on restoring as much of that world and its values as they could. One of the more steadfast among them was an experimentalist at the Physikalisch-Technische Reichsanstalt in Berlin, the national laboratory founded in 1888 by Werner von Siemens and whose first director had been Hermann von Helmholtz. This younger classical physicist was Ernst Gehrcke.

5. THE ANTI-RELATIVITY CAMPAIGN OF 1920

Einstein's "Dialogue" appeared in the fateful month of November 1918. Not long before, most Germans still thought that they would emerge victorious at the end of the war; bitter disappointments led to the search for scapegoats, particularly after the terms of the Versailles Treaty were announced. Soon right-wing agitators were calling for action against the so-called November criminals. Among academics, Einstein was one of the very few who had hoped for a very different outcome, namely the end of Prussian militarism. Although his pacifist and internationalist leanings were only vaguely known throughout the war, he afterward spoke out openly. Within the German scientific community, his unusually leftist views raised plenty of eyebrows, even among allies like Arnold Sommerfeld, who wrote him on 3 December 1918:

"I hear you believe in the new times and want to work for them – may God preserve you in your belief. I find everything unbelievably disgusting and stupid" (Hermann, Armin, ed., 1968, 54). The Munich physicist later told Einstein that his internationalist sentiments would never have been tolerated in France or England; Sommerfeld even claimed that if he had lived in either of those countries during the war years, he would have been thrown in prison for voicing such views.[38]

Amidst the aftermath of the war and his growing fame as the scientist whose achievements had surpassed those of Copernicus, Kepler, and Newton, Einstein became both a symbol of the new social and political order as well as an irresistible target for those who loathed it. Relativity now faced a new flurry of attacks from perpetrators whose concerns had more to do with political and ideological issues than with physics. Anti-Semitism was already thematized in Berlin newspapers in February 1920 following the disruption of Einstein's lectures.[39] In August of that year it reared its head in the form of a semi-organized onslaught aimed at bringing Einstein and his friends to their knees. By then the anti-relativists' cause had begun to attract the attention of one Paul Weyland, an engineer who gained notoriety after the war as a right-wing political journalist (Kleinert 1993). Weyland knew nothing about physics, but he did quickly recognize the political potency of Ernst Gehrcke's anti-relativist message of "Massensuggestion." After contacting Gehrcke and several other critics of relativity, Weyland set himself up as head of the "Working Association of German Natural Scientists for the Conservation of Pure Science," an organization whose only goal was to wage a "counter-campaign" against Einstein and his allies. Under its auspices, he advertised a series of anti-relativity lectures in the main auditorium of the Berlin Philharmonic Hall (Weyland 1920a, Weyland 1920c). This began on 24 August 1920 when Weyland and Gehrcke stepped to the podium before a large crowd mainly comprised of curiosity seekers.[40]

Weyland promoted this lecture series in provocative newspaper articles, the first of which was entitled "Einsteins Relativitätstheorie – eine wissenschaftliche Massensuggestion" (Weyland, 1920a), the slogan Gehrcke had coined back in 1913. Clearly, Gehrcke was the source for much, if not all, of what Weyland wrote here. The article referred to Lenard's train crash query, first posed in 1918, and to which Einstein had supposedly failed to reply. Weyland claimed further that the ultra-conservative spectroscopist Ludwig Glaser had demonstrated the unreliability of the results on gravitational redshift, thereby removing another bogus empirical support for general relativity. Nor did he forget to mention that Gehrcke had charged Einstein with plagiarism of Gerber's work, a charge Einstein had also failed to answer.

These, he went on, were but a few of the numerous examples that could be cited of the "bluff" of relativity theory. Weyland also claimed that a large number of former Einstein supporters had since come to recognize the error of their ways. But Einstein still controlled a "certain press" and a "certain band" who tried to influence public opinion.

Gehrcke later spelled out clearly what was meant by the assertion that Einstein controlled a "certain press," pointing his finger at *Die Naturwissenschaften*, whose editor had indeed taken a strong and active interest in relativity theory (Gehrcke, 1924b, 9). Gehrcke asserted that Arnold Berliner also fed the daily newspapers with information once the relativity story got hot. Insiders knew, of course, that Berliner was a Jew, and that his journal was published by the firm of Julius and Ferdinand Springer. They did not need much imagination to see that this claim fit a familiar pattern of opinion in Germany regarding Jewish influences in the press and popular culture. Whether or not Gehrcke's contention was true, Einstein was on friendly terms with reporters for the *Berliner Tageblatt* and *Vossische Zeitung* which suggests that Arnold Berliner probably knew some of these journalists as well. No one close to Einstein's circle had a sharper eye than he when it came to the leading scientific controversies of the day.

Just before Weyland and Gehrcke delivered their lectures against him, the *Deutsche Zeitung* published a new charge of plagiarism, claiming that Walter Ritz had given a derivation for the perihelion of Mercury back in 1908.[41] Two days later the same paper published an article by Weyland entitled "New Proofs for the Fallacy of Einstein's Relativity Theory" (Weyland, 1920c). This piece began as a more or less conventional newspaper account, but by the end its frenzied purpose became transparent as Weyland repeated the plagiarism charge. After duly noting that Glaser had refuted the experimental evidence for gravitational redshift and that Gehrcke had exposed the physical and epistemological errors in Einstein's theory, Weyland wrote: "the mathematical attack will now follow. In a series of lectures the Working Association of German Natural Scientists for the Preservation of Pure Science turns to the German public in order to prove how they have been taken in by the unconscionable Einstein press" (Weyland, 1920c). He then announced where and when that public should turn up in order to witness how "the Einsteinian phantasms will be totally plucked to pieces" (*ibid.*).

A letter from Berliner to Einstein provides a glimpse of the mood in the pro-relativist camp just before the curtain went up at the Philharmonic Hall.[42] The editor of *Die Naturwissenschaften* informed him that, although requests for Einstein's "Dialogue" continued to pour in, all the back issues containing it were now sold out. He also recalled how two years earlier Einstein had originally planned to publish a sequel, an idea that now seemed ripe in view of the "many intrigues" (*vielen Quertreibereien*) in the air, an obvious reference to Gehrcke's and Weyland's activities. Berliner even went so far as to present Einstein with a concrete plan for countering the anti-relativists: after publishing the new dialogue, he offered to have it bound together with the old one in a brochure. This, he counseled, would provide Einstein with a suitable armor for defending himself against the coming onslaught. Weyland had already published the names of the first four speakers in the planned series of talks against relativity theory, so presumably Berliner was referring to these people when he added: "I find that Gehrcke and his comrades make the most splendid advertisement for relativity theory" (*ibid.*). He soon had good reason to change his mind; almost before Einstein could consider Berliner's proposal, he found himself caught in a whirlwind of controversy that had him seriously contemplating escape from Berlin.

Einstein joined the crowd during the opening event at the Berlin Philharmonic Hall, along with Walther Nernst, Max von Laue, and his stepdaughter and secretary, Ilse.[43] Presumably all of them noticed the anti-Semitic literature and swastika lapel buttons on sale in the foyer, and they definitely heard Weyland accuse Einstein of everything from plagiarism to scientific Dadaism (Weyland, 1920b). By then, at the latest, it became clear that Gehrcke and other anti-relativists had joined forces with right-wing elements in an effort to promote their cause. What remained unclear on the evening of 24 August was the extent of their support within the German physics community. As Laue noted, Weyland's list of future speakers was a long one; he insinuated that German anti-relativists were plentiful and that German scientists would soon close ranks against Einstein to condemn his "methods."[44] Weyland's own techniques included lucrative offers to scientists who agreed to join his campaign.[45]

Paul Weyland orchestrated the anti-Einstein campaign for political purposes, but he could never have done so without Gehrcke's prior efforts. In fact, Weyland took over Gehrcke's overriding theme—that relativity was not a scientific theory but rather a mathematical dogma spread by means of "mass suggestion." This became the watchword in Weyland's opening lecture, in which he claimed that Einstein and his clique were perpetrating a propaganda campaign that made use of their privileged positions within the German physics community as well as connections with the popular press. Taking full advantage of their power, the Einsteinians rode roughshod over their opponents, whose arguments were drowned out, neglected, or suppressed.

Weyland thus claimed that the proponents of relativity theory were mere propagandists whose writings and claims were devoid of scientific value.

Gehrcke had been hinting at these things for a long time, but Weyland brought them out in the open in a blatantly political context. His critique was based on suggestive parallels he found between the Einstein conspiracy and a larger one in which the German people were being duped by a "certain press." This claim was filled with innuendo, as most educated Berliners knew that the local liberal press was owned and operated by two prominent Jewish families, the Mosses and Ullsteins. Einstein's theory had received favorable coverage in both the *Vossische Zeitung* and *Berliner Tageblatt*, but especially in the *Berliner Illustrirte Zeitung*. Lamentations about the so-called Jewish press were a standard theme among anti-Semitic propagandists, but Weyland fervently denied any political motives when Einstein and others leveled this charge. Gehrcke was also insistent that his criticisms were of a purely scientific nature. Yet Weyland evidently had his own press connections, as he published his ferocious attacks on Einstein and relativity theory in the ultra-right-wing *Deutsche Zeitung*. Gehrcke, too, published statements against relativity in this far-right newspaper.

Three days later, Einstein answered his critics in the leading liberal paper, *Berliner Tageblatt*, with an article entitled "My Reply. On the Anti-Relativity Company" ("Meine Antwort. Über die anti-relativitätstheoretische G. m. b. H."). It began as follows: "Under the pretentious name 'Syndicate of German Scientists,' a motley group has joined together to form a company with the provisional purpose of denigrating the theory of relativity and me as its author in the eyes of non-physicists" (Einstein, 1920b, 345). Referring to Weyland and Gehrcke, Einstein wrote: "I am fully aware that both speakers are unworthy of a reply from my pen; for I have good reason to believe that there are other motives behind this undertaking than the search for truth. (Were I a German national, whether bearing swastika or not, rather than a Jew of liberal international bent...). I only respond because I have received repeated requests from well-meaning quarters to have my view made known" (*ibid.*). These words resounded loudly, but Einstein soon afterward refused to ally himself with the Berlin Society for the Repulsion of Anti-Semitism, believing that Jews should not feel compelled to defend themselves against blatant anti-Semitism.[46]

Historians have often asserted that it was Einstein, and not his critics, who first raised the issue of anti-Semitism as a motive for attacks on relativity, thereby fanning the flames of controversy surrounding him and his work.[47] Yet, events quickly proved that the organizer and first speaker, Paul

Weyland, was nothing but a demagogue and rabid anti-Semite. Einstein dispensed with him by remarking that he said "nothing of pertinence" and that he merely "broke out in course abuse and base accusations." The brunt of his counterattack was directed at the second speaker, Ernst Gehrcke, whom Einstein knew very well as an experimental physicist who clung to an old-fashioned ether theory. He also knew that Gehrcke's animus against relativity theory precluded any chance for fruitful debate. Predictably enough, Gehrcke published a denial of all charges. Indeed, all his life he denied that his crusade against Einstein was motivated by anything other than scientific concerns. Thus he blamed Einstein's counterattack in *Berliner Tageblatt* for politicizing the debate between proponents and opponents of relativity theory (Gehrcke, 1924b, 12).

Shortly after Einstein's response appeared, he received a letter from Arnold Sommerfeld, the then presiding officer of the German Physical Society. Sommerfeld was intent on damage control, but he was also anxious to show Einstein that he sympathized fully with his plight. He therefore suggested to Einstein that he consider answering his critics in the *Süddeutsche Monatshefte*, a conservative paper with reactionary tendencies. "The *Berliner Tageblatt*," he added, "does not appear to me the right place to settle accounts with the brawling anti-Semites."[48] Sommerfeld, too, apparently had misgivings about the so-called Jewish press.

Statements of support for Einstein were published in a number of Berlin newspapers. Perhaps the most forceful of these came from his colleagues Laue, Nernst, and Rubens, who emphasized that beyond relativity theory "Einstein's other works ensure him an eternal place in the history of our science ... so that his influence on the scientific life not only of Berlin but all of Germany can hardly be overestimated."[49] Regarding his character, they added that those close to Einstein "know that no one surpasses him when it comes to recognizing others' personal intellectual property, his personal humility, and his distaste for self-promotion" (*ibid.*). A day later the *Berliner Tageblatt* published Einstein's "Reply" along with an article claiming that he was planning to leave Berlin due to the campaign against him. Soon a number of newspapers began spreading the rumor that Einstein's days in Berlin were numbered. This caused the Prussian Minister of Education, the Social Democrat Konrad Haenisch, considerable concern, enough so that he issued a widely circulated open letter to Einstein expressing his solidarity with Berlin's leading physicist and hoping that the published rumors were untrue (Grundmann, 1998, 161–162). Einstein replied that "I have experienced that Berlin is the place to which I am most closely bound by personal and scientific relationships. I would only follow a call to a foreign country in the event that external conditions compel me to do so" (*ibid.,* 164).

Given the volatility of the early Weimar Republic with its widespread cultural pessimism coupled with a fascination for new lifestyles and forms of expression, one can easily understand why pro- and anti-relativists began to hurl invectives at one another. Einstein was hardly to blame for this, but he did express regret about having reacted with such caustic words in the face of these provocations. In a letter to Max Born, he ruefully proclaimed that "everyone has to offer his sacrifice at the altar of stupidity from time to time, to humor God and man. And I did a thorough job of it with my article."[50] Yet he also recognized that the ground rules for civil discourse had suddenly changed in Weimar Germany. To his good friend Paul Ehrenfest, he explained how he had had no choice but to defend himself against the steady stream of charges publicly leveled against him, including dishonest self-promotion, literary theft, and outright plagiarism: "I had to do this if I wanted to stay in Berlin, where every child knows me from the photographs. If one is a democrat, then one must grant the public this much right as well."[51] By the early 1920s Einstein's name had appeared often in the popular press, and many knew what he looked like from the near life-size photo that appeared on the cover of the *Berliner Illustriter Zeitung*. Einstein clearly appreciated the fact that his life in Berlin had suddenly changed with the appearance of that issue in December 1919.

6. THE LENARD-EINSTEIN DEBATE IN BAD NAUHEIM

Lurking in the background of the Gehrcke-Weyland campaign was Philipp Lenard, who had sided with Gehrcke in suggesting that Gerber's work on the perihelion of Mercury deserved serious consideration as an alternative to Einstein's. This was rather desperate, but Lenard nevertheless raised some significant scientific points. In answer to Einstein's "Dialogue," he also published a revised version of his earlier paper for Stark's journal, also available as a separate publication. Once again he did not echo Gehrcke's reckless charge that Einstein plagiarized Gerber's formula, but neither did he distance himself from this assertion (Lenard, 1920). As a former student of Heinrich Hertz, Lenard was inclined to see an analogy between Maxwell's theory and Einstein's field-theoretic approach to gravitation (Schönbeck, 2003, 330–342). He thus contended that general relativity would be confirmed, if one could demonstrate the existence of gravitational fields devoid of a center which, according to Einstein, simulate a centrifugal force, just as Maxwell's theory was first confirmed when Hertz found in electrical waves electronic fields without centers.[52] He also raised questions about the purely formal

nature of Einstein's principle of general relativity, an issue that had also led
Erich Kretschmann to criticize Einstein's physical interpretation of the principle of general covariance (Kretschmann, 1917).

Lenard's new essay appeared just before the anti-Einstein campaign broke
out in Berlin. The timing was not entirely fortuitous, as both Weyland and
Gehrcke cited Lenard's criticisms of general relativity at the anti-relativist
meeting held in the Philharmonic Hall. Weyland even interrupted his lecture
so that the audience could purchase a copy of (Lenard, 1920), the newest
version of Lenard's pamphlet containing his critique of Einstein's theory of
gravitation. This was on sale in the foyer of the auditorium alongside as-
sorted anti-Semitic literature.[53] Even more significantly, Weyland quoted a
passage taken directly from Lenard's new text in which the Heidelberg physicist implied that Einstein's theorizing lacked the manly virtues associated
with German physics. In a clear allusion to general relativity, Lenard conceded that natural scientists should be free to "frame hypotheses"; but these,
he quickly added, must always be tested against experience. Indeed, he took
it as a kind of moral imperative for the scientist that he be reserved in treating a hypothesis as established truth, a message he conveyed by means of a
Goethean image ("das, was ursprünglich Hypothese, Dichtung des Geistes,
war, als Wahrheit auszugeben"). Regarding those who chose to flout this
dictate, Lenard had this to say:

> The more daring a natural scientist shows, the more places
> appear in his publications which fail to stand up over time;
> one can demonstrate this with examples from the distant and
> more recent past (especially easy to find for the latter). For
> this reason the daring natural scientist by no means deserves
> the high esteem accorded the daring warrior. For with his dar-
> ing the latter sets his life on the line, whereas the mistakes of
> the former usually bring only a mild rebuke before they are
> forgotten. At times it would appear that the daring ascribed to
> natural scientists consists, in reality, in a quite unscrupulous
> reckoning that no personal harm will come from inferior con-
> tributions to the scientific literature. This kind of daring is not
> a German characteristic (Lenard, 1920, 2).

Clearly Paul Weyland read this inflammatory passage in order to arouse his
audience, but his performance seems to have fallen flat for most of those who
heard him speak.[54]

These circumstances help account for why Einstein went out of his way
to attack the views of Lenard, who had not been present at the Berlin meeting, as the only "outspoken critic of relativity theory of international renown"

(Einstein, 1920b, 345). Conceding that Lenard was a "master of experimental physics," he implied that his Heidelberg colleague had moved in over his head, as he had "yet to accomplish something in theoretical physics" (*ibid.*). Indeed, Einstein thought Lenard's objections to the general theory of relativity "so superficial that [he] had not deemed it necessary until now to return to them in detail" (*ibid.*). That was all he wrote *apropos* Lenard's critique of relativity, but he promised to say more at a later opportunity. Such a chance, he pointedly remarked, would be provided four weeks later in Bad Nauheim, as he had taken the initiative to make arrangements for a discussion session on relativity theory. To his critics he then extended the following gruff invitation: "anyone willing to confront a professional forum can present his objections there" (Einstein, 1920b, 347).

Lenard, not surprisingly, was incensed when he read this.[55] Nevertheless, he was hardly innocent regarding what transpired on August 24 at the Berlin Philhamronic Hall. For Weyland had already visited him on August 1 in Heidelberg to inform him of his plans. Evidently pleased by this turn of events, Lenard wrote to Stark the following day, describing Weyland as "very enthusiastic in our direction, the fight against un-German influences."[56] He also advised Weyland to contact Stark, in part to maintain solidarity in view of the latter's forthcoming plans to bolt from the German Physical Society (Beyerchen, 1977, 106–110). That very same day, Lenard also wrote to Willy Wien, calling Weyland a "very enthusiastic supporter of our reforms, who wants in particular to fight systematically against Einstein's exaggerated machinations and the whole manner of his doings – as un-German."[57] He noted further that Weyland had the support of two Berlin physicists, Gehrcke and Glaser, and that he hoped to sign up Lenard as well. But the latter was reluctant to do so without knowing whether Wien would do the same. He therefore asked his colleague whether he was willing to place his signature on a declaration prepared by Weyland. Apparently Wien declined, causing Lenard to get cold feet. He thus decided to support Weyland's cause with moral encouragement, but to avoid any direct involvement at this stage. Whether or not he ever conveyed these tactical considerations to Weyland, the latter felt sufficiently emboldened to place Lenard's name on the list of future speakers.[58]

Einstein knew nothing, of course, about all this behind-the-scenes plotting when he went after Lenard in the *Berliner Tageblatt*. From a tactical standpoint, his mistake was not just that he attacked the ultra-sensitive Heidelberg physicist but that he did so by openly challenging him and other anti-relativists to a public debate at the forthcoming meeting of the Society of German Natural Scientists and Physicians in Bad Nauheim. Presumably few had taken notice of the special joint session on relativity on the program, a quiet initiative of the mathematician Robert Fricke, then President of the Deutsche Mathematiker-Vereinigung (see Appendix). Once Einstein issued his public challenge, however, the congress organizers realized that they suddenly had a potentially explosive situation on their hands.

Politics had indeed pervaded German academic life with a vengeance, placing further strains on the already delicate state of personal relations within Germany's physics community. These largely hinged upon a long-standing tension between leading experimental physicists – Lenard, Willy Wien, and Johannes Stark, all of whom worked outside the Prussian universities – and the theoreticians in Berlin, particularly Einstein, Planck, and Laue. Throughout the war, these experimental physicists had been longing to break Berlin's dominant power. Together they contemplated plans for the formation of a new splinter organization, headed by Stark, which would challenge the Berlin-dominated Deutsche Physikalische Gesellschaft at the Bad Nauheim meeting in September 1920. The DPG's President, Arnold Sommerfeld, had done his best to throw cold water on Stark's plans, fearing that they would shatter the community (Beyerchen, 1977, 106–109). With the "Einstein affair" he now faced a true crisis. After learning of reports that Einstein might leave Germany in the wake of the Weyland campaign, he pleaded with him not to desert ("flee the flag") with reassurances that the physics community backed him wholeheartedly.[59] Sommerfeld had since heard how Max Wolf's name had been misused by the Weyland-Gehrcke coalition, and he suggested to Einstein that the same surely applied in the case of Philipp Lenard.

When Sommerfeld wrote to Lenard, however, he discovered that his overtures to a peaceful settlement had accomplished nothing. After receiving an acidic response, he tried in vain to persuade Einstein to write Lenard directly to apologize for having associated him with the Berlin tumult.[60] In doing so, Sommerfeld hoped to mollify Einstein by asserting that "Lenard had expressed himself very properly" (*ibid.*) in the new edition of his brochure "Relativität, Äther und Gravitation" (Lenard, 1920). Einstein surely thought otherwise, having heard the key passage that Weyland had cited during his tirade against the relativity theorists. Still, Einstein hoped that some kind of agreement might be reached that would allow the controversy to subside.

Planck and Sommerfeld were especially concerned that the earlier attacks against Einstein and relativity theory staged in the Berlin Philharmonic Hall might spill over into the Bad Nauheim meeting. They therefore persuaded the society's presiding officer, Friedrich von Müller, to take preemptive action against the anti-relativists in his opening address. Müller kindly obliged. Referring directly to the special session on relativity theory, he emphatically stated that:

> ...it will be treated in an entirely different spirit than that of the tumultuous gatherings in Berlin. *Scientific questions of such difficulty and great significance* as the *theory of relativity* cannot be *brought to a vote in popular assemblies with demagogic slogans, nor can they be decided by personal attacks in the political press. They will receive here the objective appreciation that their brilliant creator deserves.*[61]

This counter-offensive evidently produced the desired effect, as Müller's statement was greeted with thunderous applause.[62]

Three days later, on the morning of 23 September, bath house number eight in Bad Nauheim was jammed with spectators, most of them anticipating real fireworks. Einstein's friends and foes showed up *en masse*, the latter hoping to witness a spectacular debate between Einstein and Lenard. Max Planck presided over this special session, a delicate undertaking given the pre-conference publicity. Six lectures were scheduled, but time constraints left room for just four speakers: Hermann Weyl, Gustav Mie, Max von Laue, and Leonhard Grebe.[63] A special correspondent for the *Berliner Tageblatt* found the lectures "a hailstorm of differentials, coordinate invariants, elementary action quanta, transformations, vectorial systems, etc."[64] Some left the sweltering bathhouse, but most waited in anticipation of the battle to come. Only fifteen minutes remained for the general discussion, most of which was taken up by the Lenard-Einstein debate (*CPAE*, vol. 7, 350-359).

This turned out to be their one and only direct encounter, and both men were intent on saving face; consequently, neither had anything new to say. Afterward, Einstein and a few of his friends tried to pacify Lenard, but he refused to shake hands with his rival or to accept Einstein's public apology for attacking him in *Berliner Tageblatt* (Schönbeck, 2003, 353). The press coverage of the confrontation was generally fair to both sides, though many reporters sensed that what they had witnessed was really a clash of worldviews rather than a scientific debate.[65] Among those who witnessed the encounter, partisan reactions prevailed. *DMV* President Robert Fricke, who hoped to use the session to showcase Germany's accomplishments in mathematical physics, was delighted by the outcome. Writing to his uncle, Felix Klein, he was certain that "even the lay people could feel Einstein's superiority over Lenard."[66] Over a year afterward, Hermann Weyl wrote a detailed report on what transpired for the *Deutsche Mathematiker-Vereinigung*. Einstein found it both "excellent and very interesting,"[67] though he must have noticed that

Weyl's views with regard to what Einstein dubbed Mach's principle had begun to waver (Weyl, 1922, 61–62).

Behind the scenes Planck and the Freiburg physicist, Friedrich Himstedt, entered into lengthy negotiations aimed at restoring civil relations between Einstein and Lenard. This resulted in a statement that they communicated to the press at the conclusion of the conference, and which read:

> In an article entitled "Meine Antwort über die antirelativistische G.m.b.H." that appeared in the "Berliner Tageblatt" on 27 August, Professor Einstein defended himself against the "Working Association of German Natural Scientists for the Conservation of Pure Science," in whose first meeting Herr Weyland attacked him in a personal and malicious manner. In this article he also turned against Professor Lenard, whose name appeared on the list of speakers alongside those of other physicists. During the recently held conference of natural scientists held in Bad Nauheim we could determine that Herr Lenard's name had been placed on the list of speakers without his authorization. In view of this fact, Herr Einstein has authorized us to express his deep regret that his article contained criticisms of his highly esteemed colleague, Herr Lenard.[68]

Himstedt realized that Lenard would not feel this public statement constituted a satisfactory apology.[69] Einstein had wounded the latter's pride by calling his objections to the general theory of relativity "so superficial that [he] had not deemed it necessary until now to return to them in detail." But Lenard laid the blame for this not just on Einstein alone: he saw this outburst as a shocking display of the arrogant attitude displayed by the Berlin clique that tried to stage-manage the relativity revolution. Bad Nauheim was, for Lenard, merely the affirmation of this conspiracy within the German Physical Society, and he soon thereafter resigned his membership (Schönbeck, 2003, 353).

Lenard was not yet the fanatical racist who would later emerge as the father figure of Deutsche Physik. Two years later, however, a series of events took place that put him on that path. When he refused to close his Heidelberg Institute on the day of mourning for Walther Rathenau, the assassinated foreign minister, a confrontation took place that later would have great symbolic significance in Lenard's mind. He came to see this incident as an episode not unlike Hitler's Beer Hall Putsch of 1923. As the former corporal sat in Landsberg prison writing *Mein Kampf*, Lenard and Stark issued a statement of complete solidarity with his cause (Beyerchen, 1977, 95–96). Back in 1920, Einstein could not have foreseen how far or how fast this would go, but he certainly saw the direction in which Lenard was moving. In fact, Lenard seems to have been just as eager to expose Einstein and his theory as was Gehrcke; he simply lacked the courage to stand up on his own.

For Paul Weyland, who reported on the Bad Nauheim meeting for the *Deutsche Zeitung*, this dramatic showdown marked the end of his campaign to discredit Einstein and the "Einstein press." He accused Planck and the Einstein clique of having throttled the opposition, who, except for Lenard, were never given a chance to speak (Weyland, 1920d). In Weyland's view, the relativity debate had been nothing but a sham with only one redeeming virtue: it revealed the deep division within the German physics community. On the one hand there were those who, "under the leadership of Lenard, rejected the rape of physics by mathematical dogmas, whereas the Einsteinophiles on the other side cling to their standpoint and try to climb the Parnassus of their rubbish of formulas... before they will fall precipitously from their icy heights" (*ibid.*). As for "the art and manner of free research, as understood by the German Physical Society," Weyland called it "a scandal without example in the history of German science," and suggested that it was "high time that fresh air enter this rat's nest of scientific corruption" (*ibid.*). Soon afterward, Gehrcke and Lenard severed their ties with Weyland, who then left the arena of relativity to pursue a crusade against other "Jewish influences" that he claimed were poisoning German culture (Kleinert, 1993).

Ernst Gehrcke, who stalked Einstein from the time he first set foot in Berlin, consistently denied that his attacks had anything to do with politics. After Bad Nauheim he continued his campaign while debating with leading relativists like Hermann Weyl and Hans Thirring. In 1924 he published "Die Massensuggestion der Relativitätstheorie" (Gehrcke, 1924b), a documentary history intended to demonstrate the wisdom of his prophetic claim that Einstein's theory of relativity was nothing but a phenomenon of mass psychology. In the meantime, Lenard and Stark had temporarily retreated, but their early attacks on relativity were by no means forgotten. These, in fact, helped lay the groundwork for the Aryan physics movement after 1933, whose followers looked up to Lenard as their revered *Führer*, precisely as Weyland had predicted (Beyerchen, 1977).

7. ON RELATIVITY IN WEIMAR GERMANY

In discussing the resistance to relativity in Germany immediately after World War I, Philipp Frank identified three distinct groups among the anti-relativists (Frank, 1949, 270–271): 1) right-wing propagandists and anti-Semites; 2) unimaginative experimental physicists; and 3) traditional philosophers (like his nemesis Oskar Kraus). According to Frank, the first group saw Einstein as the leader of a pseudo-scientific movement that aimed to undermine traditional values by proclaiming that all is relative. "They knew absolutely nothing about Einstein and his theories," writes Frank, "except that he was of Jewish extraction, a "pacifist," that he was highly regarded in England, and that he also seemed to be gaining prestige among the German public" (Frank, 1949, 270). Frank deemed the latter two groups politically naive, though they often harbored the view that Jewish scientists represented a different kind of mentality that was "un-German." Frank thus portrayed the politically motivated anti-Semites as manipulating members of the other two groups. Anti-relativists like Gehrcke and Kraus always insisted that their criticisms of Einstein's theory were purely scientific and entirely apolitical, and Frank thought so too.

As a prominent pro-relativist, Philipp Frank was hardly an unbiased observer. Moreover, his own philosophical interests reinforced an inclination to draw a sharp line between scientific and political issues. These subjective factors undoubtedly colored his interpretation of the motivations behind the anti-relativist movement. In fact, Frank presented scarcely any hard evidence for his claim that the physicists and philosophers opposed to relativity were simple-minded fellows whose political naïveté made them susceptible to anti-Semitic influences. Nevertheless, the same basic approach to the relativity revolution has been adopted by numerous Einstein scholars.

Thus one finds a similarly sharp distinction between science and politics in Klaus Hentschel's *Interpretationen und Fehlinterpretationen der speziellen und der allgemeinen Relativitätstheorie durch Zeitgenossen Albert Einsteins* (Hentschel, 1990), by far the most thorough study of the debates over relativity during the early Weimar period. The title reflects Hentschel's agenda, which primarily aims to sort out legitimate attempts to grasp the philosophical import of relativity from those which failed to do so. Regarding the philosophers who promoted and attacked relativity, Hentschel shows that within German-speaking countries these debates were part of a larger clash that pitted positivists and logical empiricists against neo-Kantians and idealists. However, his study analyzes these conflicts in the *wake* of Einstein's fame and hence overlooks several key factors that shaped the events that followed. Hentschel's project was anything but narrowly conceived; consequently he brings many other facets of the debates over relativity to light. Still, his book primarily documents the rivalries between various philosophical schools rather than focusing on other dimensions of the conflict, such as the rift that developed within the German physics community or the prominent role played by mathematicians in the relativity debates.

Political issues were, of course, traditionally taboo in German scientific discourse, but they clearly played a major role when it came to promoting careers and building alliances (Forman, 1974). Thus, relativity theory was not merely a controversial scientific hypothesis; it also carried ideological overtones that reflected the political currents of the immediate post-war culture.[70] As documented in (Jungnickel and McCormmach 1986), around the turn of the century theoretical physics emerged as a new subdiscipline in Germany. Relying heavily on sophisticated mathematics, the new breed of physicists sometimes had closer ties to the mathematics community than to their colleagues in experimental physics. Two of Germany's most influential theoreticians, Arnold Sommerfeld and Max Born, were in fact originally trained as mathematicians. Since many of these young theoreticians were of Jewish background, experimentalists often saw them as representing a new approach to physics that reflected an abstract, Semitic *Denkweise*, a view that was quite pervasive among both philo- and anti-Semites (Rowe, 1986). These disciplinary developments contributed to growing tensions within the larger physics community that led to open dissension within the German Physical Society, an organization long dominated by the physicists in Berlin. For several leading experimental physicists located in southern German states, Einstein symbolized that dominance and their growing sense of marginalization within their own community. Seen from this vantage point, Einstein's rise to fame as an international superstar shortly after the collapse of the *Kaiserreich* was bound to provoke a reaction with strong political overtones.

Like other commentators, Hentschel also distinguishes sharply between explicitly anti-Semitic arguments against relativity and so-called "scientific issues," such as the numerous priority claims made by anti-relativists in order to cast doubt on the originality of Einstein's ideas. The foregoing analysis suggests that a broader perspective is required in order to understand what was at stake. Ernst Gehrcke's persistent efforts to characterize relativity theory as a phenomenon of modern mass psychology clearly went beyond the realm of traditional scientific discourse. Indeed, his views reflect many of the same anti-modernist attitudes of the virulent anti-Semites who surfaced during the early years of the Weimar Republic. Thus, for Gehrcke and several other anti-relativists who followed his lead, one cannot draw a clear cut line between their arguments purporting to show that Einstein's theory of gravitation was either fallacious or unoriginal (or even both!) and parallel arguments that were meant to suggest he was part of a conspiracy led by influential Jews and their allies. Both types of arguments stemmed from the German analogue of the Dreyfus phenomenon in France alluded to earlier. Thus the accounts of Frank and Hentschel tend to overlook the degree to which Einstein's enemies within the scientific community were motivated by a deeply anti-modernist mentality that went hand-in-hand with the currents of anti-Semitism that swept through Berlin after World War I. By looking carefully at the events leading up to the dramatic confrontations of August and September 1920, a quite different picture emerges, one suggesting that the worlds of politics and science interpenetrated each other in a variety of ways.

8. APPENDIX

Robert Fricke an Einstein 16.5.1920 Einstein Archive [43 725]

Als derzeitiger Vorsitzende der Deutschen Mathematiker-Vereinigung habe ich mich kürlich mit Schoenfliess ... über die Disposition der [Nauheimer] Versammlung ins Einvernehmen gesetzt. Ich habe hierbei in Vorschlag gebracht, dass der Donnerstag Vormittag in der Abt. I, verbunden mit der Abt. für mathematischen Physik, ausschließlich Vorträgen über die Relativitätstheorie vorbehalten bleiben möchte. Ich nehme als sicher an, dass Sie in einer am Montag oder Dienstag stattfindenden gemeinsamen Sitzung der naturwissenschaftlichen Hauptgruppe sprechen werde. Dies würde gewiss nicht ausschließen, dass Sie im engen Kreise der eigentlichen Fachgenossen am Donnerstag nochmals das Wort ergriffen. Ich wollte neben Ihnen mich mit dergleichen Bitte an die Herren v. Laue, Hilbert, Sommerfeld, Weyl u. Born wenden. Sollte es gelingen, einen solchen Vormittag zu Stande zu bringen, so würde ich glauben, dass derselbe einen der größten Erfolge der Versammlung darstellen könnte und der Welt zeigen könnte, was Deutschland selbst in einer so tief unglücklicher Zeit auf wissenschaftlichem Gebiete zu leisten vermochte.

Robert Fricke an Klein 29.9.1920 Nachlass Klein IX, 286F

Die sensationelle Relativitätssitzung nahm einen überaus glänzenden Verlauf, der mich in die größte Begeisterung versetzt hat. Die Entwicklung wurde zu einem Triumph Einsteins, der wirklich ein überlegener Geist ist. Ich bin stolz darauf, zu dieser Sitzung den Anstoß gegeben zu haben, und freue mich nach der Sitzung Einstein noch persönlich meine Empfindungen haben aussprechen zu können. Nächst Einstein machte Weyl den tiefsten Eindruck, aber auch der vierte Vortrag über die Rothverschiebung des Spektrums wirkte außerordentlich und erregte sichtlich auch bei Einstein selbst großes Interesse. Bei der Diskussion war die Überlegenheit Einsteins über Lenard selbst der Laien fühlbar.

University of Mainz
Germany

NOTES

[1] See for example Corry 1997, Gray 1999, Rowe 2001, and Scholz 2001.

[2] That part of the story that involves Göttingen mathematicians is described in Rowe 1999, Rowe 2001, and Rowe 2004.

[3] The broader repercussions of cultural pessimism on German science are explored in Forman 1971 and Stern 1999.

[4] See Hentschel 1990, 55–129, and Goenner 1992.

[5] For background on Einstein's role, see the editorial note "Einstein's Encounters with German Anti-Relativists," *CPAE*, vol. 7, 101-113.

[6] Torsten Palmer u. Hendrik Neubauer, eds., *Die Weimarer Zeit in Pressefotos und Fotoreportagen*, Cologne: Könemann, 2000, pp. 14–17.

[7] Kayser described this photograph and its impact in detail in (Reiser, 1930, 160).

[8] Einstein Digital Archive [28 581].

[9] Lindemann to Einstein, 23 November 1919 (*CPAE*, vol. 7, 211). He received similar news in a letter from Paul Ehrenfest, 24 November 1919.

[10] Einstein's tribute to Berliner was reprinted in Einstein 1954, 68–70.

[11] A competing publication was *Naturwissenshaftliche Wochenschrift*, which published a number of articles by prominent anti-relativists.

[12] The following account is based on information in Sarkowski 1996.

[13] Courant's various wheelings and dealings later made him known to some in the United States as "dirty Dick" (Reid, 1976, 230).

[14] On Freundlich, see Hentschel 1994.

[15] On Eddington's *Report on the Relativity Theory of Gravitation* (London: Fleetway Press, 1918), *Nature*, 6 March 1919.

[16] Eddington discussed this three-pronged test in Eddington 1920, 110–122.

[17] For details on the technical problems that Eddington and his colleagues had to be overcome to produce the desired result, namely confirmation of general relativity, see Earman and Glymour 1980.

[18] Philipp Lenard lamented that German newspapers had left the impression that Germany's scientists were dependent on the expertise of their English counterparts when it came to the confirmation of general relativity. He noted, however, that such a scandalous situation was to be expected until such time as the country obtained a "truly German press" (Lenard, 1921, 427).

[19] Several months later *Die Naturwissenschaften* published an article by Erwin Freundlich which discussed the results of the report prepared by Dyson, Eddington, and Davidson (Freundlich, 1920).

[20] The photo was printed with Einstein's handwritten dedication to Ferdinand Springer.

[21] Felix Klein, who carried on a lengthy correspondence with Einstein, found him personally delightful, this "in complete contrast with the foolish promotional efforts set in motion to honor him" (Klein to Wolfgang Pauli, 8 March 1921, in *Wolfgang Pauli. Wissenschaftlicher Briefwechsel*, vol. I, ed. A. Hermann, et al., New York; Springer, 1979, p. 27).

[22]Berliner to Reichenbach, 6 October 1920, quoted in (Hentschel, 1990, 55–56).

[23]On Gehrcke's career, see Goenner 1993, 114–115.

[24]Gehrcke published his collected writings on relativity theory in Gehrcke 1924a.

[25]Born wrote: „Auffällig wäre bei der Durchsicht einer solchen Liste vielleicht die große Zahl der Mathematiker, jener so sehr zum Zweifel neigenden Gesellen, deren kritischen Betrachtungen mancher experimentelle Physiker als überflüssige Tüfteleien beiseite zu schieben liebt; die Relativitätstheorie hat nicht nur der mathematischen Kritik standgehalten, sondern durch einen der ersten deutschen Mathematiker unserer Zeit, Minkowski, ihr eigentliches formales Gewand erhalten" (Born, 1913, 92).

[26]Born added: "Dass die Theorie tatsächlich logisch widerspruchsfrei ist, lässt sich mathematisch beweisen mit Hilfe von Minkowskis geometrischer Darstellung in der vier-dimensionalen, aus Raum und Zeit gebildeten Mannigfaltigkeit, die er 'Welt' nennt" (Born, 1913, 93–94). He also referred the reader to the more intuitive model of SRT set forth by Emil Cohn in support of this point (E. Cohn, *Physikalisches über Raum und Zeit.* Leipzig: Teubner, 1911.)

[27]Einstein to Stern, 4 June 1914, *CPAE*, vol. 8A, 29.

[28]Die Uhr B, welche bewegt wurde, geht deshalb nach, weil sie im Gegensatz zu der Uhr A Beschleunigungen erlitten hat. Diese Beschleunigungen sind zwar für den Betrag der Zeitdifferenz beider Uhren belanglos, ihr Vorhandensein bedingt jedoch das Nachgehen gerade der Uhr B, und nicht der Uhr A. Beschleunigte Bewegungen sind in der Relativitätstheorie absolute (Gehrcke, 1924a, 35).

[29]Ich bin durchaus dagegen, dass diese Notiz publiziert wird, weil sie – aus dem Zusammenhange herausgenommen – nur Verwirrung stiften kann, trotzdem sie in gewissem Grade richtig ist (quoted in Gehrcke 1924a, 34).

[30]Einstein to Wien, 17 October 1916, *CPAE*, vol. 8A, p. 344.

[31]For a detailed account of Lenard's struggles with relativity theory, see Schönbeck 2003.

[32]Lenard to Stark, 16 July 1917, cited in Kleinert and Schönbeck 1978, 323.

[33]Lenard to Stark, 20 October 1917; 9 February 1918 cited in Kleinert and Schönbeck 1978, 323–324.

[34]"Die Gerbersche Arbeit ist ernstlich bemängelt worden; ich möchte auf der anderen Seite zeigen, dass auch das verallgemeinerte Relativitätsprinzip als Ausgangspunkt nicht ohne weiteres befriedigt. Es scheint mir nämlich die Notwendigkeit einer Einschränkung des verallgemeinerten Relativitätsprinzips vorzuliegen im Gegensatz zu der oft wiederholten Betonung seiner ganz allgemeinen Gültigkeit. Gleichzeitig soll hervorgehoben werden, dass das Relativitätsprinzip keineswegs den Äther ausschließt, was anscheinend als eine besondere, umstürzende Eigenschaft dieses Prinzips hingestellt wird, und dass die Mechanik des Äthers einschließlich der Elektrodynamik und der Gravitation keineswegs als aussichtslos gelten müsse" (Lenard, 1918, 117–118).

[35]"Wenn hierbei durch Trägheitswirkung alles im Zuge zu Trümmern geht, während draußen alles unbeschädigt bleibt, so wird, meine ich, kein gesunder Verstand einen anderen Schluss ziehen wollen, als den, dass es eben der Zug war, der mit

Ruck seine Bewegung geändert hat, und nicht die Umgebung" (Lenard, 1918, 122–123).

[36]Einstein's booklet was an immediate success: one year later it went into a third edition and after the war it emerged as a best-seller within the genre of semi-popular scientific literature, reaching its fourteenth edition by 1922. It appeared in English translation in Einstein 1920.

[37]For an account of competing electrodynamical theories ca. 1905, see Darrigol 2000, 372–394.

[38]Sommerfeld to Einstein, 3 September 1920 (Hermann, Armin, ed., 1968, 68).

[39]Einstein explained the circumstances that led to a tense discussion with students enrolled in his lecture course. Some of these sought to have unauthorized auditors expelled from the course, a measure Einstein found unjustified. He therefore cancelled this course, but planned to resume his lectures in a forum that would allow free access, a decision that surely disappointed several students. Regarding the "Uproar in the Lecture Hall," Einstein wrote: "Should an incident like the one yesterday occur again, I will cease my lecturing altogether. What happened yesterday cannot be called a scandal, even though some remarks that were made demonstrated a certain animosity toward me. *Anti-Semitic remarks* per se did not occur, but the *undertone* could be interpreted that way." (*8-Uhr Abendblatt* 73 (Berlin), no. 38, 13 February 1920, pp. 2 and 3; *CPAE*, vol. 7, pp. 285-286).

[40]Accounts of the meeting were published in various Berlin newspapers, including *Berliner Tageblatt*,

Vossische Zeitung, 8-Uhr Abendblatt, and *Vorwärts* (reprinted in Weyland 1920b).

[41]H. Reinhardt, "Ritz gegen Einstein," *Deutsche Zeitung*, 21 August 1920.

[42]Berliner to Einstein, 19 August 1920.

[43]Accounts of this episode can be found in Fölsing 1993, 520–529, and Hermann 1994, 246–249.

[44]Laue to Sommerfeld 25 August, 1920 (Hermann, Armin, ed., 1968, 65); Weyland had announced twenty speakers for his forum.

[45]Felix Ehrenhaft received a letter from Weyland dated 23 July 1920 in which he was assured a sum of 10,000 to 15,000 marks for his participation (*Berliner Tageblatt*, 4 September 1920, Abendausgabe, p. 3).

[46]For a discussion of Einstein's evolving interest in Jewish affairs, see the editorial note "Einstein and the Jewish Question" (*CPAE*, vol. 7, 221-236).

[47]Fölsing 1993, Hentschel 1990, 133–134; Goenner 1993, 111–112.

[48]Sommerfeld to Einstein, 3 September 1920 (Hermann, Armin, ed., 1968, 68).

[49]"'Wissenschaftliche' Kampfmethoden," *Berliner Tageblatt,* Abendausgabe, 26 August 1920.

[50]Einstein to Born, 9 September 1920 (Born, Max ed., 1969, 58).

[51]"Dies musste ich, wenn ich in Berlin bleiben wollte, wo mich jedes Kind von den Photographien her

kennt. Wenn man Demokrat ist, muss man der Öffentlichkeit auch so viel Recht geben." Einstein to

Paul Ehrenfest, before 10 September 1920.

[52] See Lenard 2003, 439–440.

[53] *8-Uhr-Abendblatt*, 25 August 1920, (reprinted in Weyland 1920b, pp. 8–9); *Vorwärts* (25 August 1920),

Abendausgabe, p. 2. Two days later a parody of the atmosphere was published in the form of a long chant

led by nationalist professors and echoed by a chorus of fraternity students ("Die Einstein Hetz–In der

Philharmonie zu singen"; *Vorwärts*, 27 August 1920; reprinted in Grundmann 1998, 155–157).

[54] *Vorwärts*, 25 August 1920, Abendausgabe.

[55] Lenard's reaction can be found in a letter to Willy Wien, 8 September 1920, Wien Nachlass, Deutsches Museum, Munich.

[56] Lenard to Stark, 2 August 1920: "Ein Herr Weyland – sehr begeistert in unserer Richtung, zur Bekämpfung undeutscher Einflüsse – war gestern bei mir und will einen Verein ,,Arbeitsgemeinschaft deutscher Naturforscher zur Erhaltung reiner Wissenschaft" gründen. Ich habe ihm geraten, vor Allem mit Ihnen sich in Verbindung zu setzen, damit nicht unnötig viele Neugründungen stattfinden und keine Zersplitterung unsere Nauheimer Absichten hindert." (Kleinert and Schönbeck 1978, 327).

[57] Lenard to Wien, 2 August 1920, Wien Nachlass, Deutsches Museum, Munich.

[58] Others named were L. Glaser, M. Wolf, O. Kraus, and M. Palagyi. Glaser alone, however, spoke in Weyland's forum.

[59] Sommerfeld to Einstein, 3 September 1920 (Hermann, Armin, ed., 1968, 68). Einstein replied that he had, indeed, thought earnestly of "Fahnenflucht" for two days, but had since decided he would stay in Berlin (6 September 1920, *ibid.*, 69).

[60] Sommerfeld to Einstein, 11 September 1920 (Hermann, Armin, ed., 1968, 71).

[61] *Berliner Tageblatt*, 20 September 1920, Abendausgabe, p. 4.

[62] Max Planck reminded Einstein of this response one year later (Planck to Einstein, 22 October 1921).

[63] Hugo Dingler and F. P. Liesegang spoke the following day, when Einstein was no longer present. Dingler was the only opponent of relativity theory on the program. The official transcript of the lectures and ensuing discussions at Bad Nauheim was prepared by Peter Debye for *Physikalische Zeitschrift* (see *CPAE*, vol. 7, 350-359). Although this is the most complete documentary account available, contemporary participants realized that it had several problematic features. This probably accounts for why Hermann Weyl later decided to publish another version of what transpired in Weyl 1922.

[64] *Berliner Tageblatt*, 24 September 1920, Abendausgabe, p. 3.

[65] See ,,Ein neuer Beweis für die Einstein-Theorie. Das Rededuell Einstein-Lenard," *Berliner Tageblatt* 24 September 1920, Abendausgabe.

[66] Fricke to Klein, 29 September 1920 (see appendix).

[67] Einstein to Weyl, 16 December 1921, referring to Weyl 1922.

[68] Press communication of F. Himstedt (Freiburg) and M. Planck (Berlin) regarding Einstein's retraction of his criticisms of Lenard in "Meine Antwort..." *Berliner*

Tageblatt, 25 September 1920, Morgenausgabe, p. 2:. Im „Berliner Tageblatt" vom 27. August hat Herr Professor Einstein unter dem Titel „Meine Antwort über die antirelativistische G.m.b.H." einen Artikel der Abwehr gegen die „Arbeitsgemeinschaft deutscher Naturforscher für Reinheit der Wissenschaft" gerichtet, in deren ersten Versammlung bekanntlich Herr Weyland ihn in persönlich gehässiger Weise angegriffen hat. In diesem Artikel hat er sich auch gegen Herrn Professor Lenard gewendet, welcher neben anderen Physikern auf der Rednerliste verzeichnet war. Bei Gelegenheit der jüngsten Tagung der Naturforscherversammlung in Bad Nauheim konnten wir feststellen, dass Herr Lenard ohne sein Zutun au die Rednerliste gekommen ist. Auf Grund dieser Tatsache hat uns Herr Einstein ermächtigt, sein lebhaftes Bedauern auszusprechen, dass er die in seinem Artikel enthaltenen Vorwürfe auch gegen den von ihm hochgeschätzten Kollegen Herrn Lenard gerichtet hat.

[69]Himstedt afterward wrote to Johannes Stark. "Es ist nach nochmaligen langen Verhandlungen eine Vereinbarung zu treffen, nach der Planck u[nd] ich eine rein sachliche Erklärung abgeben, mit der beide Paukanten sich einverstanden erklärt haben. Das ist das Erfreuliche. Das Unerfreuliche ist meiner Ansicht nach, dass die Erklärung sehr matt u[nd] die Genugtuung für Lenard zu gering [waren]. Ich habe den Eindruck gehabt, dass L. hier der Sache ein nicht zu unterschätzendes Opfer gebracht hat. Ich hatte so sehr genug von diesen unerquicklichen Hin- u[nd] Her- Verhandlungen, dass ich abgereist bin sobald die Sache erledigt war. Von Physik habe ich verdammt wenig zu hören bekommen in der ganzen Woche, aber ein großes Vergnügen ist es mir gewesen mit lieben Kollegen u. besonderes mit Ihnen zusammen sein zu können." Himstedt to Stark, 27 September 1920, Nachlass Stark, Stadtsbibliothek Berlin.

[70]The broader cultural impact of relativity is discussed in several essays in Williams 1968.

REFERENCES

Beyerchen, A. (1977). *Scientists under Hitler: Politics and the Physics Community in the Third Reich*, Yale University Press, New Haven.

Birck, O. (1917). Die Einsteinsche Gravitationstheorie und die Sonnenfinsternis im Mai 1919, *Die Naturwissenschaften* **5(46), 16 November 1917**: 689–696.

Born, M. (1913). Zum Relativitätsprinzip: Entgegnung auf Herrn Gehrckes Artikel 'Die gegen die Relativitätstheorie erhobenen Einwände', *Die Naturwissenschaften* **1 (1913)**: 92–94.

Born, M. (1920). *Die Relativitätstheorie Einsteins und ihre physikalischen Grundlagen*, Springer, Berlin.

Born, M. (1978). *My Life; Recollections of a Nobel Laureate*, Taylor & Francis, London.

Born, Max ed. (1969). *Albert Einstein/ Max Born. Briefwechsel, 1916-1955*, Nymphenburger, Munich.

Corry, L. (1997). David Hilbert and the Axiomatization of Physics (1894-1905), *Archive for History of Exact Sciences* **51**: 83–198.

Darrigol, O. (2000). *Electrodynamics from Ampère to Einstein*, Oxford University Press, Oxford.

Earman, J. and Glymour, C. (1980). Relativity and Eclipses: The British Eclipse Expeditions of 1919 and Their Predecessors, *Historical Studies in the Physical Sciences* **11**: 49–85.

Eddington, A. S. (1920). *Space, Time, and Gravitation: an Outline of the General Theory of Relativity*, Cambridge University Press, Cambridge.

Einstein (1915a). Zur allgemeinen Relativitätstheorie, *Königlich Preußische Akademie der Wissenschaften* pp. 778–786. (Berlin). *Sitzungsberichte* (1915). Reprinted in *CPAE*, vol. 6, pp. 214–224.

Einstein (1915b). Zur allgemeinen Relativitätstheorie. (nachtrag),, *Königlich Preußische Akademie der Wissenschaften* pp. 799–801. (Berlin). *Sitzungsberichte* (1915). Reprinted in *CPAE*, vol. 6, pp. 225-229.

Einstein (1915c). Erklärung der Perihelbewegung des Merkur aus der allgemeinen Relativitätstheorie, *Königlich Preußische Akademie der Wissenschaften* pp. 831–839. (Berlin). *Sitzungsberichte* (1915). Reprinted in *CPAE*, vol. 6, pp. 233-243.

Einstein (1915d). Die Feldgleichungen der Gravitation, *Königlich Preußische Akademie der Wissenschaften* pp. 844–847. (Berlin). *Sitzungsberichte* (1915). Reprinted in *CPAE*, vol. 6, 244–249.

Einstein (1916). Die Grundlage der allgemeinen Relativitätstheorie, *Annalen der Physik* **49**: 769–822. Reprinted in *CPAE*, vol. 6, 283–339.

Einstein (1917). *Über die spezielle und die allgemeine Relativitätstheorie. (Gemeinverständlich.)*, Vieweg, Braunschweig. Reprinted in *CPAE*, vol. 6, 420–539.

Einstein (1918a). Dialog über Einwände gegen die Relativitätstheorie, *Die Naturwissenschaften* **6**: 697–702.

Einstein (1918b). Bemerkung zu E. Gehrckes Notiz 'Über den Äther', *Deutsche Physikalische Gesellschaft. Verhandlungen* **20**: 261.

Einstein (1920a). *Relativity: The Special and the General Theory*, Methuen, London. Trans. Robert W. Lawson.

Einstein (1920b). Meine Antwort: Über die anti-relativitätstheoretische G.m.b.H, *Berliner Tageblatt* **27 August 1920, Morgenausgabe**: pp. 1–2. Reprinted in *CPAE*, vol. 7, 345–347.

Einstein (1954). *Ideas and Opinions*, Bonanza Books, New York.

Einstein (1996). The Berlin Years: Writings, 1914-1917, *in* A. J. Kox et al. (ed.), *Collected Papers of Albert Einstein (CPAE)*, Vol. 6, Princeton University Press, Princeton.

Einstein (1998a). The Berlin Years: Correspondence, 1914-1917, *in* Robert Schulmann, et al. (ed.), *Collected Papers of Albert Einstein (CPAE)*, Vol. 8A, Princeton University Press, Princeton.

Einstein (1998b). The Berlin Years: Correspondence, 1918, *in* Robert Schulmann, et al. (ed.), *Collected Papers of Albert Einstein (CPAE)*, Vol. 8B, Princeton University Press, Princeton.

Einstein (2002). The Berlin Years: Writings, 1918-1921, *in* Michel Janssen et al. (ed.), *Collected Papers of Albert Einstein (CPAE)*, Vol. 7, Princeton University Press, Princeton.

Fölsing, A. (1993). *Albert Einstein. Eine Biographie*, Suhrkamp, Frankfurt am Main.

Forman, P. (1971). Weimar Culture, Causality, and Quantum Theory, 1918-1927: Adaptation by German Physicists to a Hostile Intellectual Environment, *Historical Studies in the Physical Sciences* **3**: 1–115.

Forman, P. (1974). The Financial Support and Political Alignment of Physicists in Weimar Germany, *Minerva* **12**: 39–66.

Frank, P. (1947). *Einstein, his Life and Times*, Alfred Knopf, New York.

Frank, P. (1949). *Einstein, sein Leben und seine Zeit*, Paul List, München.

Freundlich, E. (1920). Ein Bericht der englischen Sonnenfinsternisexpedition über die Ablenkung des Lichtes im Gravitationsfelde der Sonne, *Die Naturwissenschaften* **8**: 667–673.

Gehrcke, E. (1913a). Die gegen die Relativitätstheorie erhobenen Einwände, *Die Naturwissenschaften* **1**: 62–66.

Gehrcke, E. (1913b). Einwände gegen die Relativitätstheorie, *Die Naturwissenschaften* **1**: 170.

Gehrcke, E. (1916). Zur Kritik und Geschichte der neueren Gravitationstheorien, *Annalen der Physik* **51**: 119–124. Reprinted in Gehrcke 1924a, 40–44.

Gehrcke, E. (1919). Berichtigung zum Dialog über die Relativitätstheorien, *Die Naturwissenschaften* **7**: 147–148. Reprinted in Gehrcke 1924a, 48–50.

Gehrcke, E. (1924a). *Kritik der Relativitätstheorie. Gesammelte Schriften über absolute und relative Bewegung*, Hermann Meusser, Berlin.

Gehrcke, E. (1924b). *Die Massensuggestion der Relativitätstheorie. Kulturhistorisch-psychologische Dokumente*, Hermann Meusser, Berlin.

Gerber, P. (1898). Die räumliche und zeitliche Ausbreitung der Gravitation, *Zeitschrift für Mathematik und Physik* **43 (1898)**: 93–104.

Gerber, P. (1902). Die Fortpflanzungsgeschwindigkeit der Gravitation, *(Programmabhandlung des städtischen Realgymnasiums zu Stargard in Pommern)* . Republished in *Annalen der Physik* **52 (1917)**: 415–441.

Goenner, H. (1992). The Reception of Relativity in Germany as Reflected by Books Published between 1908 and 1945, *in* J. Eisenstaedt and A. J. Kox (eds), *Studies in the History of General Relativity, Einstein Studies*, Vol. 3, Birkhäuser, Boston, pp. 15–38.

Goenner, H. (1993). The Reaction to Relativity Theory I: The Anti-Einstein Campaign in Germany in Germany, *Science in Context* **6**: 107–133.

Gray, Jeremy, ed. (1999). *The Symbolic Universe. Geometry and Physics, 1890-1930*, Oxford University Press, Oxford.

Grundmann, S. (1998). *Einsteins Akte*, Springer-Verlag, Berling.

Hentschel, K. (1990). *Interpretationen und Fehlinterpretationen der speziellen und der allgemeinen Relativitätstheorie durch Zeitgenossen Albert Einsteins*, Birkhäuser, Basel.

Hentschel, K. (1994). Erwin Finlay Freundlich and Testing Einstein's Theory of Relativity, *Archive for History of Exact Sciences* **47**: 143–201.

Hermann, Armin, ed. (1968). *Albert Einstein/ Arnold Sommerfeld Briefwechsel. Sechzig Briefe aus dem goldenen Zeitalter der modernen Physik*, Schwabe Verlag, Basel.

Hermann, Armin, ed. (1994). *Einstein, der Weltweise und sein Jahrhundert: eine Biographie*, Piper, München.

Hilbert, D. (1992). *Natur und mathematisches Erkennen. Vorlesungen, gehalten 1919-1920 in Göttingen*, ed. David E. Rowe. Birkhäuser, Basel.

Jungnickel, C. and McCormach, R. (1986). *Intellectual Mastery of Nature: Theoretical Physics from Ohm to Einstein, vol. 2 The Now Mighty Theoretical Physics, 1870-1925*, University of Chicago Press, Chicago.

Kleinert, A. (1993). Paul Weyland, der Berliner Einstein-Töter, *in* H. Albrecht (ed.), *Naturwissenschaft und Technik in der Geschichte. 25 Jahre Lehrstuhl für Geschichte der Naturwissenschaften und Technik am Historischen Institut der Universität Stuttgart*, Verlag für Geschichte der Naturwissenschaften und der Technik, Stuttgart, pp. 198–232.

Kleinert, A. and Schönbeck, C. (1978). Lenard und Einstein. Ihr Briefwechsel und ihr Verhältnis vor der Nauheimer Diskussion von 1920, *Gesnerus* **35**: 318–333.

Kostro, L. (2000). *Einstein and the Ether*, Apeiron, Montreal.

Kretschmann, E. (1917). Über den physikalischen Sinn der Relativitätspostulate. A. Einsteins neue und seine ursprüngliche Relativitätstheorie, *Annalen der Physik* **53**: 575–614.

Lenard, P. (1918). Über Relativitätsprinzip, Äther, Gravitation, *Jahrbuch der Radioaktivität und Elektronik* **15**: 117–136.

Lenard, P. (1920). *Über Relativitätsprinzip, Äther, Gravitation*. *2nd ed*, Hirzel, Leipzig.

Lenard, P. (1921). *Über Relativitätsprinzip, Äther, Gravitation*. *3rd ed, mit einem Zusatz, betreffend der Nauheimer Diskussion*, Hirzel, Leipzig. Reprinted with Lenard's comments in Lenard 2003, 423–460.

Lenard, P. (2003). *Wissenschaftliche Abhandlungen*, Band 4, ed. Charlotte Schönbeck, Diepholz, Berlin.

Levenson, T. (2003). *Einstein in Berlin*, Bantam Books, New York.

Mach, E. (1904). *Die Mechanik in ihrer Entwickelung*. *Historisch-kritisch dargestellt*. *5th rev. ed*, Brockhaus, Leipzig.

McCormmach, R. (1982). *Night Thoughts of a Classical Physicist*, Harvard University Press, Cambridge, Mass.

Pais, A. (1982). *'Subtle is the Lord... ' The Science and the Life of Albert Einstein*, Oxford University Press, Oxford.

Reid, C. (1976). *Courant in Göttingen and New York*, Springer-Verlag, New York.

Reiser, A. (1930). *Albert Einstein: a Biographical Portrait*, Albert & Charles Boni, New York.

Rowe, D. E. (1986). 'Jewish Mathematics' at Göttingen in the Era of Felix Klein, *Isis* **77**: 422–449.

Rowe, D. E. (1999). The Göttingen Response to General Relativity and Emmy Noether's Theorems, *in* J. Gray (ed.), *The Symbolic Universe. Geometry and Physics, 1890-1930*, Oxford University Press, Oxford, pp. 189–234.

Rowe, D. E. (2001). Einstein meets Hilbert: At the Crossroads of Physics and Mathematics, *Physics in Perspective* **3**: 379–424.

Rowe, D. E. (2004). Making Mathematics in an Oral Culture: Göttingen in the Era of Klein and Hilbert, *Science in Context* **17(1/2)**: 85–129.

Rowe, D. E. (2006). Publicity, Politics, and the Price of Fame: Einstein in Berlin, *in* M. Janssen and C. Lehner (eds), *The Cambridge Companion to Einstein*, Cambridge University Press. Scheduled to appear in 2006.

Sarkowski, H. (1996). *Springer-Verlag. History of a Scientific Publishing House, Part I, 1842-1945*, Springer-Verlag, Heidelberg. Trans. Gerald Graham.

Scholz, E. (ed) (2001). *Hermann Weyl's Raum-Zeit-Materie and a General Introduction to his Scientific Work*, Birkhäuser, Basel.

Schönbeck, C. (2003). Philipp Lenard und die frühe Geschichte der Relativitätstheorien, *in Lenard 2003, 323-375*, GNT-Verlag.

Stachel, J. (2002). Einstein from 'B' to 'Z', *Einstein Studies*, Vol. 9, Birkhäuser, Boston.

Stern, F. (1999). *Einstein's German World*, Princeton University Press, Princeton.

Warwick, A. (2003). *Masters of Theory. Cambridge and the Rise of Mathematical Physics*, University of Chicago Press, Chicago.

Weyl, H. (1918). *Raum. Zeit. Materie. Vorlesungen über allgemeine Relativitätstheorie*, Springer, Berlin, 1918.

Weyl, H. (1922). Die Relativitätstheorie auf der Naturforscherversammlung in Bad Nauheim, *Jahresbericht der Deutschen Mathematiker-Vereinigung* **31**: 51–63.

Weyl, H. (1949). Relativity Theory as a Stimulus in Mathematical Research, *Proceedings of the American Philosophical Society* **93 (1949)**: 535–541.

Weyland, P. (1920a). Einsteins Relativitätstheorie – eine wissenschaftliche Massensuggestion, *Tägliche Rundschau* . 2 August, Unterhaltungsbeilage, Evening Edition, reprinted in Weyland 1920b, pp. 21-24.

Weyland, P. (1920b). Betrachtungen über Einsteins Relativitätstheorie und die Art ihrer Einführung. Vortrag gehalten am 24. August 1920 im großen Saal der Philharmonie zu Berlin, *Schriften aus dem Verlage der Arbeitsgemeinschaft deutscher Naturforscher zur Erhaltung reiner Wissenschaft e. V* . Heft 2. Berlin: Arbeitsgemeinschaft deutscher Naturforscher zur Erhaltung reiner Wissenschaft e. V./Köhler, 1920.

Weyland, P. (1920c). Neue Beweise für die Unrichtigkeit der Einsteinschen Relativitätstheorie, *Deutsche Zeitung* . 23 August 1920.

Weyland, P. (1920d). Die Naturforschertagung in Nauheim. Erdrosselung der Einsteingegner, *Deutsche Zeitung* . 26 September 1920.

Williams, L. P. (ed) (1968). *Relativity Theory: its Origins and Impact on Modern Thought*, John Wiley, New York.

CONTRIBUTIONS RELATED TO GENERAL RELATIVITY PUBLISHED IN *DIE NATURWISSENSCHAFTEN*, 1916-1922

1916

Freundlich, E. Die Grundlagen der Einsteinschen Gravitationstheorie, *Die Naturwissenschaften* **4 (1916)** 363-372, 386-392. Reprinted as *Die Grundlagen der Einsteinschen Gravitationstheorie.* Berlin: Springer, 1916.

1917

Frank, P. Die Bedeutung der physikalischen Erkenntnistheorie Machs für das Geistesleben der Gegenwart, *Die Naturwissenschaften* **5** (**1917**): 65–72.

Schlick, M. Raum und Zeit in der gegenwärtigen Physik. Zur Einführung in das Verständnis der allgemeinen Relativitätstheorie, *Die Naturwissenschaften* **5** (**1917**): 161–167, 177–186. Reprinted as *Raum und Zeit in der gegenwärtigen Physik. Zur Einführung in das Verständnis der allgemeinen Relativitätstheorie*, Springer, Berlin 1917.

A. M. Bestätigung des Relativitätsprinzips. (Bericht über Versuch von de Haas), *Die Naturwissenschaften* **5** (**1917**): 388.

Berliner, A. Die Gegenseitige Induktion zweier Massen, *Die Naturwissenschaften* **5** (**1917**): 614.

Birck, O. Die Einsteinsche Gravitationstheorie und die Sonnenfinsternis im Mai 1919, *Die Naturwissenschaften* **5** (**1917**): 689–696.

Berliner, A. Perihelbewegung des Merkur, *Die Naturwissenschaften* **5** (**1917**): 711.

1918

Born, M. Review of: Albert Einstein, *Über die spezielle und die allgemeine Relativitätstheorie. (Gemeinverständlich.)* Braunschweig: Vieweg, 1917. *Die Naturwissenschaften* **6** (**1918**): 82.

Einstein, A. Review of: Hermann Weyl, *Raum – Zeit – Materie. Vorlesungen über allgemeine Relativitätstheorie.* Springer, Berlin 1918. *Die Naturwissenschaften* **6** (**1918**): 373.

Lense, J. Kosmologische Betrachtungen zur allgemeinen Relativitätstheorie, *Die Naturwissenschaften* **6** (**1918**): 663–664.

Einstein, A. Dialog über Einwände gegen die Relativitätstheorie, *Die Naturwissenschaften* **6** (**1918**): 697–702.

1919

Gehrcke, E. Berichtigung zum Dialog über die Relativitätstheorien, *Die Naturwissenschaften* **7** (**1919**): 147–148.

Berliner, A. Zur Sonnenfinsternis am 29. Mai, *Die Naturwissenschaften* **7** (**1919**): 368–369.

Freundlich, E. Zur Prüfung der allgemeinen Relativitätstheorie, *Die Naturwissenschaften* **7** (**1919**): 629–636.

Haas, A. Die Axiomatik der Physik, *Die Naturwissenschaften* **7** (**1919**): 745–750.

Einstein, A. Prüfung der allgemeinen Relativitätstheorie, *Die Naturwissenschaften* **7 (1919)**: 776.

1920

Haas, A. Die Physik als geometrische Notwendigkeit, *Die Naturwissenschaften* **8 (1920)**: 121–127.

Berliner, A. Report on Grebe and Bachem, Die Rotverschiebung der Spektrallinien zur Prüfung der Einsteinschen Gravitationstheorie, *Die Naturwissenschaften* **8 (1920)**: 199.

Freundlich, E. Zu dem Aufsatze, Die Physik als geometrische Notwendigkeit" von Arthur Haas, *Die Naturwissenschaften* **8 (1920)**: 234–235.

Kries, J. v. Über die zwingende und eindeutige Bestimmtheit des physikalischen Weltbildes, *Die Naturwissenschaften* **8 (1920)**: 237–247.

Laue, M. v. Zur Prüfung der allgemeinen Relativitätstheorie an der Beobachtung, *Die Naturwissenschaften* **8 (1920)**: 390–391.

Schlick, M. Naturphilosophische Betrachtungen über das Kausalitätsprinzip, *Die Naturwissenschaften* **8 (1920)**: 461–474.

Silberstein, L. Zur Prüfung der allgemeinen Relativitätstheorie an der Beobachtung, *Die Naturwissenschaften* **8 (1920)**: 474–475.

Laue, M. v. Bemerkung hierzu, *Die Naturwissenschaften* **8 (1920)**: 475.

Freundlich, E. Der Bericht der englischen Sonnenfinsternisexpedition über die Ablenkung des Lichtes im Gravitationsfelde der Sonne, *Die Naturwissenschaften* **8 (1920)**: 667–673.

Laue, M. v. Historisch-kritisches über die Perihelbewegung des Merkur, *Die Naturwissenschaften* **8 (1920)**: 735–76.

Guthnick, P. Jährliche Refraktion und Lichtablenkung in der Nähe der Sonne, *Die Naturwissenschaften* **8 (1920)**: 814.

Reichenbach, H. Review of: Harry Schmidt, *Das Weltbild der Relativitätstheorie.* Hartung, Hamburg 1920. *Die Naturwissenschaften* **8 (1920)**: 925.

Reichenbächer, E. Inwiefern lässt sich die moderne Gravitationstheorie ohne die Relativität begründen? *Die Naturwissenschaften* **8 (1920)**: 1008–1010.

Einstein, A. Antwort auf vorstehende Betrachtung, *Die Naturwissenschaften* **8 (1920)**: 1010–1011.

1921

Kopff, A. Das Rotationsproblem in der Relativitätstheorie, *Die Naturwissenschaften* **9 (1921)**: 9–15.

Freundlich, E. Sonnenatmosphäre und Einsteineffekt, *Die Naturwissenschaften* **9 (1921)**: 103–104.

Hopmann, J. Zur Ablenkung der Lichtstrahlen im Gravitationsfeld der Sonne, *Die Naturwissenschaften* **9 (1921)**: 192.

Thirring, H. Über das Uhrenparadoxon in der Relativitätstheorie, *Die Naturwissenschaften* **9 (1921)**: 209–212.

Einstein, A. Zur Abwehr, *Die Naturwissenschaften* **9 (1921)**: 219.

Jacob, M. Review of: Max Born, *Die Relativitätstheorie Einsteins und ihre physikalischen Grundlagen*. Springer, Berlin 1920. *Die Naturwissenschaften* **9 (1921)**: 371–373.

Thirring, H. Review of: Caspar Isenkrahe, *Zur Elementaranalyse der Relativitätstheorie*. Vieweg, Braunschweig 1921. *Die Naturwissenschaften* **9 (1921)**: 373.

Schlick, M. Geometrie und Erfahrung. Zu Einsteins Vortrag vom 27.1.1921, *Die Naturwissenschaften* **9 (1921)**: 435–436.

Gehrcke, E. Über das Uhrenparadoxon in der Relativitätstheorie, *Die Naturwissenschaften* **9 (1921)**: 482.

Thirring, H. Erwiederung, *Die Naturwissenschaften* **9 (1921)**: 482–483.

Gehrcke, E. Erörterung des Uhrenparadoxon in der Relativitätstheorie, *Die Naturwissenschaften* **9 (1921)**: 550–551.

Thirring, H. Erwiederung, *Die Naturwissenschaften* **9 (1921)**: 551.

1922

Kienle, H. Die Bewegung des vier inneren Planeten mit besonderer Berücksichtigung der Bewegung des Merkurperihels, *Die Naturwissenschaften* **10 (1922)**: 217–224, 246–254.

Westphal, W. Die Möglichkeit einer Prüfung des Satzes von der Gleichheit der trägen und der schweren Masse auf astronomischen Grundlage, *Die Naturwissenschaften* **10 (1922)**: 261.

Laue, M. v. u. and Pringsheim, P. St. Johns und Babcocks Beobachtungen über die Rotverschiebung der Spektrallinien auf der Sonne, *Die Naturwissenschaften* **10 (1922)**: 330.

Laue, M. v. Zum Einstein-Film, *Die Naturwissenschaften* **10 (1922)**: 434.

Bauschinger, J. Die astronomische Festlegung des Trägheitssystems, *Die Naturwissenschaften* **10 (1922)**: 1005–1010.

ERHARD SCHOLZ

THE CHANGING CONCEPT OF MATTER IN H. WEYL'S THOUGHT, 1918–1930

ABSTRACT

During the "long decade" of transformation of mathematical physics between 1915 and 1930, H. Weyl interacted with physics in two highly productive phases and contributed to it, among others, by his widely read book on *Space – Time – Matter (Raum – Zeit – Materie)* (1918–1923) and on *Group Theory and Quantum Mechanics (Gruppentheorie und Quantenmechanik)* (1928–1931). In this time Weyl's understanding of the constitution of matter and its mathematical description changed considerably. At the beginning of the period he started from a "dynamistic", classical field theoretic and geometrical conception of matter, following and extending the Mie-Hilbert approach, but gave it up during the year 1920. After transitional experiments with a singularity (and in this sense topological) approach in 1921/22, he developed an open perspective of what he called an "agency theory" of matter. The idea for it was formulated already before the advent of the new quantum mechanics in 1925/26. It turned out to be well suited to be taken over to the quantum view as a kind of heritage from the first half of the decade. At the end of the period, Weyl completely renounced his earlier belief in the possibility to "construct matter" from a geometrically unified field theory. He now posed the possibility of a geometrization of the mathematical forms underlying the rising quantum physical description of matter as a completely open problem for future research.

1. INTRODUCTION

It may appear a strange question to ask for the changing views of a mathematician on the concept of matter. Why not pose it for a natural scientist or a philosopher? But Hermann Weyl was, as we know, all of them; and in this convergence of interests and competences he was rather unique. His views on mathematics and their foundations made it impossible for him to separate mathematics and its meaning from broader contexts of its use as a conceptual form and as a symbolic tool for the understanding of nature (or at least some aspects of it).

During the "long decade" of transformation of mathematical physics, as we may call the time between 1915 and 1930, with the rise of the general theory of relativity (GRT) and the origin of the new quantum mechanics (QM), H. Weyl interacted with physics in two highly productive phases and

V.F. Hendricks, K.F. Jørgensen, J. Lützen and S.A. Pedersen (eds.), Interactions: Mathematics, Physics and Philosophy, 1860-1930, pp. 281–306.
© 2006 Springer.

contributed to the development of both, theoretical physics and the mathematical concepts and methods in it. The first phase lasted from 1916 to 1923 and had as main out-spring his widely read book on *Space – Time – Matter (Raum – Zeit – Materie)* (Weyl, 1918*b*), which we will also refer to by RZM. In this period the book had five successive editions with considerable extensions and/or alterations well documenting the shifts in the understanding of the subject by its author. Some of these changes were of a more technical nature for general relativity or the mathematics involved, others were of more basic nature, including in particular the changing characterization and mathematical description of matter. In the middle of the 1920s Weyl worked on the representation theory of Lie groups (1924/1925) and wrote a book on the philosophy of mathematics and the natural sciences (1925/1926), before he started to contribute actively to the rising quantum mechanics, culminating in his second book about mathematical physics, *Group Theory and Quantum Mechanics* (Weyl, 1928).

The growing awareness of the irreducible and far-reaching role of quantum properties resulted already in considerable shifts of Weyl's concept of matter during the first phase of involvement in physics. In the second half of the "long decade" his views were deeply transformed by the rise of quantum physics. This transformation was, of course, much more than a personal experience. It reflected the experience of the whole community of researchers in basic physics of the time, although seen from a specific Weylian perspective. As such, it may be illuminating for a historical and philosophical understanding of the transformation of the concept of matter, brought about by the tension resulting from the unfinished "double revolution" of GRT and QM during the 1920s.[1] In spite of the drastic difference between Weyl's concept of matter at the beginning of the period and at the end of it, we easily perceive a common thread linking both ends. This common underlying feature is a dynamistic view of matter. This characterization has to be understood in a general, philosophico-conceptual sense which *may* be related, but *need not* be, to the electrodynamical picture of matter which gave a new thrive to dynamism among physicists and mathematicians of the early 20th century.

In the history and philosophy of physics, the dynamistic view of matter in the early 20th century is often restricted to the exclusively electromagnetic approach. Such a restriction shadows off the intricate link to the quantum theoretical phase, which played a role for some of the protagonists of the period. Of course, also Weyl started from Mie's electrodynamical theory of matter when he first looked for an adequate "modern" mathematical expression of such a dynamistic view. From this basis he developed his program of a geometrically unified field theory in the first phase of his involvement in mathematical physics.[2] The impact of quantum physics replaced classical field pictures by quantum stochastical descriptions of the "agency nature" of matter (*Agenstheorie der Materie*), as Weyl liked to call it. At the end of the period discussed here, he completely renounced his earlier belief in the possibility to "construct matter" from a geometrically unified field theory. He now posed the possibility of a geometrization of the mathematical forms underlying the rising quantum physical description of matter as a completely open problem for future research.

In this article I present a kind of longitudinal section through the long decade, observed along the trajectory of a single person, who was partially a contributor and partially a well informed observer of the development.[3] We start with Weyl's turn towards Mie's theory of matter, his own contribution to it, and his rather early detachment, which was related to the influences of early quantum mechanics, without being a necessary conclusion from it. After a short phase of relaxation of classical explanations of matter, by a combination of metrical and topological aspects (matter characterized by singularities in space-time), Weyl developed an open perspective of what he called an "agency theory" of matter. The idea for it was formulated already before the advent of the new quantum mechanics in 1925/26. It turned out to be well suited to be taken over to the quantum view as a kind of heritage from the first half of the decade.

2. ADHERENCE TO MIE'S DYNAMISTIC APPROACH TO MATTER

After Weyl came back to neutral Switzerland from his war duty in the German army in May 1916, he started a completely new phase of his research, which was imbued by a longing for a sounder basis of knowledge.[4] For him, this meant to work in a broad and interconnected set of fields comprising the foundations of analysis, differential geometry, general relativity, unified field theory and the basic structures of matter. Only if we take this broad range of intellectual activities into account, we can get an adequate sense of Weyl's conceptual and theoretical moves inside the single fields. Let us have a look at some points of such interconnections:

- In the foundations of mathematics our author shifted from his own constructive-arithmetical approach for a characterization of the concept of continuum (Weyl, 1918*a*) to a kind of Brouwerian intuitionism (Weyl, 1921*b*). For a while, he believed Brouwer's approach to possess an intimate connection to his ideas in purely infinitesimal geometry. Weyl could well characterize "purely infinitesimal" structures on the level of differential geometry by his generalization of a Riemannian metric by combining a conformal structure with a *length* *connection* $\varphi = \sum_i \varphi_i dx^i$. It was, however, much more difficult to give them a mathematical meaning on the foundational (and topological) level. Here a precise conceptual characterization was lacking. Weyl was well aware of this deficiency which contributed to tensions and shifts inside his foundational contributions. For a while, Brouwer's "revolutionary" approach to the continuum (as Weyl called it in 1920) appeared him to offer a promising road.[5]

- For some years, Weyl considered his gauge geometrical generalization of the Riemannian metric as the proper approach for a unified field theory of gravitation and electromagnetism and, moreover, a field theory of matter based upon it.[6]
- Rising doubts with respect to the physical feasibility of this immediate physical interpretation of gauge geometry contributed to a turn towards a more basic philosophico-conceptual analysis of the principles of congruence geometry in Weyl's *mathematical analysis of the problem of space*.[7]
- The necessity, or at least usefulness, to accept classical logical principles (excluded middle) in the proof of the main theorem of the analysis of the space problem contributed to rethink his radical position in the foundations of mathematics.

In the second point indicated above, the field theoretic approach to matter constitution, Weyl was deeply influenced by Mie's electromagnetic theory of matter which he got to know through Hilbert's modification during the autumn 1915.[8] Hilbert attempted to arrive at a kind of mathematical synthesis of Mie's and Einstein's ideas on electromagnetism (Mie) and gravitation (Einstein). He indicated how to find a common Hamiltonian for gravity and electromagnetism in a generally covariant setting.[9] He was convinced, that in such a classical united field theory the riddles of the grainy structure of matter should be solvable. H. Weyl and F. Klein were not convinced that Hilbert's attempted synthesis of Mie and Einstein was acceptable as a physical theory. They argued for a broader understanding of Hilbert's approach. E. Noether's mathematical analysis of Hilbert's invariance conjectures (later "Noether theorems") contributed an essential mathematical stepping stone for it.[10]

In addition to the unclear role of energy conservation in Hilbert's approach (and general relativity more broadly), which was clarified only step by step, Weyl was not convinced that Hilbert's approach was able to lead to a unification of gravitation and electromagnetism, in which matter structures were better derivable than in Mie's original version. Probably these were the main reasons for him to discuss the field theoretic matter concept in the first edition of his book essentially like in Mie's original purely electromagnetic approach (Weyl, 1918*b*, §25). Hilbert's generalization was only mentioned in passing, in the section which treated the modification of the Hamiltonian principle for electromagnetism by gravitation (Weyl, 1918*b*, §32). Of course, such a presentation did not give sufficient credit to Hilbert's seminal step towards a unified approach to gravity and electromagnetism in the context of relativistic physics.[11] On the other hand, Weyl presented Mie's approach in such a convinced rhetoric form that the reader might easily get the impression that Mie's research goal was already nearly achieved. The desired result (derivation of a "granular" structure from field laws) seemed close to be sure. After a comparison of Mie's theory with Maxwell-Lorentz's, Weyl stated:

> The theory of Maxwell and Lorentz cannot hold for the interior of the electron; therefore, from the point of view of the ordinary theory of the electrons we must treat the electrons as something given *a priori*, as a foreign body to the field. A more general theory of electrodynamics has been proposed by *Mie*, by which it seems possible to derive the matter from the field .... (Weyl, 1918*b*, 165)

This formulation was kept unchanged by Weyl in the next three editions.[12] He only changed it during the last revision for the fifth edition (1923). Then he clearly expressed the open status of Mie's attempt and presented it, in an essentially didactical approach, as nothing but an *example* of a physical theory "which agrees completely with the recent ideas about matter" (Weyl, 1918*b*, [5]1923, 210).

Mie's proposal fitted beautifully with Einstein's discovery of the energy-mass equivalence of special relativity and seemed to extend it. In a passage commenting the equivalence $E = mc^2$, Weyl argued:

> We have thus attained a new, purely dynamical view of matter. (Footnote: Even Kant in his "Metaphysische Anfangsgründe der Naturwissenschaft" teaches the doctrine that matter fills space not by its mere existence but in virtue of the repulsive forces of all its parts.) Just as the theory of relativity theory has taught us to reject the belief that we can recognize one and the same point in space at different times, *so now we see that there is no longer a meaning in speaking of the same position of matter at different times.* (Weyl, 1918*b*, 162), (Weyl, 1922, 202)

Already here, in the context of special relativity, he described an electron as a kind of "energy knot" which "propagates through empty space like a water wave across the sea", and which could no longer be considered as element of some self-identical substance. Then, of course, there arose the problem to understand both, this kind of propagation of energy, and the stability of the "energy knot". Weyl stated the new challenge of (special) relativity to field theory, which arose from a dynamical understanding of matter/energy:

> The theory of fields has to explain why the field is granular in structure and why these energy-knots preserve themselves permanently from energy and momentum in their passage to and fro (. . .); therein lies the *problem of matter*. (Weyl, 1918*b*, 162, emphasis in original) (Weyl, 1922, 203)

Like the dynamists of the early 19th century, Weyl now insisted that atoms could not be considered as invariant fundamental constituents of matter:

> *Atoms and electrons are not*, of course, *ultimate invariable elements*, which natural forces seize from without, pushing them hither and thither, but they are themselves distributed continuously and subject to minute changes of a fluid character in their smallest pieces. It is not the field that requires matter as its carrier in order to be able to exist itself, but *matter* is, on the contrary, *an offspring of the field*. (ibid.).[13]

It seems worthwhile to remark that these general passages on the dynamistic outlook on the problem of matter were *not changed* by Weyl until (and including) the fifth edition of his book in 1923. On the other hand, the special role attributed to Mie's theory, or to his own unified field theoretic approach, underwent considerable changes during the following years. But in spite of all his enthusiasm for the new role of field theory in the understanding of matter, Weyl indicated already in 1918 after his presentation of Mie's theory, that something new was rising at the (epistemic) horizon, which might have unforeseen consequences in the future. He compared the actual status of field physics with the seemingly all-embracing character of Newtonian mass-point dynamics in the Laplacian program at the turn to the 19th century and warned:

> Physics, this time as a physics of fields, is again pursuing the object of reducing the totality of natural phenomena to *a single physical law:* it was believed that this goal was almost within reach when the mechanical physics of mass points, founded upon Newton's *Principia*, was celebrating its triumphs. But also today, provision is taken that our trees do not grow up to the sky. We do not yet know whether the state quantities underlying Mie's theory suffice for a characterization of matter,

whether it is in fact purely "electrical" in nature. Above all, the dark cloud of all those appearances that we are provisionally seeking to deal with by the quantum of action throws its shadow upon the land of of physical knowledge, threatening no one knows what new revolution. (Weyl, 1918b, 170).[14]

3. A GEOMETRICAL EXTENSION OF MIE'S THEORY

A few months after his book manuscript was finished, Weyl developed his concept of a generalized *Weylian metric* on a differentiable manifold. In technical terms his metric was given, and still can be characterized, by an equivalence class of pairs, $[(g, \varphi)]$, consisting of a (semi)Riemannian metrics $g = \sum_{i,j} g_{ij} dx_i dx_j$ and a differential form representing a *length connection*, $\varphi = \sum_i \varphi_i dx_i$, up to equivalence by conformal factors in the Riemannian component of the metric and so-called "gauge transformations" of the length connection form.[15] This generalization allowed a seemingly natural interpretation of the potential of the electromagnetic field by the length connection and thus a metrical unification of the main physical fields known at the time, gravity (g) and electromagnetism (φ). Weyl considered this structure as an important step forward for the Mie program of a dynamical characterization of matter. He published about it in several articles, starting in 1918. In the following year he included the approach into the third edition of his book (Weyl, 1918b, [3]1919).

The first edition had ended with a section on cosmology, "Considerations of the world as a whole" (Weyl, 1918b, §33). In the third edition two new sections were added, one on "the world metrics as the origin of the electromagnetic phenomena", containing an introduction to Weyl's unified field theory, and one on "matter, mechanics and the presumable (mutmaßliches) law of the world", in which Weyl's extension of the Mie program was sketched. Like in the first edition, Hilbert's extension of Mie's program was only indirectly mentioned in the section on the combined Hamiltonian principle of electromagnetism and gravitation. On the other hand, the last section culminated in Weyl's own attempt to overbid both Mie and Hilbert by a derivation of the discrete "granular" matter structures from his gauge invariant action principle. In his lecture course on mathematics and the knowledge of nature of the winter semester 1919/20, Hilbert countered by an acid remark that such a perspective would lead to a kind of "Hegelian physics", in which the "whole world process would not go beyond the limited content of a finite thought" (Hilbert, 1992, 100). He did not explain, though, why this kind of analysis should not apply to his own program just as well.

In 1919 Weyl was at the high point of enthusiasm for his new theory. The new section in the third edition of his book started with a rhetoric trumpet-blast:

> We rise to a final synthesis ... (Weyl, 1918b, [3]1919, 242)[16]

Part of his enthusiasm resulted apparently from the realization that gauge invariance with respect to the change of the length gauge led to a new invariance principle, which in Weyl's semantics of the approach could only be the invariance of electrical charge (Weyl, 1918b, [3]1919) (Weyl, 1922, 293). That was, of course, a great achievement of lasting importance, even if the specific version of gauge invariance had later to be given up.[17] But Weyl hoped for more. He expected that on the one hand the cosmological modification of the Einstein equation should be a natural result from his gauge geometry. On the other hand the stable solutions of the equations for the "problem of matter", satisfying adequate regularity conditions should lead to a discrete set of solutions depending on some parameter β. This expectation had a (formal) similarity to a set of "discrete eigenvalues" of an operator, although here the operator was not linear.

The problem was, in fact, characterized by a non-linear differential equation of great complexity. Even Weyl guessed that the available tools of analysis would probably neither suffice for a proof of their existence, nor for an approximative calculation (Weyl, 1918b, [3]1919, 260). This remark made the epistemic status of Weyl's "discrete solutions" highly problematic. It turned them rather into a symbol for a natural philosophical speculation than into an object for research in mathematical physics. In fact, the inaccessibility of Weyl's modified non-linear electromagnetic field equations seem to have contributed to his turn away from the program a little later. Weyl continued the discussion by a beautiful remark.

> The corspusculae which correspond to the possible eigenvalues had to coexist in the same world besides each other or in another, mutually enforcing on another subtle modifications of their intrinsic structure; strange consequences seem here to arise for the organization of the universe; perhaps they may make comprehensible its stillness in the large and unrest in the small. (Weyl, 1918b, [3]1919, 261)

When Weyl wrote these lines, he was at the peak of his belief in a strong unification program of forces and matter, which could be constructed on purely geometrical grounds. In the last long passage of the new added sections we find a discussion of how he now saw the relationship between geometry and physics in the light of his recent findings.

We have realized that physics and geometry coincide with each other and that the world metrics is one, and even the only one, physical reality. Thus, in the final consequence, this physical reality appears as nothing but a pure form; geometry has not been physicalized but physics has been geometrized (nicht die Geometrie ist zur Physik, sondern die Physik ist zur Geometrie geworden). (Weyl, 1918b, [3]1919)

H. Weyl was now at the apogee of the belief in a strong unification which was both, deeply reductionist and highly idealistic. In his eyes, physics seemed to be transformed to a purely formal status and was absorbed by geometry. Matter had seemingly become an epiphenomenon of the "world metrics" which started to acquire a slightly mystical flavour. In the mind of our protagonist, the physicalizing tendency of geometry among leading protagonists of the 19th century, including researchers like C.F. Gauss, N.I. Lobachevsky, and B. Riemann, appeared turned upside down—even though in this extreme form for only a year or two.

As we will see in a moment, this conviction did not hold for long. Already in the fourth edition, this extremely reductionist passage was canceled by its author. Now the book ended with another, less reductive passage on the unifying power of the mind and an éloge of the "chords from that harmony of the spheres of which Pythagoras and Kepler once dreamed" (Weyl, 1922, 312). Weyl did not hide that he had changed his mind; in a separate article written for the physical community shortly after the revisions for the fourth edition, he explained frankly:

From the first edition of RZM to the third one I took the position of (. . .) [a purely field theoretic characterization of matter, E.S.], as I was charmed by the beauty and unity of pure field theory; in the fourth edition, however, I lost confidence in the field theory of matter by striking reasons and changed to the second point of view [of a primacy of matter, irreducible to interaction fields, E.S.] . (Weyl, 1921a, 242)

Let us have a look at the "striking reasons" for this ontological shift at the beginning of the 1920s.

4. BREAK WITH MIE'S THEORY

In the year 1919 Weyl gave a talk to the Swiss *Naturforschende Gesellschaft* on the relationship between the causal and the statistical view of physics, which was published a year later (Weyl, 1920). This paper has been strongly criticized by P. Forman, in his otherwise very stimulating article on Weimar culture and its influence on the discourse among physicists (Forman, 1971), as a document of an "anti-rational" kind of "conversion to acausality". We need not take up here again the broader debate on the question how "anti-rational" the move was and what kind of "acausality" was at stake here.[18] It may suffice to add that the topic of his talk contained, for Weyl, a challenging combination of questions in the conceptual foundations of contemporary physics, including the rising "clouds" of quantum phenomena, with the question of how modern natural science can be made compatible with metaphysical considerations of the existential experience of the openness of evolving life processes and of the freedom of personal actions. A central topic of this talk was the directedness and irreversibility of time, which appeared Weyl to be linked to some process level of irreducibly statistical nature.[19]

The talk took place about the time of our protagonist's turn towards Brouwer's intuitionism. In *this respect* we could even speak of a kind of "conversion".[20] Weyl speculated that Brouwer's approach might be able to lead to a solution of several fundamental problems at one strike. In mathematics he hoped for an answer to the foundational question of the concept of the mathematical continuum and for a philosophically and mathematically sound characterization of the topological "substrate" of purely infinitesimal geometry; in physics he expected a break with the rigidity of the causality structure in classical mechanics ("Gesetzesphysik") and an access to understand the irreducible directness of time. He expected that satisfying answers to all these questions might have some intimate link to the open "process of becoming", which was inherent in Brouwer's choice sequence for the characterization of the intuitionistic continuum:

> Finally and foremost, it is inbuilt into the essence of the continuum that it cannot be treated as a rigid being, but rather only as something what is continuously evolving in an infinite, inward bound process of becoming. (Weyl, 1920, 121)

In this speculative thought, Weyl hoped to find a common thread binding together the foundations of analysis, "purely infinitesimal" geometry, the directedness of time flow in the physical world, its determinative openness, and a conjectured irreducibly stochastical nature of physical laws, which would break with the classical kind of lawfulness ("causality" in the language of the time). Such a break put into doubt any classical field theory of matter and would, probably, even put an end to it as a fundamental concept.

Other influences added salt to such doubts. In September 1920, during the discussions of the Bad Nauheim meeting of the German *Naturforscher Versammlung* and from a draft manuscript of Pauli's contribution on relativity to the *Enzyklopädie der Wissenschaften* Weyl got to know content and reason of Wolfgang Pauli's critical evaluation of his modified version of the Mie theory of matter (Pauli, 1921). Pauli gave no less than five arguments which made it unlikely that Weyl's program might ever be able to lead to a physically realistic theory (ibid., 236ff.).

(1) If there are differential equations which describe the basic constituents of matter (electron and hydrogen nucleus, at the time), they are highly complicated non-linear. A proper physical theory should, on the other hand, be able to give a simple and elementary theoretical character-ization of such elementary constituents rather than derive them as "cunning devices (Kunststücke) of analysis".

(2) The stability of energy knots in a Weylian unified field would require that the gravitational contribution counterbalances the very strong re-pulsive electric forces. Because of the huge difference of the field strengths ($\frac{e}{\sqrt{Gm}} \sim 10^{20}$, for e charge, m mass of the electron and G Newton's gravitational constant), it seemed very unlikely that such an equilibrium could ever be obtained.

(3) In the static case (which was supposed to be the theory context for the derivation of the basic solution of the electron etc.), the field equations of the unified theory are symmetric with respect to charge conjuga-tion. The masses of electron and the proton (in later terminology) strongly break this symmetry. They differ in 3 orders of magnitude.

(4) The field strength in the "interior" region of the electron is in principle unobservable. Therefore the theory may be considered as physically doubtful.

(5) In spite of strong attempts (including some of Pauli himself) no con-vincing Lagrangian had been found, which gave a strongly energy concentrated, stable, centrally symmetric solution to the field equa-tions.

This conjunction of detailed scientific criticism, coming from a person-ally close, young expert in the field, with his own most recent conceptual and metaphysical speculations, undermining the classically deterministic field structures anyhow, shattered Weyl's conviction that his program of a geo-metrically unified field theory would be able to lead to a derivation of matter structures. At the end of the year, in a letter to Felix Klein, in which he reported on his recent advances on mathematical and physical questions (in-cluded or not into the just finished fourth edition of RZM), he reported:

> Finally I thoroughly detached myself from Mie's theory and came to a different position with respect to the problem of matter. I no longer accept field physics as the key to reality. The field, the ether, appears to me only as a *transmitter* of

effects, which is completely feeble by itself; while matter is a reality lying beyond the field and causing its states .... (*Letter H. Weyl to F. Klein, December 28, 1920*, 1920).[21]

Similar phrases are to be found close to the end of the fourth edition of RZM. Here the last section, containing Weyl's version of Mie's theory was no longer announced under the emphatic title of the "presumable world law" as in the third edition. Now, the discussion was downgraded to be a presentation of the "simplest principle of action ...", a formal exercise of some methodological value only.

It contained a short discussion of some consequences of Weyl's gauge invariant quadratic action $S^2\sqrt{|det\,g|}$ for the Hamiltonian of a combined theory of gravitation and electromagnetism, with S the scalar curvature of Weyl geometry. Now he commented that this action is only the "simplest assumption for calculation", for which the author no longer wanted to "insist that it is realised in nature" (Weyl, 1922, 295). For anybody who continued to read the book until the end, Weyl made clear that he now conjectured a close interrelation between the directedness of time flow with quantum jumps as seen in the Bohr model of the atom. That was no longer compatible with the classical structures of time-invertible determinism:

> We must here state in unmistakable language that physics at its present stage can in no wise be regarded as lending support to the belief that there is a causality of physical nature founded on rigorously exact laws. The extended field, "ether" is merely the *transmitter* of effects and is, of itself, powerless; it plays a part that is in no wise different from the one which space, with its rigid Euclidean metrical structure, plays according to the old view; but now the rigid motionless character has become transformed into one which gently yields and adapts itself. ... (Weyl, 1922, 311, emphasis in original)

Now the old duality of field ("ether") and matter was again back on stage for our protagonist. That brought him closer to the perception of the problem by the majority of physicists working on the structure of matter, but also indicated a growing distance to the views held by A. Einstein.

5. A SHORT-LIVED SINGULARITY THEORY OF MATTER

As Weyl came from a strong field theoretic paradigm, it was natural for him in the years 1920/21 to characterize matter by its formal relationship to the interaction field(s). Thus in the fourth edition of RZM Weyl stated his new viewpoint clearly:

Contrary to Mie's view, *matter* now appears *as a real singularity of the field.* (Weyl, 1922, 262, emphasis in original)[22]

But then, matter had somehow to be located in a determinative boundary
structure of the field and the old question of the structures of matter was
again open. After the experience of dynamistic hopes during his period of
adherence to the Mie theory, and in the light of recent modifications coming
from experimental knowledge in microphysics, Weyl came to the conclusion:

If matter is to be regarded as a boundary singularity of the
field, our field-equations make assertions only about the *possible states of the field* and *not about the conditioning of the
states of the field by the matter.* This gap is provisionally filled
by the *quantum theory* in a manner of which the underlying
principles are not yet grasped at all. (Weyl, 1922, 303, emphasis in original)[23]

Now the task to understand matter mathematically could be approached
from different viewpoints. One was topological in nature. General relativity
offered the opportunity to consider a differential topological manifold with
boundaries, in the interior of which the fields are regular, while they are
singular on the boundaries and diverge in respective limiting processes. In
an article written for *Annalen der Physik* shortly after the publication of the
fourth edition of RZM, Weyl explained his new viewpoint more in detail
(Weyl, 1921*a*). He argued in two directions. Coming from the point of
view of special relativity and Minkowski space, the generalization for GRT
consisted not only in a deformation of the metric, but could also comprise
a topological modification of the underlying manifold. Weyl argued that in
a space-time manifold with a combined electromagnetic and gravitational
field, the subsets on which the fields obtain singular values should be cut out
and omitted.

In the general theory of relativity the world can possess arbitrary (...) connectedness: nothing excludes the assumption that in its Analysis-Situs properties it behaves like a four-
dimensional Euclidean continuum, from which different tubes
of infinite length in one dimension are cut off. (Weyl, 1921*a*,
252f.)

If the general relativistic point of view was considered as the more realistic one, it even appeared as more natural to turn the view round. One would
then have to argue in terms of pasting rather than of cutting:

The simply connected continuum from which we construct the
domain of the field by cutting off the tubes is nothing but a

mathematical fiction, although the metrical relations persisting in the field strongly propose the extension of the real space by adding such fictitious improper (erdichteter uneigentlicher) regions corresponding to the single matter particles. (ibid.)

For Weyl, this change of the mathematical construction of space went in hand with a change of the understanding of the relationship between space-time and matter:

According to [this] perception, *matter itself is nothing spatial (extensive) at all, although it is inserted in a certain spatial neighbourhood*. (Weyl, 1921*a*, 254, emphasis in original)

He must have liked this idea. One of the Fichtean motifs on the "construction" of matter and space from forces, which had impressed Weyl already at the time of his turn towards purely infinitesimal geometry, acquired here a new face and persisted in a modified form.[24]

On the other hand, there was a physical approach to the problem of matter. In addition to the proper laws of the field(s) one had to "study the laws according to which matter excites the field actions". For Weyl, matter was now turning into an irreducible originator of dynamical excitation of the interaction field(s) and was itself guided by the latter in its own spatio-temporal dynamics. He insisted that it could neither be understood as a "substance" in the sense of traditional natural philosophy, nor could it be derived from the "field" as in the Mie version of dynamicist matter explanation. Weyl preferred to characterize matter as a *dynamical agency (dynamisches Agens)*. This was not a commonly used word of the German language. It even had not been used before, to my knowledge, in the earlier discourses on the constitution of matter. In the Aristotelian and scholastic tradition, from which Weyl may have adopted it, it rather denoted the "form" as a dynamical principle *imposed upon* matter.[25] Probably it was Weyl who transferred it to *matter itself* and introduced, by this move, another semantical shift into the long dynamistic tradition of matter explanation.[26]

In the early 20th century, the dynamistic view of matter had found its clearest scientific expression in the electromagnetic world view and its classical field theoretic generalizations in the Mie – Hilbert – Weyl programs. It is interesting to see that Weyl choose a new word exactly at the time when he gave up his belief in the success of classical field theories. Apparently the choice of the new word *Agens* demarcates a cut inside this semantical field of dynamistic matter theories, between the classical field theoretic approaches and a still unknown one, with an open horizon towards quantum stochastical aspects of determination.

Different to the older field theories, Weyl considered it as an important feature of the agency view that it considered matter as something which acts *upon* spatial structures like fields, although it is *not* itself located *inside* space. Already in 1921, several years before the advent of the refined form of the quantum mechanics, Weyl stated optimistically:

> In addition to the *substance* and the *field* perceptions we have to add a third view of matter as an *agency (Agens)* effecting the field states. ... It makes place for the modern physics of matter, working with statistical concepts, besides the strictly functional physics of a classical field. (Weyl, 1921*a*, 255)

For Weyl, such a shift had nothing to do with a longing for "acausality", or even the adoration of it. He rather insisted that the view of mechanical and classically field theoretic physics had reduced causality to a purely functional mathematical relationship, while the agency perception opened a possibility to understand the causation of field states by matter in a new and deeper way.

> Here the specified direction of the passage of time: past $\rightarrow$ future, which cannot find its place in field physics, can be taken up again; in fact it is most closely related to the idea of causation. (Weyl, 1921*a*, 256)

The characterization of causality by a deterministic and time-invertible lawlike structure as in classical mechanics appeared as an inappropriate concept. The change from classical determination to a probabilistic one would therefore not at all contradict the concept of "causation". Just to the contrary, Weyl expected that it might open the path towards a more appropriate understanding of the latter. Although his most recent turn had its origin in the short-lived singularity theory of matter, the *agency paradigm* of matter was kept open for a modification in its mathematical characterization and for a future enrichment by an improved understanding of its physical properties.

6. THE AGENCY CONCEPT OF MATTER AS AN OPEN RESEARCH FIELD

The role Weyl assigned to singularities of classical fields in the fourth editionof RZM remained itself "singular" in his work. It did not appear earlier and vanished, or was at least drastically reduced in importance, nearly as fast as it appeared. In the fifth edition of his book the section on "further rigorous solutions of the statical problem of gravitation", which contained the central passages on the singularity theory of the electron, from which we quoted above, was completely reorganized. Apparently Weyl was not satisfied with the outlook on the strong interpretation of singularities as *the* mathematical clue to the solution of the "problem of matter". In the fifth edition and in his later publications on the philosophy of nature (Weyl, 1924, 1927) we find the singularity model only in a weak sense. It was mentioned only in passing, as an idea illuminating the impossibility of a direct localization of the basic agency structures of matter *inside* space.

In the fifth edition of RZM (1923), Weyl no longer gave the impression that he was already in possession of a mathematical clue to the solution of the "problem of matter". He now preferred only to characterize the terrain of investigation and discussed different approaches that had been tried up to then. Among these he mentioned, of course, Mie's theory and his own generalization as important examples. But now they were only presented as explorative theoretical models, without any claim that they might lead towards a reliable representation of reality.

In this discussion we find beautiful, nearly poetic descriptions of the actual state of knowledge as an open terrain:

> We only perceive the bounding embankment of the subtle, deep groove which is dug into the metrical face of the world by the trajectory of the electron; what is covered by the depth, remains hidden to us. It may be that the whole groove is filled by a field, qualitatively equivalent to the outer one, as Mie assumed; *but just as well the abyss may be fathomless.* Mie's perception dissolves matter into the field; the other one removes it, so to speak, from the field. According to the latter view *matter is an agency determining the field, although in itself nothing spacelike, extensional, but only located in a certain spatial neighbourhood,* from which its field effects depart. ... (Weyl, 1918b, ^{5}1923, 286)

Coming closer to the middle of the 1920s, Weyl left it open, whether it seemed more promising to smoothen the field for a mathematical representation of the basic constituents of matter (like in the Mie approach), to excise it (like in the singularity approach of 1921), or to find any other characterization which might take the statistical nature of quantum descriptions better into account than the other ones:

> Our description of the field surrounding an electron is a first, stuttering formulation of such laws. Here lies the working field for modern physics of matter, to which belong, above all, the facts and riddles of the quantum of action (...) As far as we can judge today, the lawfulness according to which matter induces effects can be described in statistical terms only, ... (RZM ^{5}1923, 286f.)

Independent of these open problems for an adequate mathematical characterization of matter, it now seemed clear to him that matter, rather than the field had to be given primacy for all experimental purposes or any practical exchange with nature.

> Our willful actions have always to grapple on matter, primarily; only thus we can change the field. In fact, we then need two kinds of laws for the explanation of natural phenomena: 1. *field laws* (...), 2. *laws regulating the excitation of the field by matter.* .... (ibid.)

For Weyl, causality returned to the status of a relation which enabled human beings to influence the course of natural processes by a willful modification of material constellations in the world. As he had come to the insight that physical knowledge of the basic matter structures was still highly restricted, he canceled those passages of the final sections of earlier editions of RZM, which appeared now much too enthusiastic. That did not exclude poetic allusions. The fifth edition, the last one revised by himself, ended with a passage which was both, sober and prophetic:

> We were unable to pursue our analysis of space and time without studying matter in detail. Here, however, we are still confronting riddles the solution of which is not to be expected from field physics. In the darkness still surrounding the problem of matter, quantum theory may perhaps be the first twinkling of light. (Weyl, 1918b, [5]1923, 317)

Weyl had entered the first phase of active intervention into mathematical physics (the "RZM-phase", as we might call it) with a strong program of reductionist unification; at the end of it, he clearly saw the necessity to distinguish ontologically and mathematically between interaction fields and matter. While for the first class the classical field theories could be considered as very successful, the problem of matter had turned back into a riddle.

7. A VIEW BACK IN 1930

Only two years after these lines were written, the "first twinkling of light" was stabilized by the establishment of quantum mechanics in the form of wave mechanics and operator theory in Hilbert spaces. The core of this development was the product of a new generation of physicists (W. Heisenberg, W. Pauli, P. Jordan, P.A.M. Dirac, E. Schrödinger, e.a.) who stood in close communication with outstanding figures of the earlier period (N. Bohr, M. Born, A. Sommerfeld, P. Ehrenfest, e.a.). Although Weyl was no member of this group, he was close enough to several of the participants that he was immediately drawn into the turn to "the new" quantum theory at the middle of the 1920s. In oral and written exchange with E. Schrödinger, W. Pauli, M. Born and P. Jordan he even contributed in certain respects to it.[27] In his lecture course in winter semester 1927/28 on *Group Theory and Quantum Mechanics*, he took up Schrödinger wave functions and Pauli spinors (in later terminology) as new mathematical forms to represent a stochastically determining matter "agency". In the book arising from it (Weyl, 1928) he could already include Dirac spinors for the characterization of a relativistic matter field of a new type. The second edition (1931) entered into the complex and irritating discussion of "second quantization" of these new provisional symbolic systems.

Knowing well about the provisional character of the quantum mechanical characterizations of matter, Weyl was deeply impressed by its successes already on the level of spectroscopy and the first steps into the quantum chemical theory of valence bonds. An invitation to the 1930 Rouse Ball lecture at Cambridge gave Weyl the opportunity to review the whole development of matter concepts which had taken place during the long decade just coming to an end.

Even from hindsight, he still considered the attempts of the early 1920s to geometrize "the whole of physics" as very comprehensible at its time, because they had tried to follow up on Einstein's successful geometrization of gravity (Weyl, 1931, 338). In this historizing perspective, he saw no reason to distance himself from his own attempts of 1918. He summarized its critical reception by physicists and reviewed Eddington's approach to unification by affine connections, including Einstein's later support for that program. Comparing the latter with his own "metrical" unification of 1918 he concluded that from hindsight both theory types appeared as "merely geometrical dressings (geometrische Einkleidungen) rather than as proper geometrical theories of electricity". He discussed the struggle between the metrical and affine field theories (i.e., Weyl 1918 versus Eddington/Einstein) and gave the whole story a smilingly ironic turn:

> ...there is no longer the question which of the two theories
> will prevail in life, but only whether the two have to be buried
> as twin brothers in the same grave or in two different graves.
> (Weyl, 1931, 343)

In the light of his changed view on the problem of matter, he could find just as little arguments in favour of the more recent brands of unification attempts proposed at the end of the decade, Einstein's distant parallelism approach or the Kaluza-Klein approach.[28] Weyl completely rejected Einstein's new theory, not only because of their completely diverging conceptions of matter, but also by a strong mathematical reason. In his opinion, Einstein's latest theory would break with important features of the infinitesimal geometric point of view, which lay at the base of general relativity. He warned:

> The result [of pursuing Einstein's *Fernparallelismus* approach, E.S.] is to give away nearly all what has been achieved in the transition from special to general relativity. The loss is not compensated by any concrete gain." (Weyl, 1931, 343)

Weyl perceived a nearly complete scientific devaluation of all unified field theories invented during the long decade. This devaluation resulted from the quantum theoretical insights into matter structures, which had found first well formed mathematical representations by complex scalar or spinor fields during the second part of the decade:

> In my opinion the whole situation has changed during the last 4 or 5 years by the detection of the matter field. All these geometrical leaps (geometrische Luftsprünge) have been premature, we now return to the solid ground of physical facts. (Weyl, 1931, 343)

He continued to sketch the theory of spinor fields and the new understanding of the underdetermination of phase which opened a new theoretical frame for the gauge principle. In 1929 he and V. Fock had proposed a revised gauge theory of electromagnetism in this context. He insisted that the new principle of phase gauge "has grown from experience and resumes a huge treasury of experimental facts from spectroscopy" (ibid. 344). That stood in marked contrast to the purely speculative principles on which all the classical unified field theories had been built, his own one from 1918 included. Now he no longer expected to achieve knowledge on natural processes by geometric speculation, but tried to anchor it in more solid grounds, the observation of matter processes and their mathematization:

> By the new gauge invariance, the *electromagnetic field now becomes a necessary appendix of the matter field, just as it had been attached to gravitation in the old theory.* (Weyl, 1931, 345, emphasis in original)

In short, Weyl had turned from his speculative and strongly idealist approach to matter, pursued at the turn to the 1920s, *to a mathematically empiristic and moderately materialistic* one at the end of the decade. He was well aware that great difficulties had still to be surmounted to come to grips with a quantization of the semiclassical fields (complex scalar or spinor wave functions), which had recently been invented for a provisional and partial representation of the quantum properties of matter. That gave geometry an outlook which was completely different to the one in the classical field the-ories. On the other hand, Weyl did not want to exclude that some day a geometrization might become possible on a new level. But if one wanted to continue along this path, he was sure that "one had to set out in search of a geometrization of the matter field" itself. If one would try to do without an improved mathematization of the agency structures of matter themselves, the geometrical theories would fall back to the methodological status of the unification attempts of the 1920s. He now considered these as immature, although comprehensible first attempts, as *Luftsprünge* (leaps into the air).

It may be appropriate to add that the German word "Luftsprünge" not only connotes unrealistic first attempts, but also the joy of youthfulness. Weyl has had both, the joy of the youthful speculation that he was close to the goal of a reduction of physics to geometry and the awareness, as a mature natural scientist, that the difficult practices of experimentation and closely related symbolical practices of the mathematics of quantum physics opened the path towards a much more reliable comprehension of the agency structures of matter.

8. ACKNOWLEDGMENT

I thank Tilman Sauer for careful reading and crucial hints to the first version of this paper.

University of Wuppertal
Germany

NOTES

[1] A. Pais' description of the change of matter concepts by the rising quantum theory as "the end of the game of pebbles" (Pais, 1986, 324) fits already well to this shift, although Pais used it as a header for the rise of second quantized fields starting in the late 1920s.

[2] For Weyl's first phase of involvement in mathematical physics compare (Sigurdsson, 1991; Scholz, 2001), for broader views on unified field theories see (Vizgin, 1994; Goldstein/Ritter, 2003; Cao, 1997; Goenner, 2004).

[3] With respect to unified field theories (UFT) a complementary view at several "transversal" sections (in time) with a broad evaluation of authors and approaches is presented in (Goldstein/Ritter, 2003). The perspective of UFT's was characteristic for all pre-quantum 20th century dynamistic approaches to the concept of matter and the first half of Weyl's trajectory.

[4] See (Sigurdsson, 1991, 64ff.), (Schappacher, 2003).

[5] Compare (Hesseling, 2003, 121ff.), (Scholz, 2000).

[6] See footnote 2.

[7](Scholz, 2004).

[8]Compare (Corry, 1999*a,b*, 2004; Kohl, 2002; Sauer, 1999; Vizgin, 1994).

[9]For a discussion of Hilbert's research program building upon and extending Mie's field theoretic matter theory see (Sauer, 1999), for a critical evaluation of Hilbert's relation to Einstein's theory of general relativity (Corry, 2004, 1997; Renn, 1999).

[10]See (Brading and Brown, 2003; Kosmann-Schwarzbach, 2004; Rowe, 1999; Brading, 2002).

[11]An additional cause (not a reason) may have been Weyl's dissatisfaction with Hilbert's position in the foundations and philosophy of mathematics, which started to get into the public at the same time (1918) and may have contributed to let Weyl emphasize Mie's role in the game more strongly. Hilbert stroke back and characterized Weyl's program to derive matter from the latter's geometrically unified theory as a kind of "Hegelian physics", undermining unnoticingly his own approach as well (Hilbert, 1992) – see below.

[12]It thus appears verbally unchanged in the third edition on which H.L. Brose's English translation is based (Weyl, 1922, 206). Here, as in other cases, our English quotes from RZM are following Brose's translation, where available.

[13]Translation slightly adapted, E.S.

[14]Unchanged in all editions, last one in (Weyl, 1918*b*, [5]1923, 216). Translation from (Weyl, 1922, 212) slightly adapted by E.S.

[15](Varadarajan, 2003; Vizgin, 1994).

[16]"Wir erheben uns zu einer letzten Synthese." Brose's translation reduced the kick of enthusiasm considerably: "We now aim at a final synthesis" (Weyl, 1922, 282). Weyl did not weaken the rhetoric until and including the fifth edition, although he slightly revised its wording by adding a "nun (now)" (Weyl, 1918*b*, [5]1923).

[17]See (Vizgin, 1994; O'Raifeartaigh and Straumann, 2000; Brading, 2002) and for a more detailed (historically oriented) discussion of the underlying mathematics (Varadarajan, 2003).

[18]See the comments and critique of Forman's original presentation in (Hendry, 1984; Sigurdsson, 1991; Stöltzner, 2002) and also the modifications in (Forman, 1980).

[19]Weyl hinted at the possibility that the classical mechanical discussion on ergodicity had to be revised in the light of "some mysterious discontinuity" introduced recently by quantum theory (Weyl, 1920, 118).

[20]Compare, e.g., (Hesseling, 2003, 127).

[21]"Endlich habe ich mich gründlich von der Mie'schen Theorie losgemacht und bin zu einer anderen Stellung zum Problem der Materie gelangt. Die Feldphysik erscheint mir keineswegs mehr als der Schlüssel zu der Wirklichkeit; sondern das Feld, der Äther ist mir nur noch der in sich selbst völlig kraftlose *Übermittler* der Wirkungen, die Materie aber eine jenseits des Feldes liegende und dessen Zusände verursachende Realität. Mit dem "Weltgesetz" (Hamiltonsches Prinzip), das die Wirkungsübertragung im Äther regelt, wäre noch gar wenig für das Verständnis aller

Naturerscheinungen gewonnen." (ibid. emphasis in original); compare (Sigurdsson, 1991).

[22]In the German original (Weyl, 1918b, [4]1921, 238), no longer in the fifth edition.

[23]The translation of the last phrase has been slightly changed to adapt it closer to the German original (Weyl, 1918b, [4]1921, 276) than in Brose's translation. The passage is no longer contained in the fifth edition.

[24]Compare (Scholz, 1995, 2005).

[25]Skùli Sigurdsson has characterized this turn (smilingly) as a Weylian brand of "idealist materialism" (Sigurdsson, 2001, 30).

[26]This claim has still, to be checked by experts in the history of natural philosophy; please note the clause "to my knowledge".

[27]For the Born and Jordan part see (Scholz, 2006).

[28]For Einstein's distant parallelism see (Sauer, 2006), for Kaluza and Kaluza-Klein (Wuensch, 2003).

REFERENCES

Brading, Katherine. 2002. Which symmetry? Noether, Weyl, and conservation of electric charge. *Studies in the Philosophy of Modern Physics* **33**:3–22.

Brading, Katherine and Brown, Harvey. 2003. Symmetries and Noether's theorems. *In* E. Castellani, K. Brading (eds). *Symmetries in Physics : Philosophical Reflections*. Cambridge: University Press.

Cao, Tian Yu. 1997. *Conceptual Developments of 20th Century Field Theories*. University Press, Cambridge.

Corry, Leo. 1999a. David Hilbert between mechanical and electromagnetic reductionism (1910–1915). *Archive for History of Exact Sciences* **53**:489–527.

Corry, Leo. 1999b. From Mie's electromagnetic theory of matter to Hilbert's unified foundations of physics. *Studies in the History and Philosophy of Modern Physics* **30**:159–183.

Corry, Leo; Stachel, John; Renn Jürgen. 1997. A belated decision in the Hilbert-Einstein priority dispute. *Science* **278**:1270–1273.

Corry, Leo. 2004. *David Hilbert and the Axiomatization of Physics (1898-1918)*. From Grundlagen der Geometrie to Grundlagen der Physik. Kluwer, Dordrecht etc..

Forman, Paul. 1971. Weimar culture, causality, and quantum theory, 1918–1927: Adaptation by German physicists and mathematicians to a hostile intellectual environment. *Historical Studies in the Physical Sciences* **3**:1–116.

Forman, Paul. 1980. Kausaltät, Anschaulichkeit und Individualität, oder wie Wesen und Thesen, die der Quantenmechanik zugeschrieben, durch kulturelle Werte vorgeschrieben wurden. *Kölner Zeitschrift für Soziologie und Sozialpsychologie, Sonderheft* **22**:393–406. In (von Meyenn, 1994, 181–200).

Goenner, Hubert. 2004. On the history of unified field theories. *Living Reviews in Relativity* . [http://relativity.livingreviews.org/Articles/lrr-2004-2, visited June 2, 2004].

Goldstein, Catherine; Ritter, Jim. 2003. The varieties of unity: Sounding unified theories 1920–1930. In Asketar, Abhay e.a. (eds.). *Revisiting the Foundations of Relativistic Physics: Festschrift in Honor of John Stachel*. Dordrecht etc. Kluwer.

Gray, Jeremy (ed.). 1999. *The Symbolic Universe: Geometry and Physics 1890–1930*. University Press, Oxford.

Hendry, John. 1984. *The Bohr-Pauli Dialogue and the Creation of Quantum Mechanics*. Reidel, Dordrecht.

Hesseling, Dennis. 2003. *Gnomes in the Fog. The Reception of Brouwer's Intuitionism in the 1920s*. Birkhäuser, Basel.

Hilbert, David. 1992. *Natur und mathematisches Erkennen.Vorlesungen, gehalten 1919–1920 in Göttingen*. Nach Ausarbeitungen von P. Bernays. Ed. D. Rowe. Birkhäuser, Basel etc..

Kohl, Gunter. 2002. Relativität in der Schwebe: Die Rolle von Gustav Mie. Preprint 209, MPI History of Science, Berlin.

Kosmann-Schwarzbach, Yvette. 2004. *Les Theorème de Noether. Invariance et lois de conservation au XXe siècle*. Paris: Edition de l'Ecole Polytechnique.

Letter H. Weyl to F. Klein, December 28, 1920. 1920. Nachlass F. Klein Universitätsbibliothek Göttingen Codex Ms Klein **12**:297.

O'Raifeartaigh, Lochlainn; Straumann, Norbert. 2000. Gauge theory: Historical origins and some modern developments. *Reviews of Modern Physics* **72**:1–23.

Pais, Abraham. 1986. *Inward Bound: Of Matter and Forces in the Physical World*. Clarendon, Oxford.

Pauli, Wolfgang 1921. Relativitätstheorie. *Encyklopädie der Mathematischen Wissenschaften* **5.2**, Teubner, Leipzig, 539–775.

Renn, Jürgen; Stachel, John. 1999. Hilbert's foundations of physics: From a theory of everything to a constituent of general relativity. Preprint 118, MPI History of Science, Berlin.

Rowe, David. 1999. The Göttingen response to general relativity and Emmy Noether's theorems. *In (Gray, 1999, 189–233)*.

Sauer, Tilmann. 1999. The relativity of discovery. *Archive for History of Exact Sciences* **53**:529–575.

Sauer, Tilman. 2006. Field equations in teleparallel spacetime: Einstein's *Fernparallelismus* approach towards unified field theory. To appear in *Historia Mathematica*.

Schappacher, Norbert. 2003. Politisches in der Mathematik: Versuch einer Spuren-sicherung. *Mathematische Semesterberichte* **50**:1–27.

Scholz, Erhard. 1995. Hermann Weyl's "Purely Infinitesimal Geometry". In *Proceedings of the International Congress of Mathematicians, Zürich Switzerland 1994*. Birkhäuser, Basel etc. pp. 1592–1603.

Scholz, Erhard. 2000. Hermann Weyl on the concept of continuum. In *Proof Theory: History and Philosophical Significance*, ed. V. Hendricks; S.A. Pedersen; K.F. Jørgensen. Kluwer, Dordrecht pp. 195–220.

Scholz, Erhard (ed.). 2001. *Hermann Weyl's* Raum – Zeit – Materie *and a General Introduction to His Scientific Work*. Birkhäuser, Basel etc.

Scholz, Erhard. 2004. Hermann Weyl's analysis of the "problem of space" and the origin of gauge structures. *Science in Context* **17**:165–197.

Scholz, Erhard. 2005. Philosophy as a Cultural Resource and Medium of Reflection for Hermann Weyl. *Révue de synthèse*. **126**: 331–352.

Scholz, Erhard. 2006. The introduction of groups into quantum theory. To appear in *Historia Mathematica*.

Sigurdsson, Skúli. 1991. Hermann Weyl, Mathematics and Physics, 1900 – 1927. Cambridge, Mass.: PhD Dissertation, Harvard University.

Sigurdsson, Skúli. 2001. Journeys in spacetime. In *Scholz (2001)*. pp. 15–47.

Stöltzner, Michael. 2002. *Vienna Indeterminsm. Causality, Realism and the Two Strands of Boltzmann's Legacy (1896–1939)*. PhD Dissertation Bielefeld University.

Varadarajan, V.S. 2003. Vector bundles and conections in physics and mathematics: some historical remarks. In *A Tribute to C. S. Seshadri (Chennai, 2002)*, ed. V. Balaji, V.; Lakshmibai. Birkhäuser, Trends in Mathematics Basel etc. pp. 502–541.

Vizgin, Vladimir. 1994. *Unified Field Theories in the First Third of the 20th Century.* Translated from the Russian by J. B. Barbour. Birkhäuser, Basel etc..

von Meyenn, Karl (Ed.). 1994. *Quantenmechanik und Weimarer Republik*. Vieweg, Braunschweig.

Weyl, Hermann. 1918*a*. *Das Kontinuum. Kritische Untersuchungen über die Grundlagen der Analysis*. Veit, Leipzig.

Weyl, Hermann. 1918*b*. *Raum, – Zeit – Materie*. Springer, Berlin etc.. Later editions: ^{2}1919, ^{3}1919, ^{4}1921, ^{5}1923, ^{6}1970, ^{7}1988, ^{8}1993.

Weyl, Hermann. 1920. Das Verhältnis der kausalen zur statistischen Betrachtungsweise in der Physik. *Schweizerische Medizinische Wochenschrift* **50**:737–741. (Weyl, 1968, II, 113–122).

Weyl, Hermann. 1921*a*. Feld und Materie. *Annalen der Physik* 65:541–563. In (Weyl, 1968, II, 237–259) [47].

Weyl, Hermann. 1921*b*. Über die neue Grundlagenkrise der Mathematik. *Mathematische Zeitschrift* **10**:39–79, (Weyl, 1968 II, 143–180), [41].

Weyl, Hermann. 1922. *Space, Time, Matter.* Translated from the 4th German edition by H. Brose. Methuen, London.

Weyl, Hermann. 1924. Was ist Materie? *Die Naturwissenschaften* **12**:561–568, 585–593, 604–611. Reprint Springer 1924, Berlin. Wissenschaftliche Buchgesellschaft 1977, Darmstadt. In (Weyl, 1968, II, 486–510) [66].

Weyl, Hermann. 1927. *Philosophie der Mathematik und Naturwissenschaft,* Handbuch der Philosophie, Abt. 2A. Later editions [2]1949, [3]1966. English with comments and appendices Weyl (1949). Oldenbourg, München.

Weyl, Hermann. 1928. *Gruppentheorie und Quantenmechanik.* Hirzel, Leipzig. [2]1931, English 1931.

Weyl, Hermann. 1931. Geometrie und Physik. *Die Naturwissenschaften* **19**:49–58. Rouse Ball Lecture Cambridge, May 1930. (Weyl, 1968, III, 336–345) [93].

Weyl, Hermann. 1949. *Philosophy of Mathematics. and Natural Science.* 2nd ed. 1950. University Press, Princeton.

Weyl, Hermann. 1968. *Gesammelte Abhandlungen, 4 vols.* Ed. K. Chandrasekharan. Springer, Berlin etc..

Wuensch, Daniela. 2003. The fifth dimension: Theodor Kaluza's ground-breaking idea. *Annalen der Physik* **12**:519–542.

LAWRENCE SKLAR

WHY DOES THE STANDARD MEASURE WORK IN STATISTICAL MECHANICS?

1. THE FOUNDATIONAL PROBABILITY POSIT AND ITS JUSTIFICATION

There are two main puzzles that continue to demand resolution in work on the foundations of statistical mechanics. One of these is the explanatory ground for the success of the fundamental probabilistic posit in the theory, the posit that the micro-states of a system are to be taken as distributed uniformly with respect to standard (Lebesgue) measure in the phase space of points representing total microscopic states of the system. The other is the origin of the time asymmetry that is so characteristic of the thermodynamic world the theory of statistical mechanics is supposed to explain.

Let us put the second question to the side, noting that the generally preferred solution to it nowadays is to be found in a posited low-entropy for the spacetime structure of the world at the "Big Bang." That leaves the first question: Why should we assign probabilities to micro-states of systems in the standard way? After all an infinity of other possible probability distributions are possible. What, physically, is so special about the standard distribution?

Let us start by noting three things that bother people about the standard statistical posit:

1. The posit seems too "arbitrary." Why pick this particular measure for specifying uniformity of the probability distribution and not some other measure (one based on position and energy, say, as opposed to position and momentum)?
2. The posit seems to be one that depends on "mere matter of fact" about how micro-states are distributed. But the thermodynamic theory seems to have "lawlike" status. How could such lawlikeness be attributed to our initial probability distribution?
3. The posit of uniformity with respect to the standard measure is a very strong one. Could the foundations of the theory be redone so as to depend upon a much weaker probabilistic assumption instead?

The standard probabilistic posit over micro-states is used in both equilibrium and non-equilibrium statistical mechanics. In equilibrium theory it is used to calculate averages of functions of the micro-states. These "phase averages" are then identified with the macroscopic parameters that characterize equilibrium states of systems.

There are many subtle issues involved in justifying the identification of such phase averages with the macroscopic parameters. But there is also the

V.F. Hendricks, K.F. Jørgensen, J. Lützen and S.A. Pedersen (eds.), Interactions: Mathematics, Physics and Philosophy, 1860-1930, pp. 307–320.

issue of why it is reasonable to pick the standard probability distribution for calculating these averages in the first place. Here one is often offered a "transcendental" justification for picking the standard probability distribution. For certain idealized models of systems (say, gases as hard spheres in a box), ergodic theory can show that the standard probability distribution is the only such distribution that assigns zero probability to the collections of states assigned zero probability by the standard measure, and that is also invariant in time as the dynamics takes systems from one micro-state to another. Needless to say, this rationale for choosing the standard measure is also fraught with problematic aspects.

In non-equilibrium theory the standard probability distribution plays a number of roles. Most importantly, it is used as the probability distribution over initial micro-states for systems placed in an initial non-equilibrium condition. Without positing some such initial probability distribution, no finite time results for the behavior of systems evolving from non-equilibrium towards equilibrium can be derived from the usual practice in the theory of modeling such evolution by the evolution in time of an initially posited probability distribution over micro-states in phase space.

This dynamic evolution of a phase space distribution is usually modeled using what is called "coarse graining" of the phase space. One breaks the allowed phase space region of points corresponding to possible micro-states of the system up into small regions, and then calculates coarse grained quantities relative to that partitioning of the phase space. Dynamical considerations that show "instability" of trajectories of individual systems for idealized models, hard spheres in a box again, for example, can then be used to derive so called "mixing" results for the system. These results are combined with the initial probabilistic posit to try to derive the standard kinetic equations that describe the approach to equilibrium. This is modeled by an evolution in time of a coarse-grained quantity, usually Gibbs coarse-grained entropy.

Another very important use of the standard probability measure in non-equilibrium statistical mechanics is in characterizing the degree to which the dynamics "stirs up" the phase space. This provides a systematic and abstract characterization of the degree to which the dynamics can be expected to drive a system towards equilibrium. Here one needs to adopt a fundamental probabilistic measure in order to begin to define the relevant quantities (such as the Kolmogorov-Sinai entropy of a shift) that are used in these characterizations.

But, it might be asked, why does the standard probability measure really need some kind of "justification?" Isn't it just the posit that probability is spread uniformly over the allowable phase space region? And, given that we have no more specific information about the likelihood that the system has one micro-state rather than another, isn't such an attribution of uniformity to the probability measure the "natural" one to make? But such a justification for the posit is fraught with well-known difficulties. It amounts to relying on some principle of "symmetry" or "insufficient reason" famous from the history of probability theory. But the objections to founding probability measures on some kind of "principle of indifference" are also well known. Choose a different way of measuring or partitioning the space of possible outcomes, and a principle of indifference will result in different probability assignments for classes of those possible outcomes. These issues have been illustrated by the so-called "Bertrand's Paradoxes" that showed how one and the same principle of symmetry or uniformity for probability distribution resulted in contradictory probability assignments depending on the measure or partition chosen for the sample space.

The standard measure distributes probability uniformly with respect to Lebesgue measure (or with special variants of it depending upon the case at issue). That is, it relies upon phase volumes measured by extent in position and in momentum. Now such a measure is "special" in that it has many nice features not possessed by other measures (such as invariance under the dynamics and under canonical transformations of the dynamical variables). But other measures can easily be imagined, such as a measure given by extent in position and in energy. And it is far from obvious how to block an application of some principle of indifference that would assign uniform probability relative to one of these alternative measures, especially in the non-equilibrium case.

Of course one could, following Tolman and others, simply take the fundamental probability distribution as an otherwise unexplained posit. But it is not clear that one should or need do so. Let us look at three suggestions as to how to found the posit on some more fundamental principle. My aim here is not to explore any one of these three approaches in the detail they deserve. Rather it is to outline how very different the approaches are in their fundamental posits and in their conception as to just where one should look for a justification of the foundational probability assumptions of statistical mechanics.

2. MEASURING THE STIRRING-UP OF A SHIFT WITHOUT INTRODUCING A MEASURE

A long train of thought in non-equilibrium statistical mechanics models the change toward equilibrium of a system by exploring how the underlying dynamics will "stir up" or "mix" an initial distribution of possible initial microstates in phase space. Since Gibbs one proposal has been to "coarse grain" the phase space by partitioning it up into small cells. Quantities can then be defined relative to the partition chosen. Most important of these is what is called the "Gibbs coarse grained entropy." This is a quantity that can have its value change in time, unlike "Gibbs fine grained entropy." It might, therefore, serve as an indicator of the degree tio which a system has moved toward equilibrium.

But the value of this quantity will then depend upon the coarse graining that has been chosen. There is then a fear that this will introduce an unacceptable "subjective" or "arbitrary" element into the theory. A nice way around this problem is due to Kolmogorov and Sinai. They propose a measure of the degree to which the dynamics provides a stirring up of the system that looks at how coarse grained entropies change over time, but focuses on the maximal change possible relative to that determined by all possible coarse grainings. It turns out that this is easily calculable, since some special coarse grainings can be found that are guaranteed to generate this maximal amount.

So the use of the "Kolmogorov-Sinai entropy of a shift" eliminates the problem of the relativity of coarse grained entropy to a chosen partition. But the quantity still depends upon having chosen the standard probability measure as fundamental. Can one get around the new accusation that this choice is also "arbitrary" or "subjective" and so illegitimate in a fundamental theory?

One interesting proposal suggests measuring how much the dynamics stirs up the phase space in a way that doesn't depend upon choosing any measure at all. This proposal generates a measure of dynamical mixing called the "topological entropy of a shift."

How does it work? Crudely it goes like this: Define "shephard points" for the shift. These are points so that after one shift each point remains within some small specified distance of its shephard. Find the minimum number of shepherds for a single shift. Then ask what the least number of shepherds are required to keep the sheep (the associated phase points) within the requisite distance for n shifts, and divide that by n. Take the log of this number. Let n go to infinity. Then let the distance a point must be to its shepherd go to zero. The resulting number (that can be proven to exist) is the topological entropy of the shift. Remarkably the topological entropy of a shift is provably equal to the maximum over all Kolmogorov-Sinai measure theoretic entropies of the shift over all measures that are invariant under the shift (of which the standard measure is one).

But of course things are never simple. The topological entropy of a shift depends upon picking the standard topology. And that usually is formulated in terms of "distance" between phase space points. While distance between points in phase space along a single trajectory is naturally given by how long a system would take in time to get from one point to another, distance between points not of the same trajectory doesn't have any obvious natural constraint on its definition. And while the standard topology on phase space is a natural one, it is far from clear that we can use its existence as a knock down argument for a physically appropriate device to use in generating our model of a measure of mixing in the phase space that is appropriate to capture what we were interested in in the first place—that is, the actual approaching to equilibrium of real, individual physical systems in real time.

Once again, we cannot pursue the details here. But what we should note is this: Here is a proposal to come to grips with one of the objections to the invocation of the standard measure in non-equilibrium statistical mechanics, the objection that that measure was too "arbitrary." It works by exploring a non-measure theoretic feature of the dynamical transformation in phase space, hoping to find in topology some mathematical aspect of the model with less apparent "arbitrariness" than choosing the standard measure was accused of invoking.

3. INITIAL PROBABILITY OUT OF NON-STANDARD QUANTUM MECHANICS

In order to make non-equilibrium statistical mechanics work we must posit, for each temporarily energetically isolated system of the world, a probability distribution over its possible initial conditions that is uniform with respect to the standard measure. The result of such a posit is the possibility of deriving the usual thermodynamic, time-asymmetric, approach to equilibrium of systems. In thermodynamics we think of that time-asymmetric behavior as being "lawlike" in nature. But in ordinary statistical mechanics we seem forced to maintain that the fact that the isolated systems have their initial states distributed in the way that they do seems merely "de facto." There isn't anything in the *laws* of standard classical or quantum mechanics that forces the world to behave in this particular way. Should we just accede to this "mere matter of fact" status for thermodynamic time asymmetry? Or is there someway of letting us recover its lawlike status?

Quantum mechanics is afflicted with the so-called "measurement problem." The orthodox presentation of the theory postulates a change of state of a system upon measurement that can in no way be assimilated to an ordinary dynamical evolution. Proposed solutions to this abound. These range from taking measurement as a fundamentally distinct kind of process in the world not at all like ordinary dynamical evolution, to viewing measurement as an ordinary dynamical interaction that can be usefully *misrepresented* in the standard way.

One solution to the measurement problem is that of Ghirardi, Rimini and Weber (GRW). They propose that there is a level of physics below that treated by standard quantum mechanics. At this level the world consists of ongoing genuinely stochastic processes, where the probabilistic transitions proposed are of a lawlike nature. Ordinary quantum mechanical wavefunctions must be multiplied by GRW probability functions. These are normalized Gaussian functions centered on the points over which the wave function is defined. The probability that the wave function will be so multiplied is given by the inner product of the wave function with the GRW Gaussian. There are independent GRW functions for each particle of the system.

This leads to a probability for each isolated particle that it will, in any time interval, have its wave function collapse to a near eigenfunction of position about some point or other. For any single particle, the GRW functions have their amplitudes and widths chosen so that the probability of collapse of the wave function over quite lone periods of time is very small. But for a system of a vast number of particles the cumulative effects of the GRW functions add up so that within reasonable times there is a very large probability that the system will have its wave function as a whole collapse to one in which the particles of the system all have near eigenstates of position about some point or other. The low probability of "real collapse" of wave functions for systems with few particles is used to explain how the familiar interference effects can be demonstrated for these systems. The large probability of "real collapse" for macroscopic systems with many particle components is used to explain the results of measurement. That is, that once measurement has taken place no interference effects can any longer be discerned. Here measurement is thought of as requiring an interaction between the system being measured and some measurement apparatus, the latter being a large system with many particle components, and so subject to rapid GRW collapses.

David Albert has suggested that one could invoke this "real quantum collapse" to account as well for the initial probability distribution in statistical mechanics, thereby killing two tough birds with one speculative stone!

The argument goes something like this: There are initial micro-conditions that lead to "bad," that is, anti-thermodynamic, behavior, and there are "good" initial micro-states that lead to the expected thermodynamic behavior. A system at the moment of its preparation, having many, many micro components, will be immediately subjected to a GRW "kick" that will drive it into some near eigenstate of position for all the molecular components. Given the probability distribution inherent as a matter of law in the GRW theory, it is overwhelmingly likely that the system will be driven to a good micro-state. Why? Because in every small region about "almost every" micro-state, there will be an overwhelmingly dominant number of good micro-states.

The regions of good micro-states into which a GRW kick will drive a system with high probability will be much smaller that even very tiny regions of phase points characterized by tiny ranges in the macroscopic parameters. But they will be very large compared to the even smaller clumps of bad micro-states. So whatever the micro-state of the system before the kick, it will be highly probable that the system will almost instantly be in a good initial micro-state.

It isn't quite that simple. If the system happens from the beginning to be in a nearly eigenstate of position over a bad point, then one GRW kick will not take it, with high probability, into a good state. This is because the eigenstates over good points and those over bad points are orthogonal to one another, and the near eigenstates are "nearly orthogonal." But then, it is claimed, with a sequence of GRW kicks, with ordinary deterministic evolution of ordinary quantum states between the kicks, even starting in a near eigenstate over an anti-thermodynamic point, the system will quickly find itself, with overwhelming probability, in a thermodynamic micro-state.

There are some interesting problematic test cases with which this theory must deal. What about very small systems with few components? GRW kicks will be far apart, but the systems will still behave thermodynamically, won't they? In reply one might claim that such systems are not truly isolated, but interact with the very large environment and are driven by the kicks in that environment to thermodynamic behavior. Or you might look to the past before the system was isolated and claim that it was then that it got kicked into the good initial state it has when first isolated. Or, more implausibly, one might try to claim that thermodynamic behavior for such systems over small times isn't really to be expected.

Another problem for the GRW approach are spin-echo systems. These are systems that show apparent thermodynamic behavior, but where it is demonstrable that there has been no "randomizing" of any kind over the micro-states of the system during the time interval in which the thermodynamic behavior has been displayed. So GRW randomizing by kicks cannot be relevant to the explanation of the thermodynamic behavior. In the case of nuclei locked in a rigid lattice (the common locus of spin-echo type effects), the wave-functions are strongly localized making them impervious to GRW collapses.

One reply here is to look, once again, at the past of the system. Another more promising one is to ask what the source is of the appearance of thermodynamic behavior for the system. This is usually found in the inhomogeneity of the system's internal magnetic field. The GRW account might then look to explaining that randomness in GRW kicks, which are then the ultimate source of the thermodynamic behavior.

Again consider system over very short times, times too short for GRW kicks to play a significant role even if the system has many components. Once again one might refer to the past history of the system. Or one might claim that the system doesn't show thermodynamic behavior over such short time intervals. Or one might deny that measurements of the systems parameter values are not instantaneous but over time intervals sufficiently long for GRW randomization to matter.

Very little of this has been worked out in detail. More importantly, there isn't very much evidence in favor of the GRW account of measurement in quantum mechanics. The GRW theory would nicely account for the well known theoretical features of a measurement, the disappearance of interference and the existence of determinate values of observable quantities – at least as long as these observables are functions of position). But the GRW theory is empirically distinguishable from orthodox quantum mechanics, since over short enough times the theories predict observably different behavior. But no experimental evidence exists so far that there really is a GRW stochastic level of behavior underneath the usual quantum states of systems.

For our purposes, though, it is interesting to see how resort to the GRW theory is used to respond to one of the standard difficulties with the usual probabilistic basic posit of non-equilibrium statistical mechanics. For if the GRW account is true, it is claimed, then this posit does indeed have fully lawlike status.

4. MOLECULAR ROULETTE WHEELS AND WEAK A PRIORI POSITS

If topological entropy attempts to deal with the alleged arbitrariness of the standard measure, and if the GRW theory attempts to deal with the alleged non-lawlikeness of the standard probabilistic posit, the next approach we will look at focuses instead on the alleged problem that the standard probabilistic posit is too *strong* to take as a basic and otherwise unexplained principle in the theory.

A suggestion that one might found non-equilibrium statistical mechanics on a much weaker probabilistic posit that the usual one has been made by Michael Strevens. He relies on fascinating work initiated many years ago by Poincaré. Why does a roulette wheel work? Forget the green slots. Why do red and black come up an equal number of times on the average?

Simplify. Suppose that the slot the ball falls into depends only upon the initial tangential velocity given the ball by the croupier. Then probabilities of outcomes depend upon the probability distribution over velocities of the ball given by the croupier. But can we really assume that all croupiers will have similar such probability distributions? Poincaré shows that we don't have do. So long as the croupier's distribution meets very weak requirements, red and black will have equal chances of occurring. Why?

We may break up the range of initial velocities of the ball into small, adjacent intervals. For velocities in one interval, the ball will end up in a red slot. For those in an, approximately equally wide, adjacent interval, the ball will end up in a black slot. So long as the initial probability distribution over velocities is more-or-less flat over adjacent intervals, we will end up with a probability of outcomes that gives equal likelihood to red and black. So long as this minimal condition is met, the overall global shape of the probability distribution is irrelevant to that result!

Strevens uses a result of this kind to try and demonstrate that what is dubiously posited in Boltzmann's original derivation of the kinetic equation for approach to equilibrium can be derived as the result of a much weaker posit of the sort needed for the roulette wheel case. Boltzmann's original posit, the *Stosszahlansatz* or Hypothesis of Molecular Chaos, requires that at each moment of time there is a standard probability of molecular collision that is independent of the past history of the system and depends only upon the systems instantaneous state. It is this "perpetual rerandomization" that he uses to derive the famous Boltzmann equation. But not only is the hypothesis quite strong at one moment, assuming that it will continue to hold over time may not even be consistent with the underlying dynamical laws of the system.

Strevens takes as his prime descriptive variable the relative angle of impact of two molecules (RAI). He aims to show that:

1. The evolution function on the probability distribution over the RAI's is "microconstant." What this means is that there is a way of partitioning the range of RAI's into small intervals such that the transition function from any one such interval into the RAI's after a collision has occurred is the same for each such interval of the original RAI's.
2. For probabilistic purposes the RAI screens off other initial conditions such as the position or velocity of the colliding molecules.
3. A "macroperiodic distribution" will be maintained. That is, if we start with a distribution over the RAI's so that the probability in that initial distribution is uniform over each of the small micro-partition intervals, that condition of uniformity over the small intervals will persist through collisions.

From these he will try to show that:

4. All of the micro-variables (not just RAI's but such things as molecular positions and velocities as well) will display a random walk process that leads to microconstancy of the variables.

And, finally:

5. The probabilities of these variables will all be independent of one another and of the past history of the system.

To show (1) he argues that we can partition the range of RAI's into intervals corresponding to which next molecule will be hit after the first collision. He claims that no matter which molecule that will be, the transition function from the angle with which the first molecule is hit to the angle at which the next will be hit will be the same no matter which the second molecule is.

To show (2) he argues that the transition from angle of impact to angle of impact doesn't depend on where or how fast the relevant molecules are with respect to one another.

To show (3) he argues that the evolution of the RAI distribution is "inflationary," that is that small intervals of RAI's blow up to the full range of RAI's on the next impact. Then he argues that the transition function from RAI to RAI will be approximately linear.

He then makes his one independent probabilistic *posit*. This is that the initial distribution of RAI's is macroperiodic. That is he posits at the first collision that the probability distribution over each small region in the partition is uniform. This is the analogue of the posit needed in Poincaré's derivation of the roulette result. And it is an otherwise unjustified foundational, probabilistic posit. But it is, first of all, a posit made only at the initial time. In this it doesn't suffer from the defect of Boltzmann's *Stosszahlansatz* of needing to be continuously made throughout the dynamic evolution, with all the possibilities of contradiction to the dynamics that this might entail. Second, it is a much weaker posit that the usual one of non-equilibrium statistical mechanics, that is the posit that at the initial time the probability is distributed over the micro-variables uniformly with respect to the natural measure. Here it is only "uniformity in the small," that is over each small region of the partition of the RAI range corresponding to some specific next molecule being hit that is being posited.

Strevens uses the physical and mathematical facts about collisions to justify his claims that the collision dynamics is inflationary and linear. Here he needs results familiar from the orthodox "mixing" approach to the derivation of thermodynamic behavior that relies upon trajectory instability. From these results and the independent posit about initial macroperiodicity he gets sustained macroperiodicity. And from this the results about the independence of probability distributions over the multiplicity of collisions and the continuous reproduction of macroperiodicity will follow. This provides the substitute he needs for the dubious Boltzmannian posit. And his substitute, unlike the usual orthodox approaches using mixing and an initial postulate of uniformity of probability distribution in the large, uses only the Poincaré-like weaker posit of initial macroperiodicity.

Whether all of this can be sustained without uncovering some hidden probabilistic assumptions buried in the argument is an interesting question. But for us what is important to notice that here we have a distinctive and very inventive way of responding to the third objection to the usual probabilistic posit, that the standard posit is so very strong. For now the results about thermodynamic behavior are to be obtained from the dynamics and a much weaker initial probabilistic posit that demands only "uniformity in the small" relative to a clever choice of micro-variable (RAI) and a clever choice of partition (intervals of RAI's that lead to some specific molecule being next hit).

5. SUMMARY

It is sometimes claimed that although an account is needed of the origin of time asymmetry in statistical mechanics, no "explanation" for the basic probabilistic posit is required. Sometimes such denials that there is any explaining to do simply rest on the confusion between a desire to find a *justification for belief* in the correctness of the posit by its success with the desire to *understand why* the posit actually holds in the world. In other cases there seems to be an assumption that posits that probability is distributed uniformly have no need of justification due to their "naturalness" or to some "minimal assumption" feature of them. But, as we have noted, reliance on any kind of principle of symmetry or principle of indifference to ground a probabilistic posit is dubious indeed. Even if one accepts a kind of symmetry argument, without some grounds for picking the right characterization of the underlying event space principles of symmetry are simply empty of content.

We have seen here the outlines of three radically different ways of trying to throw some light on the reason the basic probabilistic posit of statistical mechanics works so well in the world. One scheme seeks for a way of establishing results that avoids adopting any underlying basic probability measure at all. In the case we looked at this works by relying on a distinct, non-measure-theoretic feature of the phase space, its topology. A second proposal looks for the physical explanation of the posit's success in a deeper underlying dynamics of the constituents of the system. In the version we looked at one tries to rationalize the standard posit as it is used in non-equilibrium theory by showing it a consequence of the underlying stochastic, lawlike probabilities attributed to the world by the GRW interpretation of quantum mechanics. A third proposal accepts the need for an autonomous statistical posit at the foundation of the theory, but seeks for one adequate to the explanatory task but far weaker than the standard posit. In the example we surveyed, this new posit requires only a kind of "uniformity in the small" for the posited probability distribution rather than the global uniformity demanded by the orthodox theory.

The place of the standard posit in the theory of statistical mechanics, especially in the more general non-equilibrium portion of that theory, remains a puzzle. We are tempted to think of the dynamical laws governing the micro-constituents of a macroscopic system as constituting a full list of all of the laws of nature. Yet to obtain thermodynamic behavior something more is needed. This is the basic probabilistic posit over initial conditions of isolated systems. Viewing the truth of this posit as resting merely on "de facto" features of the world seems unsatisfying. Here we have seen several approaches to trying to at least mitigate that dissatisfaction.

What is curious is how radically different these different "ways out" are. While certainly not incompatible with each other, they seek the solution in wholly distinct conceptual directions. Nothing more vividly illustrates the mystery at the heart of statistical mechanics.

6. SUGGESTED FURTHER READING

For a general survey of the nature of the basic statistical posit and its role in both equilibrium and non-equilibrium statistical mechanics see L. Sklar,

Physics and Chance (Cambridge University Press, 1993), especially chapters 2, 5, 6 and 7. For an outline of topological entropy of a shift and how it is related to the Kolmogorov-Sinai entropy see L. Sklar, "Topology Versus Measure in Statistical Mechanics," *Monist* 83(2000), 258-73. For the proposal to use the GRW theory of quantum measurement to ground the statistical posit of statistical mechanics see D. Albert, *Time and Chance* (Harvard University Press, 2000), chapter 7. For the proposal that the orthodox posit can be replaced by the weaker posit of the sort employed by Poincaré in his theory of the roulette wheel see M. Strevens, *Bigger Than Chaos*, (Philosophy Department – Stanford University, 2001), chapter 4 especially section 4.8.

The University of Michigan
USA

INDEX

$E = mc^2$, 285

Abraham, M., 65, 87
acceleration, 10
Adler, F., 107
Albert, D., 312
analysis
  foundations, 283
Archimedean axiom, 139
Arthur, R., 40
atom
  Bohr model, 292
axiom
  Archimedean, 165
axiomatic method, 157, 164, 175
axiomatics, 201, 212

Bargmann, V., 111
Bergson, H., 190
Berliner, A., 235
Bertrand's paradoxes, 309
Birck, O., 238
Blumenthal, O., 236
Bohr, N., 110, 174, 297
Boltzmann, L., 52, 55, 165, 172, 186, 315
Bolzano, B., 211
Bonola, R., 139
Borchardt, G., 47

Born, M., 67, 174, 236, 239, 297
Brouwer, L.E.J., 175, 283, 290
Brush, S., 172
Bucherer, A., 65

Carnap, R., 185, 216
Carrier, M., 2
Carus, P., 150
Cassirer, E., 196
Castelnuovo, G., 135
causality, 21, 295, 297
choice sequence, 290
Clausius, R., 2
concept, 144, 194
  higher-order, 15
congruence, 23
  concept of, 35
  relation, 31
conservation
  energy, 6
  momentum, 108
  vis viva, 6, 11, 31
constructable in intuition, 2, 28
construction
  geometrical, 23
continuum
  intuitionistic, 290
Copernicus, N., 232

321

Boston Studies in the Philosophy of Science

Editor: Robert S. Cohen, *Boston University*

1. M.W. Wartofsky (ed.): *Proceedings of the Boston Colloquium for the Philosophy of Science, 1961/1962.* [Synthese Library 6] 1963   ISBN 90-277-0021-4
2. R.S. Cohen and M.W. Wartofsky (eds.): *Proceedings of the Boston Colloquium for the Philosophy of Science, 1962/1964.* In Honor of P. Frank. [Synthese Library 10] 1965
                                                                ISBN 90-277-9004-0
3. R.S. Cohen and M.W. Wartofsky (eds.): *Proceedings of the Boston Colloquium for the Philosophy of Science, 1964/1966.* In Memory of Norwood Russell Hanson. [Synthese Library 14] 1967   ISBN 90-277-0013-3
4. R.S. Cohen and M.W. Wartofsky (eds.): *Proceedings of the Boston Colloquium for the Philosophy of Science, 1966/1968.* [Synthese Library 18] 1969   ISBN 90-277-0014-1
5. R.S. Cohen and M.W. Wartofsky (eds.): *Proceedings of the Boston Colloquium for the Philosophy of Science, 1966/1968.* [Synthese Library 19] 1969   ISBN 90-277-0015-X
6. R.S. Cohen and R.J. Seeger (eds.): *Ernst Mach, Physicist and Philosopher.* [Synthese Library 27] 1970   ISBN 90-277-0016-8
7. M. Čapek: *Bergson and Modern Physics.* A Reinterpretation and Re-evaluation. [Synthese Library 37] 1971   ISBN 90-277-0186-5
8. R.C. Buck and R.S. Cohen (eds.): *PSA 1970.* Proceedings of the 2nd Biennial Meeting of the Philosophy and Science Association (Boston, Fall 1970). In Memory of Rudolf Carnap. [Synthese Library 39] 1971   ISBN 90-277-0187-3; Pb 90-277-0309-4
9. A.A. Zinov'ev: *Foundations of the Logical Theory of Scientific Knowledge (Complex Logic).* Translated from Russian. Revised and enlarged English Edition, with an Appendix by G.A. Smirnov, E.A. Sidorenko, A.M. Fedina and L.A. Bobrova. [Synthese Library 46] 1973
                                                                ISBN 90-277-0193-8; Pb 90-277-0324-8
10. L. Tondl: *Scientific Procedures.* A Contribution Concerning the Methodological Problems of Scientific Concepts and Scientific Explanation.Translated from Czech. [Synthese Library 47] 1973   ISBN 90-277-0147-4; Pb 90-277-0323-X
11. R.J. Seeger and R.S. Cohen (eds.): *Philosophical Foundations of Science.* Proceedings of Section L, 1969, American Association for the Advancement of Science. [Synthese Library 58] 1974   ISBN 90-277-0390-6; Pb 90-277-0376-0
12. A. Grünbaum: *Philosophical Problems of Space and Times.* 2nd enlarged ed. [Synthese Library 55] 1973   ISBN 90-277-0357-4; Pb 90-277-0358-2
13. R.S. Cohen and M.W. Wartofsky (eds.): *Logical and Epistemological Studies in Contemporary Physics.* Proceedings of the Boston Colloquium for the Philosophy of Science, 1969/72, Part I. [Synthese Library 59] 1974   ISBN 90-277-0391-4; Pb 90-277-0377-9
14. R.S. Cohen and M.W. Wartofsky (eds.): *Methodological and Historical Essays in the Natural and Social Sciences.* Proceedings of the Boston Colloquium for the Philosophy of Science, 1969/72, Part II. [Synthese Library 60] 1974   ISBN 90-277-0392-2; Pb 90-277-0378-7
15. R.S. Cohen, J.J. Stachel and M.W. Wartofsky (eds.): *For Dirk Struik.* Scientific, Historical and Political Essays in Honor of Dirk J. Struik. [Synthese Library 61] 1974
                                                                ISBN 90-277-0393-0; Pb 90-277-0379-5
16. N. Geschwind: *Selected Papers on Language and the Brains.* [Synthese Library 68] 1974
                                                                ISBN 90-277-0262-4; Pb 90-277-0263-2
17. B.G. Kuznetsov: *Reason and Being.* Translated from Russian. Edited by C.R. Fawcett and R.S. Cohen. 1987   ISBN 90-277-2181-5

Boston Studies in the Philosophy of Science

18. P. Mittelstaedt: *Philosophical Problems of Modern Physics*. Translated from the revised 4th German edition by W. Riemer and edited by R.S. Cohen. [Synthese Library 95] 1976
ISBN 90-277-0285-3; Pb 90-277-0506-2

19. H. Mehlberg: *Time, Causality, and the Quantum Theory*. Studies in the Philosophy of Science. Vol. I: *Essay on the Causal Theory of Time*. Vol. II: *Time in a Quantized Universe*. Translated from French. Edited by R.S. Cohen. 1980   Vol. I: ISBN 90-277-0721-9; Pb 90-277-1074-0
Vol. II: ISBN 90-277-1075-9; Pb 90-277-1076-7

20. K.F. Schaffner and R.S. Cohen (eds.): *PSA 1972*. Proceedings of the 3rd Biennial Meeting of the Philosophy of Science Association (Lansing, Michigan, Fall 1972). [Synthese Library 64] 1974
ISBN 90-277-0408-2; Pb 90-277-0409-0

21. R.S. Cohen and J.J. Stachel (eds.): *Selected Papers of Léon Rosenfeld*. [Synthese Library 100] 1979
ISBN 90-277-0651-4; Pb 90-277-0652-2

22. M. Čapek (ed.): *The Concepts of Space and Time*. Their Structure and Their Development. [Synthese Library 74] 1976
ISBN 90-277-0355-8; Pb 90-277-0375-2

23. M. Grene: *The Understanding of Nature*. Essays in the Philosophy of Biology. [Synthese Library 66] 1974
ISBN 90-277-0462-7; Pb 90-277-0463-5

24. D. Ihde: *Technics and Praxis*. A Philosophy of Technology. [Synthese Library 130] 1979
ISBN 90-277-0953-X; Pb 90-277-0954-8

25. J. Hintikka and U. Remes: *The Method of Analysis*. Its Geometrical Origin and Its General Significance. [Synthese Library 75] 1974   ISBN 90-277-0532-1; Pb 90-277-0543-7

26. J.E. Murdoch and E.D. Sylla (eds.): *The Cultural Context of Medieval Learning*. Proceedings of the First International Colloquium on Philosophy, Science, and Theology in the Middle Ages, 1973. [Synthese Library 76] 1975   ISBN 90-277-0560-7; Pb 90-277-0587-9

27. M. Grene and E. Mendelsohn (eds.): *Topics in the Philosophy of Biology*. [Synthese Library 84] 1976   ISBN 90-277-0595-X; Pb 90-277-0596-8

28. J. Agassi: *Science in Flux*. [Synthese Library 80] 1975
ISBN 90-277-0584-4; Pb 90-277-0612-3

29. J.J. Wiatr (ed.): *Polish Essays in the Methodology of the Social Sciences*. [Synthese Library 131] 1979   ISBN 90-277-0723-5; Pb 90-277-0956-4

30. P. Janich: *Protophysics of Time*. Constructive Foundation and History of Time Measurement. Translated from German. 1985   ISBN 90-277-0724-3

31. R.S. Cohen and M.W. Wartofsky (eds.): *Language, Logic, and Method*. 1983
ISBN 90-277-0725-1

32. R.S. Cohen, C.A. Hooker, A.C. Michalos and J.W. van Evra (eds.): *PSA 1974*. Proceedings of the 4th Biennial Meeting of the Philosophy of Science Association. [Synthese Library 101] 1976   ISBN 90-277-0647-6; Pb 90-277-0648-4

33. G. Holton and W.A. Blanpied (eds.): *Science and Its Public*. The Changing Relationship. [Synthese Library 96] 1976   ISBN 90-277-0657-3; Pb 90-277-0658-1

34. M.D. Grmek, R.S. Cohen and G. Cimino (eds.): *On Scientific Discovery*. The 1977 Erice Lectures. 1981   ISBN 90-277-1122-4; Pb 90-277-1123-2

35. S. Amsterdamski: *Between Experience and Metaphysics*. Philosophical Problems of the Evolution of Science. Translated from Polish. [Synthese Library 77] 1975
ISBN 90-277-0568-2; Pb 90-277-0580-1

36. M. Marković and G. Petrović (eds.): *Praxis*. Yugoslav Essays in the Philosophy and Methodology of the Social Sciences. [Synthese Library 134] 1979
ISBN 90-277-0727-8; Pb 90-277-0968-8

Boston Studies in the Philosophy of Science

37. H. von Helmholtz: *Epistemological Writings.* The Paul Hertz / Moritz Schlick Centenary Edition of 1921. Translated from German by M.F. Lowe. Edited with an Introduction and Bibliography by R.S. Cohen and Y. Elkana. [Synthese Library 79] 1977
ISBN 90-277-0290-X; Pb 90-277-0582-8

38. R.M. Martin: *Pragmatics, Truth and Language.* 1979
ISBN 90-277-0992-0; Pb 90-277-0993-9

39. R.S. Cohen, P.K. Feyerabend and M.W. Wartofsky (eds.): *Essays in Memory of Imre Lakatos.* [Synthese Library 99] 1976 ISBN 90-277-0654-9; Pb 90-277-0655-7

40. Not published.

41. Not published.

42. H.R. Maturana and F.J. Varela: *Autopoiesis and Cognition.* The Realization of the Living. With a Preface to "Autopoiesis' by S. Beer. 1980 ISBN 90-277-1015-5; Pb 90-277-1016-3

43. A. Kasher (ed.): *Language in Focus: Foundations, Methods and Systems.* Essays in Memory of Yehoshua Bar-Hillel. [Synthese Library 89] 1976
ISBN 90-277-0644-1; Pb 90-277-0645-X

44. T.D. Thao: *Investigations into the Origin of Language and Consciousness.* 1984
ISBN 90-277-0827-4

45. F.G.-I. Nagasaka (ed.): *Japanese Studies in the Philosophy of Science.* 1997
ISBN 0-7923-4781-1

46. P.L. Kapitza: *Experiment, Theory, Practice.* Articles and Addresses. Edited by R.S. Cohen. 1980 ISBN 90-277-1061-9; Pb 90-277-1062-7

47. M.L. Dalla Chiara (ed.): *Italian Studies in the Philosophy of Science.* 1981
ISBN 90-277-0735-9; Pb 90-277-1073-2

48. M.W. Wartofsky: *Models.* Representation and the Scientific Understanding. [Synthese Library 129] 1979 ISBN 90-277-0736-7; Pb 90-277-0947-5

49. T.D. Thao: *Phenomenology and Dialectical Materialism.* Edited by R.S. Cohen. 1986
ISBN 90-277-0737-5

50. Y. Fried and J. Agassi: *Paranoia.* A Study in Diagnosis. [Synthese Library 102] 1976
ISBN 90-277-0704-9; Pb 90-277-0705-7

51. K.H. Wolff: *Surrender and Cath.* Experience and Inquiry Today. [Synthese Library 105] 1976
ISBN 90-277-0758-8; Pb 90-277-0765-0

52. K. Kosík: *Dialectics of the Concrete.* A Study on Problems of Man and World. 1976
ISBN 90-277-0761-8; Pb 90-277-0764-2

53. N. Goodman: *The Structure of Appearance.* [Synthese Library 107] 1977
ISBN 90-277-0773-1; Pb 90-277-0774-X

54. H.A. Simon: *Models of Discovery* and Other Topics in the Methods of Science. [Synthese Library 114] 1977 ISBN 90-277-0812-6; Pb 90-277-0858-4

55. M. Lazerowitz: *The Language of Philosophy.* Freud and Wittgenstein. [Synthese Library 117] 1977 ISBN 90-277-0826-6; Pb 90-277-0862-2

56. T. Nickles (ed.): *Scientific Discovery, Logic, and Rationality.* 1980
ISBN 90-277-1069-4; Pb 90-277-1070-8

57. J. Margolis: *Persons and Mind.* The Prospects of Nonreductive Materialism. [Synthese Library 121] 1978 ISBN 90-277-0854-1; Pb 90-277-0863-0

58. G. Radnitzky and G. Andersson (eds.): *Progress and Rationality in Science.* [Synthese Library 125] 1978 ISBN 90-277-0921-1; Pb 90-277-0922-X

59. G. Radnitzky and G. Andersson (eds.): *The Structure and Development of Science.* [Synthese Library 136] 1979 ISBN 90-277-0994-7; Pb 90-277-0995-5

Boston Studies in the Philosophy of Science

60. T. Nickles (ed.): *Scientific Discovery.* Case Studies. 1980
     ISBN 90-277-1092-9; Pb 90-277-1093-7
61. M.A. Finocchiaro: *Galileo and the Art of Reasoning.* Rhetorical Foundation of Logic and
     Scientific Method. 1980   ISBN 90-277-1094-5; Pb 90-277-1095-3
62. W.A. Wallace: *Prelude to Galileo.* Essays on Medieval and 16th-Century Sources of Galileo's
     Thought. 1981   ISBN 90-277-1215-8; Pb 90-277-1216-6
63. F. Rapp: *Analytical Philosophy of Technology.* Translated from German. 1981
     ISBN 90-277-1221-2; Pb 90-277-1222-0
64. R.S. Cohen and M.W. Wartofsky (eds.): *Hegel and the Sciences.* 1984   ISBN 90-277-0726-X
65. J. Agassi: *Science and Society.* Studies in the Sociology of Science. 1981
     ISBN 90-277-1244-1; Pb 90-277-1245-X
66. L. Tondl: *Problems of Semantics.* A Contribution to the Analysis of the Language of Science.
     Translated from Czech. 1981   ISBN 90-277-0148-2; Pb 90-277-0316-7
67. J. Agassi and R.S. Cohen (eds.): *Scientific Philosophy Today.* Essays in Honor of Mario Bunge.
     1982   ISBN 90-277-1262-X; Pb 90-277-1263-8
68. W. Krajewski (ed.): *Polish Essays in the Philosophy of the Natural Sciences.* Translated from
     Polish and edited by R.S. Cohen and C.R. Fawcett. 1982
     ISBN 90-277-1286-7; Pb 90-277-1287-5
69. J.H. Fetzer: *Scientific Knowledge.* Causation, Explanation and Corroboration. 1981
     ISBN 90-277-1335-9; Pb 90-277-1336-7
70. S. Grossberg: *Studies of Mind and Brain.* Neural Principles of Learning, Perception, Develop-
     ment, Cognition, and Motor Control. 1982   ISBN 90-277-1359-6; Pb 90-277-1360-X
71. R.S. Cohen and M.W. Wartofsky (eds.): *Epistemology, Methodology, and the Social Sciences.*
     1983.   ISBN 90-277-1454-1
72. K. Berka: *Measurement.* Its Concepts, Theories and Problems. Translated from Czech. 1983
     ISBN 90-277-1416-9
73. G.L. Pandit: *The Structure and Growth of Scientific Knowledge.* A Study in the Methodology
     of Epistemic Appraisal. 1983   ISBN 90-277-1434-7
74. A.A. Zinov'ev: *Logical Physics.* Translated from Russian. Edited by R.S. Cohen. 1983
     [*see also* Volume 9]   ISBN 90-277-0734-0
75. G-G. Granger: *Formal Thought and the Sciences of Man.* Translated from French. With and
     Introduction by A. Rosenberg. 1983   ISBN 90-277-1524-6
76. R.S. Cohen and L. Laudan (eds.): *Physics, Philosophy and Psychoanalysis.* Essays in Honor
     of Adolf Grünbaum. 1983   ISBN 90-277-1533-5
77. G. Böhme, W. van den Daele, R. Hohlfeld, W. Krohn and W. Schäfer: *Finalization in Science.*
     The Social Orientation of Scientific Progress. Translated from German. Edited by W. Schäfer.
     1983   ISBN 90-277-1549-1
78. D. Shapere: *Reason and the Search for Knowledge.* Investigations in the Philosophy of Science.
     1984   ISBN 90-277-1551-3; Pb 90-277-1641-2
79. G. Andersson (ed.): *Rationality in Science and Politics.* Translated from German. 1984
     ISBN 90-277-1575-0; Pb 90-277-1953-5
80. P.T. Durbin and F. Rapp (eds.): *Philosophy and Technology.* [*Also* Philosophy and Technology
     Series, Vol. 1] 1983   ISBN 90-277-1576-9
81. M. Marković: *Dialectical Theory of Meaning.* Translated from Serbo-Croat. 1984
     ISBN 90-277-1596-3
82. R.S. Cohen and M.W. Wartofsky (eds.): *Physical Sciences and History of Physics.* 1984.
     ISBN 90-277-1615-3

Boston Studies in the Philosophy of Science

83. É. Meyerson: *The Relativistic Deduction*. Epistemological Implications of the Theory of Relativity. Translated from French. With a Review by Albert Einstein and an Introduction by Milič Čapek. 1985   ISBN 90-277-1699-4

84. R.S. Cohen and M.W. Wartofsky (eds.): *Methodology, Metaphysics and the History of Science*. In Memory of Benjamin Nelson. 1984   ISBN 90-277-1711-7

85. G. Tamás: *The Logic of Categories*. Translated from Hungarian. Edited by R.S. Cohen. 1986
                                                                                 ISBN 90-277-1742-7

86. S.L. de C. Fernandes: *Foundations of Objective Knowledge*. The Relations of Popper's Theory of Knowledge to That of Kant. 1985   ISBN 90-277-1809-1

87. R.S. Cohen and T. Schnelle (eds.): *Cognition and Fact*. Materials on Ludwik Fleck. 1986
                                                                                 ISBN 90-277-1902-0

88. G. Freudenthal: *Atom and Individual in the Age of Newton*. On the Genesis of the Mechanistic World View. Translated from German. 1986   ISBN 90-277-1905-5

89. A. Donagan, A.N. Perovich Jr and M.V. Wedin (eds.): *Human Nature and Natural Knowledge*. Essays presented to Marjorie Grene on the Occasion of Her 75th Birthday. 1986
                                                                                 ISBN 90-277-1974-8

90. C. Mitcham and A. Hunning (eds.): *Philosophy and Technology II*. Information Technology and Computers in Theory and Practice. [*Also* Philosophy and Technology Series, Vol. 2] 1986
                                                                                 ISBN 90-277-1975-6

91. M. Grene and D. Nails (eds.): *Spinoza and the Sciences*. 1986   ISBN 90-277-1976-4

92. S.P. Turner: *The Search for a Methodology of Social Science*. Durkheim, Weber, and the 19th-Century Problem of Cause, Probability, and Action. 1986.   ISBN 90-277-2067-3

93. I.C. Jarvie: *Thinking about Society*. Theory and Practice. 1986   ISBN 90-277-2068-1

94. E. Ullmann-Margalit (ed.): *The Kaleidoscope of Science*. The Israel Colloquium: Studies in History, Philosophy, and Sociology of Science, Vol. 1. 1986
                                                       ISBN 90-277-2158-0; Pb 90-277-2159-9

95. E. Ullmann-Margalit (ed.): *The Prism of Science*. The Israel Colloquium: Studies in History, Philosophy, and Sociology of Science, Vol. 2. 1986
                                                       ISBN 90-277-2160-2; Pb 90-277-2161-0

96. G. Márkus: *Language and Production*. A Critique of the Paradigms. Translated from French. 1986   ISBN 90-277-2169-6

97. F. Amrine, F.J. Zucker and H. Wheeler (eds.): *Goethe and the Sciences: A Reappraisal*. 1987
                                                       ISBN 90-277-2265-X; Pb 90-277-2400-8

98. J.C. Pitt and M. Pera (eds.): *Rational Changes in Science*. Essays on Scientific Reasoning. Translated from Italian. 1987   ISBN 90-277-2417-2

99. O. Costa de Beauregard: *Time, the Physical Magnitude*. 1987   ISBN 90-277-2444-X

100. A. Shimony and D. Nails (eds.): *Naturalistic Epistemology*. A Symposium of Two Decades. 1987   ISBN 90-277-2337-0

101. N. Rotenstreich: *Time and Meaning in History*. 1987   ISBN 90-277-2467-9

102. D.B. Zilberman: *The Birth of Meaning in Hindu Thought*. Edited by R.S. Cohen. 1988
                                                                                 ISBN 90-277-2497-0

103. T.F. Glick (ed.): *The Comparative Reception of Relativity*. 1987   ISBN 90-277-2498-9

104. Z. Harris, M. Gottfried, T. Ryckman, P. Mattick Jr, A. Daladier, T.N. Harris and S. Harris: *The Form of Information in Science*. Analysis of an Immunology Sublanguage. With a Preface by Hilary Putnam. 1989   ISBN 90-277-2516-0

105. F. Burwick (ed.): *Approaches to Organic Form*. Permutations in Science and Culture. 1987
                                                                                 ISBN 90-277-2541-1

Boston Studies in the Philosophy of Science

Boston Studies in the Philosophy of Science

127. Z. Bechler: *Newton's Physics on the Conceptual Structure of the Scientific Revolution.* 1991
ISBN 0-7923-1054-3

128. É. Meyerson: *Explanation in the Sciences.* Translated from French by M-A. Siple and D.A. Siple. 1991 ISBN 0-7923-1129-9

129. A.I. Tauber (ed.): *Organism and the Origins of Self.* 1991 ISBN 0-7923-1185-X

130. F.J. Varela and J-P. Dupuy (eds.): *Understanding Origins.* Contemporary Views on the Origin of Life, Mind and Society. 1992 ISBN 0-7923-1251-1

131. G.L. Pandit: *Methodological Variance.* Essays in Epistemological Ontology and the Methodology of Science. 1991 ISBN 0-7923-1263-5

132. G. Munévar (ed.): *Beyond Reason.* Essays on the Philosophy of Paul Feyerabend. 1991
ISBN 0-7923-1272-4

133. T.E. Uebel (ed.): *Rediscovering the Forgotten Vienna Circle.* Austrian Studies on Otto Neurath and the Vienna Circle. Partly translated from German. 1991 ISBN 0-7923-1276-7

134. W.R. Woodward and R.S. Cohen (eds.): *World Views and Scientific Discipline Formation.* Science Studies in the [former] German Democratic Republic. Partly translated from German by W.R. Woodward. 1991 ISBN 0-7923-1286-4

135. P. Zambelli: *The Speculum Astronomiae and Its Enigma.* Astrology, Theology and Science in Albertus Magnus and His Contemporaries. 1992 ISBN 0-7923-1380-1

136. P. Petitjean, C. Jami and A.M. Moulin (eds.): *Science and Empires.* Historical Studies about Scientific Development and European Expansion. ISBN 0-7923-1518-9

137. W.A. Wallace: *Galileo's Logic of Discovery and Proof.* The Background, Content, and Use of His Appropriated Treatises on Aristotle's *Posterior Analytics.* 1992 ISBN 0-7923-1577-4

138. W.A. Wallace: *Galileo's Logical Treatises.* A Translation, with Notes and Commentary, of His Appropriated Latin Questions on Aristotle's *Posterior Analytics.* 1992 ISBN 0-7923-1578-2
Set (137 + 138) ISBN 0-7923-1579-0

139. M.J. Nye, J.L. Richards and R.H. Stuewer (eds.): *The Invention of Physical Science.* Intersections of Mathematics, Theology and Natural Philosophy since the Seventeenth Century. Essays in Honor of Erwin N. Hiebert. 1992 ISBN 0-7923-1753-X

140. G. Corsi, M.L. dalla Chiara and G.C. Ghirardi (eds.): *Bridging the Gap: Philosophy, Mathematics and Physics.* Lectures on the Foundations of Science. 1992 ISBN 0-7923-1761-0

141. C.-H. Lin and D. Fu (eds.): *Philosophy and Conceptual History of Science in Taiwan.* 1992
ISBN 0-7923-1766-1

142. S. Sarkar (ed.): *The Founders of Evolutionary Genetics.* A Centenary Reappraisal. 1992
ISBN 0-7923-1777-7

143. J. Blackmore (ed.): *Ernst Mach – A Deeper Look.* Documents and New Perspectives. 1992
ISBN 0-7923-1853-6

144. P. Kroes and M. Bakker (eds.): *Technological Development and Science in the Industrial Age.* New Perspectives on the Science–Technology Relationship. 1992 ISBN 0-7923-1898-6

145. S. Amsterdamski: *Between History and Method.* Disputes about the Rationality of Science. 1992 ISBN 0-7923-1941-9

146. E. Ullmann-Margalit (ed.): *The Scientific Enterprise.* The Bar-Hillel Colloquium: Studies in History, Philosophy, and Sociology of Science, Volume 4. 1992 ISBN 0-7923-1992-3

147. L. Embree (ed.): *Metaarchaeology.* Reflections by Archaeologists and Philosophers. 1992
ISBN 0-7923-2023-9

148. S. French and H. Kamminga (eds.): *Correspondence, Invariance and Heuristics.* Essays in Honour of Heinz Post. 1993 ISBN 0-7923-2085-9

149. M. Bunzl: *The Context of Explanation.* 1993 ISBN 0-7923-2153-7

Boston Studies in the Philosophy of Science

Boston Studies in the Philosophy of Science

171. M.A. Grodin (ed.): *Meta Medical Ethics*: The Philosophical Foundations of Bioethics. 1995
ISBN 0-7923-3344-6

172. S. Ramirez and R.S. Cohen (eds.): *Mexican Studies in the History and Philosophy of Science*.
1995
ISBN 0-7923-3462-0

173. C. Dilworth: *The Metaphysics of Science*. An Account of Modern Science in Terms of Principles, Laws and Theories. 1995
ISBN 0-7923-3693-3

174. J. Blackmore: *Ludwig Boltzmann, His Later Life and Philosophy, 1900–1906* Book Two: The Philosopher. 1995
ISBN 0-7923-3464-7

175. P. Damerow: *Abstraction and Representation*. Essays on the Cultural Evolution of Thinking.
1996
ISBN 0-7923-3816-2

176. M.S. Macrakis: *Scarcity's Ways: The Origins of Capital*. A Critical Essay on Thermodynamics,
Statistical Mechanics and Economics. 1997
ISBN 0-7923-4760-9

177. M. Marion and R.S. Cohen (eds.): *Québec Studies in the Philosophy of Science*. Part I: Logic,
Mathematics, Physics and History of Science. Essays in Honor of Hugues Leblanc. 1995
ISBN 0-7923-3559-7

178. M. Marion and R.S. Cohen (eds.): *Québec Studies in the Philosophy of Science*. Part II: Biology,
Psychology, Cognitive Science and Economics. Essays in Honor of Hugues Leblanc. 1996
ISBN 0-7923-3560-0
Set (177–178) ISBN 0-7923-3561-9

179. Fan Dainian and R.S. Cohen (eds.): *Chinese Studies in the History and Philosophy of Science
and Technology*. 1996
ISBN 0-7923-3463-9

180. P. Forman and J.M. Sánchez-Ron (eds.): *National Military Establishments and the Advancement of Science and Technology*. Studies in 20th Century History. 1996
ISBN 0-7923-3541-4

181. E.J. Post: *Quantum Reprogramming*. Ensembles and Single Systems: A Two-Tier Approach
to Quantum Mechanics. 1995
ISBN 0-7923-3565-1

182. A.I. Tauber (ed.): *The Elusive Synthesis: Aesthetics and Science*. 1996   ISBN 0-7923-3904-5

183. S. Sarkar (ed.): *The Philosophy and History of Molecular Biology: New Perspectives*. 1996
ISBN 0-7923-3947-9

184. J.T. Cushing, A. Fine and S. Goldstein (eds.): *Bohmian Mechanics and Quantum Theory: An
Appraisal*. 1996
ISBN 0-7923-4028-0

185. K. Michalski: *Logic and Time*. An Essay on Husserl's Theory of Meaning. 1996
ISBN 0-7923-4082-5

186. G. Munévar (ed.): *Spanish Studies in the Philosophy of Science*. 1996   ISBN 0-7923-4147-3

187. G. Schubring (ed.): *Hermann Günther Graßmann (1809–1877): Visionary Mathematician,
Scientist and Neohumanist Scholar*. Papers from a Sesquicentennial Conference. 1996
ISBN 0-7923-4261-5

188. M. Bitbol: *Schrödinger's Philosophy of Quantum Mechanics*. 1996   ISBN 0-7923-4266-6

189. J. Faye, U. Scheffler and M. Urchs (eds.): *Perspectives on Time*. 1997   ISBN 0-7923-4330-1

190. K. Lehrer and J.C. Marek (eds.): *Austrian Philosophy Past and Present*. Essays in Honor of
Rudolf Haller. 1996
ISBN 0-7923-4347-6

191. J.L. Lagrange: *Analytical Mechanics*. Translated and edited by Auguste Boissonade and Victor
N. Vagliente. Translated from the *Mécanique Analytique, novelle édition* of 1811. 1997
ISBN 0-7923-4349-2

192. D. Ginev and R.S. Cohen (eds.): *Issues and Images in the Philosophy of Science*. Scientific
and Philosophical Essays in Honour of Azarya Polikarov. 1997   ISBN 0-7923-4444-8

Boston Studies in the Philosophy of Science

193. R.S. Cohen, M. Horne and J. Stachel (eds.): *Experimental Metaphysics.* Quantum Mechanical Studies for Abner Shimony, Volume One. 1997   ISBN 0-7923-4452-9
194. R.S. Cohen, M. Horne and J. Stachel (eds.): *Potentiality, Entanglement and Passion-at-a-Distance.* Quantum Mechanical Studies for Abner Shimony, Volume Two. 1997
          ISBN 0-7923-4453-7; Set 0-7923-4454-5
195. R.S. Cohen and A.I. Tauber (eds.): *Philosophies of Nature: The Human Dimension.* 1997
          ISBN 0-7923-4579-7
196. M. Otte and M. Panza (eds.): *Analysis and Synthesis in Mathematics.* History and Philosophy. 1997   ISBN 0-7923-4570-3
197. A. Denkel: *The Natural Background of Meaning.* 1999   ISBN 0-7923-5331-5
198. D. Baird, R.I.G. Hughes and A. Nordmann (eds.): *Heinrich Hertz: Classical Physicist, Modern Philosopher.* 1999   ISBN 0-7923-4653-X
199. A. Franklin: *Can That be Right?* Essays on Experiment, Evidence, and Science. 1999
          ISBN 0-7923-5464-8
200. D. Raven, W. Krohn and R.S. Cohen (eds.): *The Social Origins of Modern Science.* 2000
          ISBN 0-7923-6457-0
201. Reserved
202. Reserved
203. B. Babich and R.S. Cohen (eds.): *Nietzsche, Theories of Knowledge, and Critical Theory.* Nietzsche and the Sciences I. 1999   ISBN 0-7923-5742-6
204. B. Babich and R.S. Cohen (eds.): *Nietzsche, Epistemology, and Philosophy of Science.* Nietzsche and the Science II. 1999   ISBN 0-7923-5743-4
205. R. Hooykaas: *Fact, Faith and Fiction in the Development of Science.* The Gifford Lectures given in the University of St Andrews 1976. 1999   ISBN 0-7923-5774-4
206. M. Fehér, O. Kiss and L. Ropolyi (eds.): *Hermeneutics and Science.* 1999 ISBN 0-7923-5798-1
207. R.M. MacLeod (ed.): *Science and the Pacific War.* Science and Survival in the Pacific, 1939-1945. 1999   ISBN 0-7923-5851-1
208. I. Hanzel: *The Concept of Scientific Law in the Philosophy of Science and Epistemology.* A Study of Theoretical Reason. 1999   ISBN 0-7923-5852-X
209. G. Helm; R.J. Deltete (ed./transl.): *The Historical Development of Energetics.* 1999
          ISBN 0-7923-5874-0
210. A. Orenstein and P. Kotatko (eds.): *Knowledge, Language and Logic.* Questions for Quine. 1999   ISBN 0-7923-5986-0
211. R.S. Cohen and H. Levine (eds.): *Maimonides and the Sciences.* 2000   ISBN 0-7923-6053-2
212. H. Gourko, D.I. Williamson and A.I. Tauber (eds.): *The Evolutionary Biology Papers of Elie Metchnikoff.* 2000   ISBN 0-7923-6067-2
213. S. D'Agostino: *A History of the Ideas of Theoretical Physics.* Essays on the Nineteenth and Twentieth Century Physics. 2000   ISBN 0-7923-6094-X
214. S. Lelas: *Science and Modernity.* Toward An Integral Theory of Science. 2000
          ISBN 0-7923-6303-5
215. E. Agazzi and M. Pauri (eds.): *The Reality of the Unobservable.* Observability, Unobservability and Their Impact on the Issue of Scientific Realism. 2000   ISBN 0-7923-6311-6
216. P. Hoyningen-Huene and H. Sankey (eds.): *Incommensurability and Related Matters.* 2001
          ISBN 0-7923-6989-0
217. A. Nieto-Galan: *Colouring Textiles.* A History of Natural Dyestuffs in Industrial Europe. 2001
          ISBN 0-7923-7022-8

Boston Studies in the Philosophy of Science

218. J. Blackmore, R. Itagaki and S. Tanaka (eds.): *Ernst Mach's Vienna 1895–1930*. Or Phenomenalism as Philosophy of Science. 2001   ISBN 0-7923-7122-4
219. R. Vihalemm (ed.): *Estonian Studies in the History and Philosophy of Science*. 2001
                                                                                        ISBN 0-7923-7189-5
220. W. Lefèvre (ed.): *Between Leibniz, Newton, and Kant*. Philosophy and Science in the Eighteenth Century. 2001   ISBN 0-7923-7198-4
221. T.F. Glick, M.Á. Puig-Samper and R. Ruiz (eds.): *The Reception of Darwinism in the Iberian World*. Spain, Spanish America and Brazil. 2001   ISBN 1-4020-0082-0
222. U. Klein (ed.): *Tools and Modes of Representation in the Laboratory Sciences*. 2001
                                                                                        ISBN 1-4020-0100-2
223. P. Duhem: *Mixture and Chemical Combination*. And Related Essays. Edited and translated, with an introduction, by Paul Needham. 2002   ISBN 1-4020-0232-7
224. J.C. Boudri: *What was Mechanical about Mechanics*. The Concept of Force Betweem Metaphysics and Mechanics from Newton to Lagrange. 2002   ISBN 1-4020-0233-5
225. B.E. Babich (ed.): *Hermeneutic Philosophy of Science, Van Gogh's Eyes, and God*. Essays in Honor of Patrick A. Heelan, S.J. 2002   ISBN 1-4020-0234-3
226. D. Davies Villemaire: *E.A. Burtt, Historian and Philosopher*. A Study of the Author of The Metaphysical Foundations of Modern Physical Science. 2002   ISBN 1-4020-0428-1
227. L.J. Cohen: *Knowledge and Language*. Selected Essays of L. Jonathan Cohen. Edited and with an introduction by James Logue. 2002   ISBN 1-4020-0474-5
228. G.E. Allen and R.M. MacLeod (eds.): *Science, History and Social Activism: A Tribute to Everett Mendelsohn*. 2002   ISBN 1-4020-0495-0
229. O. Gal: *Meanest Foundations and Nobler Superstructures*. Hooke, Newton and the "Compounding of the Celestiall Motions of the Planetts". 2002   ISBN 1-4020-0732-9
230. R. Nola: *Rescuing Reason*. A Critique of Anti-Rationalist Views of Science and Knowledge. 2003   Hb: ISBN 1-4020-1042-7; Pb ISBN 1-4020-1043-5
231. J. Agassi: *Science and Culture*. 2003   ISBN 1-4020-1156-3
232. M.C. Galavotti (ed.): *Observation and Experiment in the Natural and Social Science*. 2003
                                                                                        ISBN 1-4020-1251-9
233. A. Simões, A. Carneiro and M.P. Diogo (eds.): *Travels of Learning*. A Geography of Science in Europe. 2003   ISBN 1-4020-1259-4
234. A. Ashtekar, R. Cohen, D. Howard, J. Renn, S. Sarkar and A. Shimony (eds.): *Revisiting the Foundations of Relativistic Physics*. Festschrift in Honor of John Stachel. 2003
                                                                                        ISBN 1-4020-1284-5
235. R.P. Farell: *Feyerabend and Scientific Values*. Tightrope-Walking Rationality. 2003
                                                                                        ISBN 1-4020-1350-7
236. D. Ginev (ed.): *Bulgarian Studies in the Philosophy of Science*. 2003   ISBN 1-4020-1496-1
237. C. Sasaki: *Descartes Mathematical Thought*. 2003   ISBN 1-4020-1746-4
238. K. Chemla (ed.): *History of Science, History of Text*. 2004   ISBN 1-4020-2320-0
239. C.R. Palmerino and J.M.M.H. Thijssen (eds.): *The Reception of the Galilean Science of Motion in Seventeenth-Century Europe*. 2004   ISBN 1-4020-2454-1

Boston Studies in the Philosophy of Science